The Second-Order Adjoint Sensitivity Analysis Methodology

Advances in Applied Mathematics

Series Editor: Daniel Zwillinger

Published Titles

Advanced Engineering Mathematics with MATLAB, Fourth Edition
Dean G. Duffy

CRC Standard Curves and Surfaces with Mathematica®, Third Edition
David H. von Seggern

CRC Standard Mathematical Tables and Formulas, 33rd Edition
Dan Zwillinger

Dynamical Systems for Biological Modeling: An Introduction
Fred Brauer and Christopher Kribs

Fast Solvers for Mesh-Based Computations *Maciej Paszyński*

Green's Functions with Applications, Second Edition *Dean G. Duffy*

Handbook of Peridynamic Modeling *Floriin Bobaru, John T. Foster, Philippe H. Geubelle, and Stewart A. Silling*

Introduction to Financial Mathematics *Kevin J. Hastings*

Introduction to Radar Analysis, Second Edition *Bassem R. Mahafza*

Linear and Complex Analysis for Applications *John P. D'Angelo*

Linear and Integer Optimization: Theory and Practice, Third Edition
Gerard Sierksma and Yori Zwols

Markov Processes *James R. Kirkwood*

Pocket Book of Integrals and Mathematical Formulas, 5th Edition
Ronald J. Tallarida

The Second-Order Adjoint Sensitivity Analysis Methodology
Dan Gabriel Cacuci

Stochastic Partial Differential Equations, Second Edition *Pao-Liu Chow*

Quadratic Programming with Computer Programs *Michael J. Best*

The Second-Order Adjoint Sensitivity Analysis Methodology

Dan Gabriel Cacuci

CRC Press
Taylor & Francis Group
Boca Raton London New York

CRC Press is an imprint of the
Taylor & Francis Group, an **informa** business

CRC Press
Taylor & Francis Group
6000 Broken Sound Parkway NW, Suite 300
Boca Raton, FL 33487-2742

International Standard Book Number-13: 978-1-4987-2648-1 (Hardback)

Library of Congress Cataloging-in-Publication Data

Names: Cacuci, Dan Gabriel, author.
Title: The second-order adjoint sensitivity analysis methodology / Dan Gabriel Cacuci.
Description: Boca Raton : CRC Press, 2018. | Includes bibliographical references and index.
Identifiers: LCCN 2017040197 | ISBN 9781498726481
Subjects: LCSH: Sensitivity theory (Mathematics) | Large scale systems. | Nonlinear systems.
Classification: LCC QA402 .C2987 2018 | DDC 003/.71--dc23
LC record available at https://lccn.loc.gov/2017040197

Visit the Taylor & Francis Web site at
http://www.taylorandfrancis.com

and the CRC Press Web site at
http://www.crcpress.com

MIX
Paper from responsible sources
FSC www.fsc.org FSC® C013056

Printed and bound in Great Britain by
TJ International Ltd, Padstow, Cornwall

Contents

Preface

The material presented in this book uses certain generic terms for defining the objects or phenomena under study. Thus, a *system* is the object or phenomena under study; it can refer to a process, mechanical device, biological mechanism, society, economy, or any other conceivable object that is under study. The *output* or *response* of a system is a result that can be quantitatively measured and/or computed. A *model* is a mathematical idealization that is used as an approximation to represent the system and its output. Models can be quite simple or highly complex; they can be expressed in terms of a single variable, many variables, or sets of nonlinear integro-differential equations. Regardless of its complexity, a model is an idealization of the system, so it cannot be exact: usually, the more complex the system, the less exact the model, particularly since the ability to solve exactly mathematically highly complex expressions diminishes with increasing complexity. In other words, the simpler the model, the easier it is to analyze but the less precise the results. A *parameter of a model* is a quantity that expresses a characteristic of the system; a parameter could be constant or variable. Observations of physical phenomena tend to fall somewhere between total unpredictability, at the one extreme, and statistically well-behaved processes, at the other extreme.

The combination of computational and experimental data from different sources involves a weighted propagation of parameter uncertainties, using the local functional derivatives of the system responses with respect to the input model parameters. These response derivatives are customarily called *response sensitivities* or simply *sensitivities*. At a fundamental level, the need for computing such sensitivities stems from the impossibility of measuring or computing any physical quantity with absolute precision, as has been discussed in the foregoing. Response sensitivities with respect to the model's parameters are also needed for the following purposes:

1. Determining the effects of parameter variations on the system's behavior.
2. Understanding the system by highlighting important data.
3. Eliminating unimportant data.
4. Prioritizing introduction of data uncertainties.
5. Quantification of response uncertainties due to parameter uncertainties.
6. Performing *forward* "predictive modeling," including data assimilation and model calibration, for the purpose of obtaining best-estimate predicted results with reduced predicted uncertainties.

7. Performing *inverse* "predictive modeling."
8. Improving the system design, possibly reducing conservatism, and redundancy.
9. Prioritizing possible improvements for the system under consideration.
10. Designing and optimizing the system (e.g., maximize availability/minimize maintenance).
11. Reducing overdesign.
12. Performing *code verification*, which addresses the question, "Are you solving the mathematical model correctly?"
13. Performing *code validation*, which uses comparisons between computations and experiments to addresses the question, "Does the model represent reality?" Code verification and validation can be performed only by taking into account all of the sensitivities and uncertainties that affect the respective computations and experiments.

In 2003, the author published a book entitled *Sensitivity and Uncertainty Analysis: Theory*, Volume 1 (Chapman & Hall/CRC), indicating that the adjoint sensitivity analysis procedure (ASAP) developed by the author in 1980–1981 is the most efficient deterministic method for computing exactly and exhaustively the local first-order response sensitivities to model parameters. This sensitivity analysis methodology requires a single large-scale computation for computing all of the first-order derivatives of a scalar-valued response with respect to all of the model's parameters. The remarkable efficiency of the ASAP stems from the fact that the adjoint sensitivity system is linear in the adjoint function and is independent of any parameter variations. The adjoint functions are used in simple quadrature of definite integrals to obtain the sensitivities to all system parameters, without needing to repeatedly solve large systems of differential and/or integral equations. Building upon this work, the author recently conceived the generally applicable *Second-Order Adjoint Sensitivity Analysis Methodology* (2nd-ASAM) for computing exactly and most efficiently *all* of the second-order functional derivatives of system responses to system parameters.

The theoretical underpinnings underlying the 2nd-ASAM are presented in this book, along with illustrative examples from various fields. These illustrative examples are sufficiently simple to admit exact solutions, most of them in closed form, in order to facilitate the understating of the theory underlying the 2nd-ASAM, yet they comprise a sufficient number of model parameters so as to make them representative of practical problems that could not be analyzed exactly by any other method of sensitivity analysis. These illustrative examples will be carried beyond just "sensitivity analysis" to include uncertainty quantification, model calibration and validation, data assimilation, and predictive modeling for reducing the uncertainties in predicted model responses and parameters. These illustrative paradigm problems underscore the fact that the 2nd-ASAM is especially useful for the computation of response sensitivities for realistic systems that are characterized by a very large number of parameters. The 2nd-ASAM applies to any model (deterministic, statistical, etc.); the larger the number of model parameters, the more efficient the 2nd-ASAM becomes for computing arbitrarily high-order response sensitivities. Computing efficiently and exactly functional derivatives is paramount to all activities that rely on models for predictions, including uncertainty quantification, model validation, optimization, data assimilation, model calibration, sensor fusion, reduced-order modeling, inverse problems, and predictive modeling.

The essential roles played by response sensitivities in the fields of uncertainty quantification and predictive modeling are briefly described in Chapter 1. In particular, the first- and second-order response sensitivities play the following roles in the quantification of uncertainties in the model responses due to uncertainties in the model parameters:

1. The first-order sensitivities provide the leading terms to the second-, third-, and fourth-order moments of the response distribution (responses variance and covariances, skewness, and kurtosis). As is well known, the expected values quantify the location of the respective distribution, while standard deviations provide a measure for the dispersion of the respective distribution around the mean values. The skewness of a distribution quantifies the departure of the subject distribution from symmetry. A distribution with a long right tail would have a positive skewness, while a distribution with a long left tail would have a negative skewness. The *kurtosis* quantifies, in a location- and scale-free sense, the dispersion of probability mass away from the shoulders, towards either the distribution's center or the tails or both, thus quantifying the degree of sharpness in the peaking of the subject distribution.
2. The second-order sensitivities contribute the leading correction terms to the response's expected value (causing it to differ from the response's computed value). The second-order sensitivities also contribute to the response variances and covariance; however, since these contributions are multiplied by the fourth power of the parameters' standard deviations, the total of these contributions is expected to be relatively smaller than the contributions stemming from the first-order response sensitivities. If the parameters follow a normal (Gaussian) multivariate distribution, the second-order sensitivities contribute the leading terms to the response's third-order moment. Thus, neglecting the second-order response sensitivities to normally distributed parameters would nullify the third-order response correlations and hence would nullify the *skewness* of a response.

Consequently, any events occurring in a response's long and/or short tails, which are characteristic of rare but decisive events (e.g., major accidents and catastrophes), would likely be missed if the second-order sensitivities of responses to normally distributed parameters were unavailable.

3. The illustrative examples presented in this book also underscore the fact that erroneous conclusions are likely to be drawn based on partial information, that is, conclusions based on only some, rather than all, sensitivities. This is because the sensitivities can be, as one would intuitively expect, of varying magnitudes, negative or positive, so cancellations and/or enhancement of contributions to the total uncertainties (covariance and variances, skewness, etc.) in the respective response distributions are likely to occur. It is therefore very important to compute *all*, rather than just some, response sensitivities. This argument underscores, once again, the practical importance and usefulness of the 2nd-ASAM, which is the only practical methodology for computing exactly all of the first- and second-order response sensitivities for physical systems involving many parameters.

The salient theoretical features of the 2nd-ASAM are anticipated in Chapter 2 by illustrating them using a simple evolution model that admits a closed-form analytical solution, so that all of the results are exact and easily tractable. On the other hand, this paradigm illustrative model comprises sufficiently many parameters (over a thousand), just as typically found in realistic physical systems, with very many of them having equally important relative sensitivities, to preclude the use of statistical and or other deterministic methods for computing exactly all of the first- and second-order response sensitivities. This paradigm model also applies the formulas introduced in Chapter 1 to illustrate the main applications of the first- and second-order response sensitivities for quantifying the uncertainties in model response, as well as for performing predictive modeling by combining computational with experimental results to reduce the uncertainties in predicted results.

As shown in Chapters 3 and 7, the 2nd-ASAM works in linearly increasing higher-dimensional Hilbert spaces, as opposed to exponentially increasing parameter-dimensional spaces. In particular, for one scalar-valued response associated with a nonlinear model comprising N_α parameters, the 2nd-ASAM requires at most N_α adjoint computations for computing *exactly* all of the first- and second-order sensitivities, as opposed to the $\left(N_\alpha^2/2+3N_\alpha/2\right)$ large-scale computations that would be required for obtaining all of the first- and second-order sensitivities by using forward deterministic or statistical methods based on recomputations with perturbed parameter values.

Chapter 4 presents an illustrative application of the 2nd-ASAM to a heat transport benchmark problem modeling the steady-state radial conduction in a heated rod coupled to the axial heat convection in a coolant surrounding the rod and flowing along it in a turbulent steady state. The development of this benchmark problem is motivated by the need to verify the numerical results produced by the commercially available thermal-hydraulics computational tools. This paradigm benchmark problem has six thermal-hydraulics parameters. Therefore, for each functional type response, such as the temperature at a specific location within the rod and/or on its surface, there are *6* first-order sensitivities and *21* second-order sensitivities. Furthermore, there are an additional *6* first-order sensitivities and an additional *21* second-order sensitivities also for the coolant temperature response. By applying the 2nd-ASAM, all of these first- and second-order sensitivities are obtained using a total of only *five independent adjoint computations*, would correspond to "large-scale computations," as follows:

1. *Two computations* using the first-level adjoint sensitivity system for obtaining the 18 first-order sensitivities, as follows: six first-order sensitivities of temperature distribution inside the heated rod, six first-order sensitivities of the rod's surface temperature, and six first-order sensitivities of the coolant temperature distribution
2. *Three computations* using the second-level adjoint sensitivity system for obtaining all 63 second-order sensitivities of the temperature distribution in the heated rod, on the rod's surface, and within the coolant temperature distribution.

Chapter 5 presents an illustrative application of the 2nd-ASAM developed in Chapter 3 to a paradigm neutron diffusion problem, in which all of the first-order relative response sensitivities to the model parameters have significantly large values, of order unity. Furthermore, most of the second-order relative detector sensitivities to cross sections turn out to be actually larger than the corresponding first-order relative sensitivities, contrary to previously held beliefs. Chapter 6 presents the application of the 2nd-ASAM to a problem of monoenergetic radiation transport through a heterogeneous non-multiplying three-dimensional medium that does not scatter the respective radiation more than once. Such radiation transport phenomena are modeled mathematically by a simplified form of the Boltzmann transport equation, customarily called the "ray-tracing" form, in which the seven-dimensional integro-differential Boltzmann transport equation is simplified to a partial differential equation in three spatial independent variables. Albeit much simplified, the ray-tracing form of the transport equation nevertheless comprises hundreds of model parameters (number densities, cross sections, and sources characterizing the medium), many having relative sensitivities of very similar magnitudes, and can only be solved numerically.

Chapter 7 presents the 2nd-ASAM for nonlinear systems. Just as has been the case for the 2nd-ASAM for linear systems (presented in Chapter 3), the 2nd-ASAM for nonlinear systems also works in linearly increasing higher-dimensional Hilbert spaces so that for one scalar-valued response associated with a nonlinear model comprising N_α parameters, the 2nd-ASAM requires at most N_α adjoint computations for computing *exactly* all of the first- and second-order sensitivities, as opposed to $\left(N_\alpha^2/2 + 3N_\alpha/2\right)$ large-scale computations that would be required for obtaining all of the first- and second-order sensitivities by using forward deterministic or statistical methods based on recomputations with perturbed parameter values. The fundamental difference between the 2nd-ASAM for nonlinear versus linear systems is that the adjoint sensitivity systems for linear systems are independent of the forward original state functions, whereas this is not true for nonlinear systems, for which the respective adjoint sensitivity systems depend on the forward original state functions. Of course, the results obtained in Chapter 7 for the 2nd-ASAM for nonlinear systems include, as particular cases, the results obtained in Chapter 3 for the 2nd-ASAM for linear systems.

Chapter 8 presents an illustrative application of the 2nd-ASAM for nonlinear systems formulated in Chapter 7 to a paradigm nonlinear heat conduction benchmark problem that simulates the steady state temperature distribution in a conceptual design of a test section comprising heated rods immersed in lead–bismuth eutectic. This benchmark problem comprises five model parameters and is sufficiently simple to admit an exact closed-form solution, thereby enabling the verification of numerical results provided by commercially available computer codes for modeling nonlinear heat conduction phenomena. For this illustrative nonlinear heat conduction benchmark, the computations of the first- and second-order sensitivities underscore that

1. Six adjoint-model computations suffice to compute all 5 of the first-order sensitivities and the 15 distinct second-order sensitivities.
2. The magnitudes of the second-order sensitivities are generally of the same order as the magnitudes of the first-order sensitivities.

Since real Hilbert spaces provide the natural mathematical setting for computational purposes, the 2nd-ASAM presented in this book is set in *real*, as opposed to complex, Hilbert spaces. This setting does not restrict, in any way, the generality of the concepts presented here, since the 2nd-ASAM can be readily set in complex Hilbert spaces by simply changing some terminology, without affecting its substance. The author plans to generalize the 2nd-ASAM to enable the efficient and exact computation of response sensitivities of arbitrarily high order, ultimately overcoming the so-called curse of dimensionality in the computation of functional derivatives, by performing the computations of these derivatives in linearly increasing higher-dimensional Hilbert spaces, as opposed to exponentially increasing parameter-dimensional spaces. In addition, the generalization also addresses the computation of sensitivities of *operator-valued responses,* which cannot be handled by current methods when the underlying problem comprises many parameters.

Dan Gabriel Cacuci
University of South Carolina

Acknowledgments

The material presented in this book has greatly benefited from the critical reviews provided by Dr. Mihaela Ionescu-Bujor, Nuclear Fusion Program at the Karlsruhe Institute of Technology, Germany, whose contributions are hereby gratefully acknowledged. Last but not least, the publication of this book would not have been possible without the kind understanding and guidance provided by Senior Editor Robert (Bob) Ross of CRC Press/Taylor & Francis Group.

Author

Professor Dan Gabriel Cacuci received his PhD degree in applied physics and nuclear engineering from Columbia University, New York, in 1978. Prior to his PhD, he had received Master of Science and Master of Philosophy degrees, also from Columbia University. Since 2012, Dr. Cacuci is holding an appointment as SmartState chaired professor and director of the Center for Nuclear Science and Energy at the University of South Carolina. His scientific expertise encompasses the following areas: predictive best-estimate analysis of large-scale physical and engineering systems, large-scale scientific computations, and nuclear engineering (reactor multi-physics, dynamics, and safety). Professor Cacuci's career encompasses both the academia and largescale multidisciplinary research centers. His research and management experience at leading national research centers includes employments as senior section head at Oak Ridge National Laboratory (1976–1988), institute director at the Nuclear Research Center Karlsruhe in Germany (1993–2004), and scientific director of the Nuclear Energy Directorate at the Commissariat a l'Energie Atomique in France (2004–2009).

Simultaneously, Professor Cacuci served as a tenured full professor at the following academic institutions: University of California at Santa Barbara (1988–1990), University of Illinois at Urbana-Champaign (1990–1993), the University of Karlsruhe (ordinarius chaired professor and department director, 1992–2009), and Karlsruhe Institute of Technology (2009-2012) in Germany. Professor Cacuci also held visiting or adjunct professorships at the following academic institutions: University of Virginia (1993–2000), University of Michigan (1995–2000), University of California at Berkeley (2001–2006), Royal Institute of Technology Stockholm (2003), the French National Institute for Nuclear Sciences and Technologies in Paris (2006–2009), and Imperial College of London (2013-2016).

Since 1986, Professor Cacuci has been the editor of *Nuclear Science and Engineering; The Research Journal of the American Nuclear Society* (ANS). He is a member of the European Academy of Arts and Sciences (2004), honorary member of the Romanian Academy (1994), and ANS Fellow (1986), and he has received many prestigious awards, including four titles of Doctor Honoris Causa, the E. O. Lawrence Award and Gold Medal (US DOE, 1998), the Eugene P. Wigner Reactor Physics Award (ANS, 2003), the Glenn Seaborg Medal (ANS, 2002), and the Alexander von Humboldt Prize for Senior Scholars (Germany, 1990). Professor Cacuci has served on numerous international committees, and was the founding coordinator of the EURATOM-Integrated Project NURESIM (European Platform for Nuclear

Reactor Simulation, 2004–2008), and the Coordinated Action for establishing a Sustainable Nuclear Fission Technology Platform (SNF-TP) in Europe (2004–2007). He has given over 600 presentations worldwide, authored 5 books, 7 book chapters, 250 peer-reviewed articles, and he has edited the comprehensive *Handbook of Nuclear Engineering* (5 volumes, 29 chapters, 3,580 pages, Springer, 2010).

1

Motivation for Computing First- and Second-Order Sensitivities of System Responses to the System's Parameters

Experience shows that it is practically impossible to measure exactly the true value of a physical quantity. This is because of various imperfections that occur at various stages involved in a measurement, including uncontrollable experimental errors, inaccurate standards, and other uncertainties arising in the data measurement and interpretation (reduction) process. Therefore, around any reported experimental value, there always exist a certain ranges of plausible values that may also be true. A "measurable" or "physical" quantity is a property of phenomena, bodies, or substances that can be defined qualitatively and can be expressed quantitatively. *Measurement* is the process of finding experimentally the value of a physical quantity, with the help of devices called *measuring instruments*. A measurement has three features:

1. The result of a measurement must always be a number expressed in sanctioned units of measurements. The purpose of a measurement is to represent a property of an object by a number.
2. A measurement is always performed with the help of a measuring instrument; measurement is impossible without measuring instruments.
3. A measurement is always an experimental procedure.

The true value of a measurable quantity would ideally reflect, both qualitatively and quantitatively, the corresponding property of the object. The theory of measurement relies on the following postulates:

1. The true value of the measurable quantity exists.
2. The true value of the measurable quantity is constant relative to the conditions of the measurement.
3. The true value cannot be found.

Since measuring instruments are imperfect, and since every measurement is an experimental procedure, the results of measurements cannot be absolutely accurate. This unavoidable imperfection of measurements is characterized quantitatively by *measurement uncertainty* or *measurement error*, which can be expressed in absolute or relative form. Consequently, repeated measurements of the same physical quantity can never yield identical results. Legitimate errors, as opposed to blunders, are categorized either as *random errors* or as *systematic errors*. If the results of separate measurements of the same quantity differ from one another, and the respective differences cannot be predicted individually, then the error owing to this scatter of the results is called *random error*. Random errors can be identified by repeatedly measuring the same quantity under the same conditions. The scatter in results cannot always be tested in practice, particularly in large-scale modern experiments, where it may be impractical to provide sufficient repetition in order to satisfy the explicit needs for quantifying the random errors based on strict statistical requirements. Nevertheless, reasonable estimates of random errors can often be made, particularly when the nature of the underlying probability distribution can be inferred from previous experience. In contradistinction to a random error, a *systematic error* is defined as a measurement error that remains constant or changes in a regular fashion when the measurements of that quantity are repeated. Such errors arise because of inherent flaws in the investigative process itself, and they lead to bias. Although systematic errors are difficult to distinguish from blunders, particularly when the impact of a blunder is small, the most reliable way to uncover systematic errors is by using a more accurate measuring instrument and/or by comparing a given result with a measurement of the same quantity but performed by a different method. Each distinct approach leads to results that differ somewhat from those obtained in other ways. These differences exhibit a pattern (i.e., are "systematic") no matter how many times each approach is repeated, because the inherent systematic deficiencies of each method cannot be avoided by mere repetition. When the errors are truly systematic, statistical regularity will emerge from the ensemble of all measurements. Such a statistical regularity will not emerge when the data sets are afflicted with blunders, since blunders are generally one-time occurrences that can be detected if a particular procedure is repeated. Consequently, redundancy within a given investigative procedure is desirable not only to improve precision but also to purge the results of blunders.

The quality of measurements that reflects the closeness of the results of measurements of the same quantity performed under the same conditions is called the *repeatability* of measurements. Good repeatability indicates that the random errors are small. On the other hand, the quality of measurements that reflects the closeness of the results of measurements of the same quantity performed under different conditions (e.g., in different laboratories, at different locations, and/or using different equipment) is called the *reproducibility* of measurements. Good reproducibility indicates that both the random and systematic errors are small.

Since the true value of the measured quantity is unknown, a measurement error cannot be found by using its definition as an algorithm. The measurement error must be evaluated by identifying its underlying sources and reasons and by performing computations based on the estimates of all components of the respective measurement inaccuracy. The smallest of the measurement errors are customarily referred to as *elementary errors* (of a measurement) and are defined as those components of the overall measurement error that are associated with a single source of inaccuracy for the respective measurement. In turn, the total measurement error is computed by using the estimates of the component elementary errors. Even though it is sometimes possible to correct, partially, certain elementary errors (e.g., systematic ones), no amount or combination of corrections can produce an accurate measurement result; there will always remain a residual error. In particular, the corrections themselves cannot be absolutely accurate, and, even after they are implemented, there remain residuals of the corresponding errors, which cannot be eliminated and which later assume the role of elementary errors.

Since a measurement error can only be calculated indirectly, based on models and experimental data, it is important to identify and classify the underlying elementary errors. This identification and classification is subsequently used to develop mathematical models for the respective elementary errors. Finally, the resulting *overall measurement error* is obtained by synthesizing the mathematical models of the underlying elementary errors.

The term *uncertainty* is customarily used to express the inaccuracy of measurement results when the numerical value of the respective inaccuracy is accompanied by a corresponding confidence probability. The term *error* is customarily used for all components of uncertainty, while the term *limits of error* is used in cases in which the measurement inaccuracy is caused by the intrinsic error of the measuring instrument and when a corresponding level of confidence cannot be stated. Measurements must be reproducible, since otherwise they lose their objective character and become meaningless. The *limits of measurement error or uncertainty* estimated by the experimenter provide a measure of the nonreproducibility of a measurement permitted by the experimentalist. The validity of the uncertainty computed for every measurement is based on the validity of the estimates of errors underlying the respective measurement. A correctly estimated measurement uncertainty permits comparisons of the respective result with results obtained by other experimenters. In many cases, it is of interest to know not only the limiting values of the total measurement error but also the characteristics of the random and systematic error components separately, in order to analyze discrepancies between results of measurements of the same quantity performed under different conditions. Knowing the error components separately is particularly important when the result of a measurement is to be used for computations together with other data that are not precise. Furthermore, the main sources of errors should also be described together with estimates of their contributions to the total measurement uncertainty. For a random error, for example, it is of interest to indicate the form and parameters of the distribution function underlying the observations, the method employed for testing the hypothesis regarding the form of the distribution function, the significance level used in the testing, and so on.

Uncertainties are broadly grouped into two categories: (1) *stochastic uncertainties*, which arise because the system under investigation can behave in many different ways, and (2) *subjective* or *epistemic uncertainties*, which arise from the inability to specify an exact value for some model parameter that is assumed to have a constant value in the respective investigation. In the risk analysis of a nuclear power reactor plant, for example, stochastic uncertainties arise due to the several hypothetical accident scenarios that are considered in the risk analysis of the respective plant, while epistemic uncertainties arise because of the many uncertain parameters that underlie the estimation of the probabilities and consequences of each of the respective hypothetical accident scenarios.

The methodology that will be presented in this book uses certain generic terms for defining the objects or phenomena under study. These generic terms will be briefly described next. Thus, a *system* is the object or phenomena under study. It represents the largest unit being considered. A system can refer to a nuclear reactor, corporation, chemical process, mechanical device, biological mechanism, society, economy, or any other conceivable object that is under study. The *output* or *response* of a system is a result that can be quantitatively measured and/or computed. The power of a nuclear reactor, the yield of a process, life span of cell, and the atmospheric temperature and pressure are all examples of system outputs. A *model* is a mathematical idealization that is used as an approximation to represent the system and its output. Models can be quite simple or highly complex; they can be expressed in terms of a single variable, many variables, or sets of nonlinear integro-differential equations. Regardless of its complexity, the model is an idealization of the system, so it cannot be exact: usually, the more complex the system, the less exact the model, particularly since the ability to solve exactly mathematically highly complex expressions diminishes with increasing complexity. In other words, the simpler the model, the easier it is to analyze but the less precise the results.

A *statistical model* comprises mathematical formulations that express the various outputs of a system in terms of probabilities. A statistical model is usually used when the system's output cannot be expressed as a fixed function of the input variables. Statistical models are particularly useful for representing the behavior of a system based on a limited number of measurements and for summarizing and/or analyzing a set of data obtained experimentally or numerically. For example, there exist families of probability distributions, covering a wide range of shapes, which can be used for representing experimentally obtained data sets. The resulting statistical model can then be used to perform extrapolations or interpolations, for determining the probability of occurrence of some event in some specific interval of values, and so on. Statistical models should be constructed based on the physical properties of the system. When the physical properties and, hence, the principles of operation of the system are well understood, the model will be derived from these underlying principles. Most often, though, the basic physical properties of the system under consideration are known incompletely. In simple cases, and when sufficient data are available, it is possible to select a general yet flexible family of statistical models and fit one member of this family to the observations. The validity of such fits is then verified by performing a *sensitivity analysis* to determine the sensitivity of the system's output to the selection of a specific model to represent the system. In situations where the physical properties of the system under investigation are poorly understood but measurements are nevertheless available, the model for the system is selected on a trial-and-error basis, usually based not only on objective criteria but also on subjective expert opinion. Finally, when no data are available, it is possible to study the conceptual model on a computer to generate numerically synthetic observations. The previously mentioned statistical methods can then be applied to analyze the synthetic observations generated numerically, just as if they had been physical observations.

The group to be measured or counted, or the conceptual entity for which predictions are to be made, is called *population*. A *parameter of a model* is a quantity that expresses a characteristic of the system; a parameter could be constant or variable. A *sample* is a subset of the population selected for study. An *experiment* is a sequence of a limited (or, occasionally, unlimited) number of trials. An *event* is an outcome of an experiment. The set of all events (i.e., the set that represents all outcomes of an experiment) is called the *event space* or *sample space* of the experiment; the respective events are also referred to as *sample points*.

When a process operates on an event space such that the individual outcomes of the observations cannot be controlled, the respective process is called a *random process* and the respective events are called *random events*. Although the exact outcome of any single *random trial* is unpredictable, a sample of random trials of reasonable size is expected to yield a *pattern of outcomes*. In other words, *randomness implies lack of deterministic regularity but existence of statistical regularity*. Thus, random phenomena must be distinguished from *completely unpredictable* phenomena, for which no pattern can be constructed, even for exceedingly large samples. Observations of physical phenomena tend to fall somewhere between total unpredictability, at the one extreme, and statistically well-behaved processes, at the other extreme. For this reason, experimentalists must invest considerable effort to identify and eliminate, as much as possible, statistically unpredictable effects (e.g., malfunctioning equipment), in order to perform experiments under controlled conditions conducive to generating statistically meaningful results.

The combination of data from different sources involves a weighted propagation of parameter uncertainties, using the local functional derivatives of the system responses with respect to the input model parameters. These response derivatives are customarily called *response sensitivities* or simply *sensitivities*. At a fundamental level, the need for computing such sensitivities stems from the impossibility of measuring or computing any physical quantity with absolute precision, as has been discussed in the foregoing.

Response sensitivities with respect to the model's parameters are also needed for the following purposes:

1. Determining the effects of parameter variations on the system's behavior.
2. Understanding the system by highlighting important data.
3. Eliminating unimportant data.
4. Prioritizing introduction of data uncertainties.
5. Quantification of response uncertainties due to parameter uncertainties.
6. Performing *forward* "predictive modeling," including data assimilation and model calibration, for the purpose of obtaining best-estimate predicted results with reduced predicted uncertainties.
7. Performing *inverse* "predictive modeling."
8. Improving the system design, possibly reducing conservatism and redundancy.
9. Prioritizing possible improvements for the system under consideration.
10. Designing and optimizing the system (e.g., maximize availability/minimize maintenance).
11. Reducing overdesign.
12. Performing "code verification," which addresses the question "are you solving the mathematical model correctly?"
13. Performing "code validation," which uses comparisons between computations and experiments to addresses the question "does the model represent reality?" Code verification and validation can be performed only by taking into account all of the sensitivities and uncertainties that affect the respective computations and experiments.

1.1 The Fundamental Role of Response Sensitivities for Uncertainty Quantification

The system (or model) parameters will be denoted as $\alpha_i(\mathbf{x})$, $i = 1, \ldots, N_\alpha$, and will be considered, for convenience, to be components of the column vector $\boldsymbol{\alpha}(\mathbf{x}) = \left[\alpha_1(\mathbf{x}),\ldots,\alpha_{N_\alpha}(\mathbf{x})\right]$, defined in an N_α-dimensional normed linear space denoted as $\mathcal{E}_\alpha$, that is, $\boldsymbol{\alpha} \in \mathcal{E}_\alpha$. Since they are affected by uncertainties, the components of $\boldsymbol{\alpha}$ are considered to be variates following a multivariate probability distribution, denoted as $p(\boldsymbol{\alpha})$, which is seldom known in practice.

Considering that $p(\boldsymbol{\alpha})$ is defined on a domain D_α, the various moments (e.g., mean values and covariance and variances) of the distribution $p(\boldsymbol{\alpha})$ are formally defined as follows:

1. The *expected* (or mean) value of a model parameter α_i, denoted as α_i^0, defined as

$$\alpha_i^0 \triangleq \int_{D_\alpha} \alpha_i p(\boldsymbol{\alpha}) d\boldsymbol{\alpha} \, ; \tag{1.1}$$

2. The *covariance*, $\text{cov}(\alpha_i,\alpha_j)$, of two parameters, α_i and α_j, defined as

$$\mu_2^{ij}(\boldsymbol{\alpha}) \triangleq \text{cov}(\alpha_i,\alpha_j) \triangleq \int_{D_\alpha} \left(\alpha_i - \alpha_i^0\right)\left(\alpha_j - \alpha_j^0\right) p(\boldsymbol{\alpha}) d\boldsymbol{\alpha}, \quad i, j = 1, \ldots, N_\alpha \, ; \tag{1.2}$$

3. The *variance*, $\text{var}(\alpha_i)$, of a parameter α_i, defined as

$$\text{var}(\alpha_i) \triangleq \int_{D_\alpha} \left(\alpha_i - \alpha_i^0\right)^2 p(\boldsymbol{\alpha}) d\boldsymbol{\alpha}, \quad i = 1, \ldots, N_\alpha \, ; \tag{1.3}$$

4. The *standard deviation*, σ_i, of α_i, defined as $\sigma_i \triangleq \sqrt{\text{var}(\alpha_i)}$;
5. The correlation, ρ_{ij}, between two parameters α_i and α_j, defined as

$$\rho_{ij} \triangleq \text{cov}(\alpha_i,\alpha_j) / \left(\sigma_i \sigma_j\right); \quad i, j = 1, \ldots, N_\alpha. \tag{1.4}$$

The covariances $\mu_2^{ij}(\boldsymbol{\alpha})$ constitute the elements of a symmetric, positive-definite parameter covariance matrix, denoted here as $\mathbf{C}_{\alpha\alpha}$, of dimension $N_\alpha \times N_\alpha$. Since the elements of $\mathbf{C}_{\alpha\alpha}$ are obtained, in practice, by a variety of experimental, empirical, and/or theoretical methods, it is imperative to ensure, at the outset, that the covariance matrix $\mathbf{C}_{\alpha\alpha}$ fulfills the physical and mathematical properties required of a covariance matrix. Thus, $\mathbf{C}_{\alpha\alpha}$ must be positive definite and the *correlation coefficients* ρ_{ij} must satisfy the Cauchy–Schwartz inequality

$$-1 \le \rho_{ij} \le 1. \tag{1.5}$$

As is well known, *the trace, the determinant, and the principal minors of a positive definite matrix are positive* so that these properties must be satisfied, in particular, by $\mathbf{C}_{\alpha\alpha}$. As is also well known, an equivalent necessary and sufficient condition for $\mathbf{C}_{\alpha\alpha}$ to be positive definite is that it admits the *Cholesky decomposition*

$$\mathbf{C}_{\alpha\alpha} = \mathbf{L}_{\alpha\alpha}\left(\mathbf{L}_{\alpha\alpha}\right)^\dagger = \left(\mathbf{U}_{\alpha\alpha}\right)^\dagger \mathbf{U}_{\alpha\alpha}, \tag{1.6}$$

where $\mathbf{L}_{\alpha\alpha}$ (respectively, $\mathbf{U}_{\alpha\alpha}$) is a nonsingular lower (respectively, upper) triangular matrix with positive entries on its diagonal, and where the dagger (†) will be used in this book to denote *transposition*. If the parameter covariance matrix $\mathbf{C}_{\alpha\alpha}$ is nearly singular, the Cholesky decomposition usually fails, in which case it is essential to compute the singular value decomposition

$$\mathbf{C}_{\alpha\alpha} = \mathbf{A}_{\alpha\alpha}\,\Sigma_{\alpha\alpha}\left(\mathbf{B}_{\alpha\alpha}\right)^{\dagger}, \tag{1.7}$$

where $\mathbf{A}_{\alpha\alpha}$ and $\mathbf{B}_{\alpha\alpha}$ are unitary matrices, and the diagonal matrix $\Sigma_{\alpha\alpha}$ contains the nonsingular eigenvalues (usually in decreasing magnitude values) of $\mathbf{C}_{\alpha\alpha}$. The numerical rank of $\mathbf{C}_{\alpha\alpha}$ must be known to enable the removal of spurious colinearities, thereby ensuring that $\mathbf{C}_{\alpha\alpha}$ is a bona fide positive definite matrix that makes physical and mathematical sense for subsequent processing.

6. The third-order moment, μ_3^{ijk}, of the multivariate parameter distribution function $p(\boldsymbol{\alpha})$, and the third-order parameter correlation, t_{ijk}, respectively, defined as follows:

$$\begin{aligned}\mu_3^{ijk}(\boldsymbol{\alpha}) &\triangleq \int_{D_\alpha}\left(\alpha_i-\alpha_i^0\right)\left(\alpha_j-\alpha_j^0\right)\left(\alpha_k-\alpha_k^0\right)p(\boldsymbol{\alpha})\,d\boldsymbol{\alpha}\\ &\triangleq t_{ijk}\sigma_i\sigma_j\sigma_k, \quad i,j,k=1,\ldots,N_\alpha;\end{aligned} \tag{1.8}$$

7. The fourth-order moment, μ_4^{ijkl}, of the multivariate parameter distribution function $p(\boldsymbol{\alpha})$, and the fourth-order parameter correlation, q_{ijkl}, respectively, defined as follows:

$$\begin{aligned}\mu_4^{ijkl}(\boldsymbol{\alpha}) &\triangleq \int_{D_\alpha}\left(\alpha_i-\alpha_i^0\right)\left(\alpha_j-\alpha_j^0\right)\left(\alpha_k-\alpha_k^0\right)\left(\alpha_l-\alpha_l^0\right)p(\boldsymbol{\alpha})\,d\boldsymbol{\alpha}\\ &\triangleq q_{ijkl}\sigma_i\sigma_j\sigma_k\sigma_l, \quad i,j,k,l=1,\ldots,N_\alpha.\end{aligned} \tag{1.9}$$

The notations used in Equations 1.1 through 1.9 are not universal. For univariate distributions, the third and fourth central moments $\mu_3(\alpha_i) \equiv \left\langle\left(\alpha_i-\alpha_i^0\right)^3\right\rangle$ and $\mu_4(\alpha_i) \equiv \left\langle\left(\alpha_i-\alpha_i^0\right)^4\right\rangle$ are related, respectively, to the so-called skewness, $\gamma_1(\alpha_i)$, and kurtosis, $\beta_2(\alpha_i)$, or excess kurtosis, $\gamma_2(\alpha_i)$, which are defined as follows:

$$\gamma_1(\alpha_i) \triangleq \frac{\mu_3(\alpha_i)}{\sigma_i^3} \triangleq \sqrt{\beta_1(\alpha_i)}; \tag{1.10}$$

$$\beta_2(\alpha_i) \triangleq \frac{\mu_4(\alpha_i)}{\sigma_i^4}; \quad \gamma_2(\alpha_i) \triangleq \beta_2(\alpha_i) - 3. \tag{1.11}$$

For a univariate distribution of a variate α_i, the expected (or mean) value, α_i^0, is a measure of the *location* of the respective distribution, while the standard deviation, σ_i, provides a measure for the *dispersion* of the respective distribution around the mean value. In the same sense, the mean value can be interpreted as the distribution's center of gravity, while the variance can be interpreted as the distribution's moment of inertia (which in mechanics linearly relates the applied torque to the induced acceleration). The *skewness* of a distribution quantifies the departure of the subject distribution from symmetry. Symmetric univariate distributions are characterized by $\gamma_1(\alpha_i) = 0 = \beta_1(\alpha_i)$ A distribution with a long right tail would have a positive skewness, while a distribution with a long left tail would have a negative skewness. Thus, if $\gamma_1(\alpha_i) < 0$, the respective distribution is skewed toward the left of the mean α_i^0, favoring lower values of α_i relative to α_i^0. On the contrary, if $\gamma_1(\alpha_i) < 0$, the respective distribution is skewed toward the right of the mean α_i^0, favoring higher values of α_i relative to α_i^0. The *kurtosis*, $\beta_2(\alpha_i)$, is always nonnegative and quantifies the degree of sharpness in the peaking of the subject distribution. For univariate distributions, the standard for comparison of kurtosis is the Gaussian distribution, for which $\beta_2(\alpha_i) = 3$. Distributions characterized by $\beta_2(\alpha_i) < 3$ are called *platykurtic* distributions, indicating that they are generally more "flat-topped" compared with the Gaussian. Distributions characterized by $\beta_2(\alpha_i) = 3$ are called *mesokurtic* distributions, indicating that their general shape is comparable to the Gaussian, while those characterized by $\beta_2(\alpha_i) = 3$ are called *leptokurtic* distributions, indicating that they are more sharply peaked than the Gaussian. It is also well known that for any distribution with mean α_i^0 and standard deviation σ_i, the kurtosis can be written in the form $\beta_2(\alpha_i) = \text{var}\left\{\left[\left(\alpha_i - \alpha_i^0\right)/\sigma_i\right]^2\right\} + 1$, thus measuring the dispersion of α_i about the points $\left(\alpha_i^0 \pm \sigma_i\right)$, which can be interpreted as the "shoulders" of the distribution. Kurtosis thus quantifies, in a location- and scale-free sense, the dispersion of probability mass away from the shoulders, toward either the distribution's center or the tails, or both. In the important particular case when the parameters $\boldsymbol{\alpha}$ obey a multivariate Gaussian distribution, the following relations hold:

$$t_{ij\mu} \equiv 0; \quad q_{ij\mu\nu} = \rho_{ij}\rho_{\mu\nu} + \rho_{i\mu}\rho_{j\nu} + \rho_{i\nu}\rho_{j\mu}; \ \textit{for all } i, j, \mu, \nu = 1, \ldots, N_\alpha.$$

Since the system responses, $R_k(\boldsymbol{\alpha})$, $k = 1, \ldots, N_r$, depend on the system's parameters, they can also be considered as variates that follow a multivariate distribution $p(\mathbf{R})$, $\mathbf{R} \triangleq \left[R_1(\boldsymbol{\alpha}), \ldots, R_k(\boldsymbol{\alpha}), \ldots, R_{N_r}(\boldsymbol{\alpha})\right]$ defined on a domain D_r. In general, this distribution is intractable. By definition, the first four moments of distribution $p(\mathbf{R})$ are as follows:

1. The expected value, denoted as $E(R_k)$, of a response $R_k(\boldsymbol{\alpha})$, which is defined below

$$E(R_k) \triangleq \int_{D_r} R_k(\boldsymbol{\alpha}) p(\mathbf{R}) d\mathbf{R}, \quad k = 1, \ldots, N_r; \tag{1.12}$$

2. The *covariance*, $\text{cov}(R_k,R_l)$, of two responses, $R_k(\boldsymbol{\alpha})$ and $R_l(\boldsymbol{\alpha})$, which is defined below

$$\text{cov}(R_k,R_l) \triangleq \int_{D_r} [R_k - E(R_k)][R_l - E(R_l)]p(\mathbf{R})d\mathbf{R}, \quad k,l=1,\ldots,N_r; \tag{1.13}$$

3. The *variance*, $\text{var}(R_k)$, of a response R_k, which is defined below

$$\text{var}(R_k) \triangleq \int_{D_r} [R_k - E(R_k)]^2 p(\mathbf{R})d\mathbf{R}, \quad k=1,\ldots,N_r; \tag{1.14}$$

4. The third-order moment (triple correlation), $\mu_3(R_k, R_l, R_m)$, of three responses, R_k, R_l, and R_m, which is defined below

$$\mu_3(R_k,\, R_l,\, R_m) \triangleq \int_{D_r} [R_k - E(R_k)][R_l - E(R_l)][R_m - E(R_m)]p(\mathbf{R})d\mathbf{R}, \quad k,l,m=1,\ldots,N_r; \tag{1.15}$$

5. The fourth-order moment (quadruple correlation), $\mu_4(R_k, R_l, R_m, R_n)$, of four responses, R_k, R_l, R_m, and R_n, which is defined below

$$\mu_4(R_k,R_l,R_m,R_n) \triangleq \int_{D_r} [R_k - E(R_k)][R_l - E(R_l)][R_m - E(R_m)][R_n - E(R_n)]p(\mathbf{R})d\mathbf{R}, \quad k,l,m,n=1,\ldots,N_r. \tag{1.16}$$

The role of the response sensitivities in propagating the parameter uncertainties (i.e., variances and higher-order moments) to compute the uncertainties (i.e., covariances and higher-order moments) characterizing the response distribution can be readily illustrated by recalling the well-known "propagation of errors" or "propagation of moments" method. This method relies on expanding formally the response $R_k(\boldsymbol{\alpha})$ in a Taylor series around the mean parameter values $\boldsymbol{\alpha}^0$, constructing appropriate products of such Taylor series, and integrating formally these products over the unknown parameter distribution function $p(\boldsymbol{\alpha})$, to obtain the moments of the multivariate response distribution. Assuming, for illustrative purposes, that only the first- and second-order response sensitivities with respect to parameters are available, the Taylor series of a response $R_k(\boldsymbol{\alpha})$ around the mean values $\boldsymbol{\alpha}^0 \equiv \left(\alpha_1^0,\ldots,\alpha_{N_\alpha}^0\right)$ is

$$\begin{aligned} R_k(\alpha) = R_k(\alpha^0) &+ \sum_{i=1}^{N_\alpha} \left\{\frac{\partial R_k}{\partial \alpha_i}\right\}_{\alpha^0} \left(\alpha_i - \alpha_i^0\right) \\ &+ \frac{1}{2}\sum_{i,j=1}^{N_\alpha} \left\{\frac{\partial^2 R_k}{\partial \alpha_i \partial \alpha_j}\right\}_{\alpha^0} \left(\alpha_i - \alpha_i^0\right)\left(\alpha_j - \alpha_j^0\right), \end{aligned} \tag{1.17}$$

where $R_k(\boldsymbol{\alpha}^0)$ denotes the computed nominal value of the response. Assuming that just the first four moments of the parameter distribution (cf. Equations 1.1 through 1.9) are available, using Equation 1.17 in the definitions provided in Equations 1.12 through 1.16 and using the definitions provided in Equations 1.1 through 1.9 yields the following expressions for the first four moments of the multivariate response distribution:

$$E(r_k) = R_k(\boldsymbol{\alpha}^0) + \frac{1}{2}\sum_{i,j=1}^{N_\alpha}\left\{\frac{\partial^2 R_k}{\partial\alpha_i\,\partial\alpha_j}\right\}_{\boldsymbol{\alpha}^0}\mathrm{cov}(\alpha_i,\alpha_j); \tag{1.18}$$

$$\begin{aligned}
cov(R_k,R_\ell) = {} & \sum_{i=1}^{N_\alpha}\sum_{j=1}^{N_\alpha}\left(\frac{\partial R_k}{\partial\alpha_i}\frac{\partial R_\ell}{\partial\alpha_j}\right)\rho_{ij}\sigma_i\sigma_j \\
& + \frac{1}{2}\sum_{i=1}^{N_\alpha}\sum_{j=1}^{N_\alpha}\sum_{\mu=1}^{N_\alpha}\left(\frac{\partial^2 R_k}{\partial\alpha_i\partial\alpha_j}\frac{\partial R_\ell}{\partial\alpha_\mu} + \frac{\partial R_k}{\partial\alpha_i}\frac{\partial^2 R_l}{\partial\alpha_j\partial\alpha_\mu}\right)t_{ij\mu}\sigma_i\sigma_j\sigma_\mu \\
& + \frac{1}{4}\sum_{i=1}^{N_\alpha}\sum_{j=1}^{N_\alpha}\sum_{\mu=1}^{N_\alpha}\sum_{\nu=1}^{N_\alpha}\left(\frac{\partial^2 R_k}{\partial\alpha_i\partial\alpha_j}\right)\left(\frac{\partial^2 R_\ell}{\partial\alpha_\mu\partial\alpha_\nu}\right)\left(q_{ij\mu\nu} - \rho_{ij}\rho_{\mu\nu}\right)\sigma_i\sigma_j\sigma_\mu\sigma_\nu \\
& + \frac{1}{6}\sum_{i=1}^{N_\alpha}\sum_{j=1}^{N_\alpha}\sum_{\mu=1}^{N_\alpha}\sum_{\nu=1}^{N_\alpha}\left(\frac{\partial R_k}{\partial\alpha_i}\frac{\partial^3 R_l}{\partial\alpha_j\partial\alpha_\mu\partial\alpha_\nu} + \frac{\partial R_l}{\partial\alpha_i}\frac{\partial^3 R_k}{\partial\alpha_j\partial\alpha_\mu\partial\alpha_\nu}\right)q_{ij\mu\nu}\sigma_i\sigma_j\sigma_\mu\sigma_\nu;
\end{aligned} \tag{1.19}$$

$$\begin{aligned}
\mu_3(R_k,R_l,R_m) = {} & \sum_{i=1}^{N_\alpha}\sum_{j=1}^{N_\alpha}\sum_{\mu=1}^{N_\alpha}\frac{\partial R_k}{\partial\alpha_i}\frac{\partial R_l}{\partial\alpha_j}\frac{\partial R_m}{\partial\alpha_\mu}t_{ij\mu}\sigma_i\sigma_j\sigma_\mu \\
& + \frac{1}{2}\sum_{i=1}^{N_\alpha}\sum_{j=1}^{N_\alpha}\sum_{\mu=1}^{N_\alpha}\sum_{\nu=1}^{N_\alpha}\frac{\partial R_k}{\partial\alpha_i}\frac{\partial R_l}{\partial\alpha_j}\frac{\partial^2 R_m}{\partial\alpha_\mu\partial\alpha_\nu}\left(q_{ij\mu\nu} - \rho_{ij}\rho_{\mu\nu}\right)\sigma_i\sigma_j\sigma_\mu\sigma_\nu \\
& + \frac{1}{2}\sum_{i=1}^{N_\alpha}\sum_{j=1}^{N_\alpha}\sum_{\mu=1}^{N_\alpha}\sum_{\nu=1}^{N_\alpha}\frac{\partial R_k}{\partial\alpha_i}\frac{\partial^2 R_l}{\partial\alpha_j\partial\alpha_\mu}\frac{\partial R_m}{\partial\alpha_\nu}\left(q_{ij\mu\nu} - \rho_{i\nu}\rho_{j\mu}\right)\sigma_i\sigma_j\sigma_\mu\sigma_\nu \\
& + \frac{1}{2}\sum_{i=1}^{N_\alpha}\sum_{j=1}^{N_\alpha}\sum_{\mu=1}^{N_\alpha}\sum_{\nu=1}^{N_\alpha}\frac{\partial^2 R_k}{\partial\alpha_i\partial\alpha_j}\frac{\partial R_l}{\partial\alpha_\mu}\frac{\partial R_m}{\partial\alpha_\nu}\left(q_{ij\mu\nu} - \rho_{ij}\rho_{\mu\nu}\right)\sigma_i\sigma_j\sigma_\mu\sigma_\nu;
\end{aligned} \tag{1.20}$$

$$\mu_4(R_k,R_l,R_m,R_n) = \sum_{i=1}^{N_\alpha}\sum_{j=1}^{N_\alpha}\sum_{\mu=1}^{N_\alpha}\sum_{\nu=1}^{N_\alpha}\frac{\partial R_k}{\partial\alpha_i}\frac{\partial R_l}{\partial\alpha_j}\frac{\partial R_m}{\partial\alpha_\mu}\frac{\partial R_n}{\partial\alpha_\nu}\mu_4^{ij\mu\nu}(\boldsymbol{\alpha}). \tag{1.21}$$

Examining Equations 1.18 through 1.21 reveals that the first- and second-order response sensitivities play the following roles in the quantification of uncertainties in the model responses due to uncertainties in the model parameters:

1. The first-order sensitivities contribute the leading terms to the second-, third-, and fourth-order moments of the response distribution, thus providing the leading contributions to the response's variance/covariances, skewness, and kurtosis; clearly, obtaining the exact and complete set of first-order sensitivities of responses to model parameters is of paramount importance for any analysis of a computational model.
2. The second-order sensitivities contribute the leading correction terms to the response's expected value (causing it to differ from the response's computed value). The second-order sensitivities also contribute to the response variances and covariance; however, since these contributions are multiplied by the fourth power of the parameters' standard deviations, the total of these contributions is expected to be relatively smaller than the contributions stemming from the first-order response sensitivities. If the parameters follow a normal (Gaussian) multivariate distribution, the second-order sensitivities contribute the leading terms to the response's third-order moment. Thus, neglecting the second-order response sensitivities to normally distributed parameters would nullify the third-order response correlations and hence would nullify the *skewness* of a response. Consequently, any events occurring in a response's long and/or short tails, which are characteristic of rare but decisive events (e.g., major accidents and catastrophes), would likely be missed if the second-order sensitivities of responses to normally distributed parameters were unavailable.

1.2 The Fundamental Role of Response Sensitivities for Predictive Modeling

Cacuci and Ionescu-Bujor (2010) published a comprehensive *predictive modeling* methodology for large-scale nonlinear time-dependent systems that generalizes and significantly extends the "4D-VAR data assimilation" procedures in the geophysical sciences (see, e.g., Lahoz et al. 2010). The fundamental features of this predictive modeling methodology are as follows:

1. It dispenses with the need for minimizing a cost functional and generalizes the state-of-the-art "data assimilation" methodologies by using the maximum entropy principle to combine all available experimental and computational information to calibrate simultaneously

all uncertain quantities, including model parameters, initial conditions, boundary conditions, and observed responses.

2. It provides explicit formulas for the calibrated best estimate predicted values for these model responses and parameters.
3. It reduces the predicted uncertainties in these predicted model responses and parameters, providing explicit formulas for the predicted covariance matrices of responses and parameters.
4. It provides a quantitative indicator, constructed from parameter and response covariance and sensitivity matrices, for quantifying in the corresponding metric the consistency (agreement or disagreement) among the *a priori* computational and experimental data.

The time-dependent generic physical system considered within the predictive modeling methodology developed by Cacuci and Ionescu-Bujor (2010) comprises N_α^ν model parameters and N_r^ν distinct responses, respectively, at time instances ("nodes") $\nu = 1, 2, \ldots, N_t$ which are considered to be components of the following column vectors:

$$\boldsymbol{\alpha}^\nu \triangleq \left(\alpha_1^\nu,\ldots,\alpha_{N_\alpha^\nu}^\nu\right)^\dagger,\quad \mathbf{r}^\nu \triangleq \left(r_1^\nu,\ldots,r_{N_r^\nu}^\nu\right)^\dagger,\quad \nu = 1,\ldots,N_t. \tag{1.22}$$

At any time node v, the system parameters are considered to be variates with mean values $(\boldsymbol{\alpha}^0)^\nu$. Furthermore, the correlations between two parameters α_i^ν and α_j^μ, at two time nodes μ and ν, are generally defined as

$$c_{\alpha,ij}^{\nu\mu} \triangleq \left\langle \left[\alpha_i^\nu - \left(\alpha_i^\nu\right)^0\right]\left[\alpha_j^\mu - \left(\alpha_j^\mu\right)^0\right]\right\rangle.$$

The above quantities constitute the elements of symmetric covariance matrices of the form

$$\mathbf{C}_\alpha^{\mu\nu} \triangleq \left\langle \left(\alpha-\alpha^0\right)^\mu \left[\left(\alpha-\alpha^0\right)^\nu\right]^\dagger \right\rangle = \left(\mathbf{C}_\alpha^{\mu\nu}\right)^\dagger = \mathbf{C}_\alpha^{\nu\mu} = \left(\mathbf{C}_\alpha^{\nu\mu}\right)^\dagger. \tag{1.23}$$

Similarly, the *measured* responses are characterized by mean values $(\boldsymbol{r}_m)^\nu$ at a time node ν and by symmetric covariance matrices between two time nodes μ and ν defined as

$$\mathbf{C}_m^{\mu\nu} \triangleq \left\langle \left(\mathbf{r}-\mathbf{r}_m\right)^\mu \left[\left(\mathbf{r}-\mathbf{r}_m\right)^\nu\right]^\dagger \right\rangle = \left(\mathbf{C}_m^{\mu\nu}\right)^\dagger = \mathbf{C}_m^{\nu\mu} = \left(\mathbf{C}_m^{\nu\mu}\right)^\dagger. \tag{1.24}$$

In the most general case, the measured responses may be correlated to the parameters through symmetric *response-parameter uncertainty* matrices of the form

$$\mathbf{C}_{r\alpha}^{\mu\nu} \triangleq \left\langle \left(\mathbf{r}-\mathbf{r}_m\right)^\mu \left[\left(\alpha-\alpha^0\right)^\nu\right]^\dagger \right\rangle = \left(\mathbf{C}_{r\alpha}^{\mu\nu}\right)^\dagger = \mathbf{C}_{r\alpha}^{\nu\mu} = \left(\mathbf{C}_{r\alpha}^{\nu\mu}\right)^\dagger. \tag{1.25}$$

Note that the matrices $\mathbf{C}_{r\alpha}^{\mu\nu}$ are *not* bona fide variance–covariance matrices, in that they are not necessarily square positive matrices, and the elements on their respective main diagonals (if they happen to be square) are also covariances (or correlations) rather that variances.

At any given time instance ν, a response r_i^ν can be a function of not only the system parameters at time node ν but also of the system parameters at all previous time instances μ, $1 \le \mu \le \nu$. This means that $\boldsymbol{r}^\nu = \boldsymbol{R}^\nu(\boldsymbol{p}^\nu)$, where the vector $\boldsymbol{p}^\nu \triangleq (\boldsymbol{\alpha}^1, \ldots, \boldsymbol{\alpha}^\mu, \ldots, \boldsymbol{\alpha}^\nu)$ has been introduced for notational convenience, and where $\boldsymbol{R}^\nu(\boldsymbol{p}_0^\nu)$ denotes the vector of computed responses at a time node ν, at the nominal parameter values $\boldsymbol{p}_0^\nu$. The $(N_r^\nu \times N_\alpha^\mu)$-dimensional matrix of first-order sensitivities of the response $\boldsymbol{R}^\nu(\boldsymbol{p}^\nu)$ to the parameters $\boldsymbol{p}^\nu$, evaluated at the nominal values $\boldsymbol{p}_0^\nu$, will be denoted as $\boldsymbol{S}^{\nu\mu}(\boldsymbol{p}_0^\mu)$, $1 \le \mu \le \nu$ and is defined as

$$\boldsymbol{S}^{\nu\mu}(\boldsymbol{p}_0^\mu) \triangleq \begin{pmatrix} \dfrac{\partial R_1^\nu(\boldsymbol{p}_0^\mu)}{\partial \alpha_1^\mu} & \cdots & \dfrac{\partial R_1^\nu(\boldsymbol{p}_0^\mu)}{\partial \alpha_{N\alpha}^\mu} \\ \vdots & \dfrac{\partial R_i^\nu}{\partial \boldsymbol{\alpha}_n^\mu} & \vdots \\ \dfrac{\partial R_{N_r}^\nu(\boldsymbol{p}_0^\mu)}{\partial \alpha_1^\mu} & \cdots & \dfrac{\partial R_{N_r}^\nu(\boldsymbol{p}_0^\mu)}{\partial \alpha_{N\alpha}^\mu} \end{pmatrix}, 1 \le \mu \le \nu. \tag{1.26}$$

Since the response $\boldsymbol{R}^\nu(\boldsymbol{p}_0^\nu)$ at time node ν can depend only on parameters $(\boldsymbol{\alpha}^0)^\mu$ that appear up to the current time node ν, it follows that $\mathbf{S}^{\nu\mu} = \mathbf{0}$ when $\mu > \nu$.

The predictive modeling methodology of Cacuci and Ionescu-Bujor (2010) yields the following results:

1. *The expression given in Equation 1.27 for computing the best-estimate predicted nominal values, denoted as* $\boldsymbol{\alpha}^{be}$, *for the calibrated (adjusted) parameters*:

$$\boldsymbol{\alpha}^{be} = \boldsymbol{\alpha}^0 + \left(C_{\alpha r} - C_\alpha \left[S(\boldsymbol{\alpha}^0)\right]^\dagger\right)\left[C_d(\boldsymbol{\alpha}^0)\right]^{-1} \boldsymbol{d}, \tag{1.27}$$

   where

$$\boldsymbol{d} \triangleq \boldsymbol{R}(\boldsymbol{\alpha}^0) - \boldsymbol{r}_m, \boldsymbol{R}(\boldsymbol{\alpha}^0) \triangleq (\boldsymbol{R}^1, \ldots, \boldsymbol{R}^\mu, \ldots, \boldsymbol{R}^{N_t}), \tag{1.28}$$

$$S \triangleq \begin{pmatrix} S^{11} & \cdots & \mathbf{0} \\ \vdots & \ddots & \vdots \\ S^{N_t 1} & \cdots & S^{N_t N_t} \end{pmatrix}, \tag{1.29}$$

$$\boldsymbol{\alpha} \triangleq \left(\boldsymbol{\alpha}^{1},\ldots,\boldsymbol{\alpha}^{\mu},\ldots,\boldsymbol{\alpha}^{N_t}\right), \tag{1.30}$$

$$\boldsymbol{r} \triangleq \left(\boldsymbol{r}^{1},\ldots,\boldsymbol{r}^{\mu},\ldots,\boldsymbol{r}^{N_t}\right), \tag{1.31}$$

$$\begin{aligned} C_d\left(\boldsymbol{\alpha}^0\right) &\triangleq \left\langle \boldsymbol{d}\boldsymbol{d}^{\dagger}\right\rangle = \left\langle \left(\delta\boldsymbol{r} - \boldsymbol{S}\left(\boldsymbol{\alpha}^0\right)\delta\boldsymbol{\alpha}\right)\left(\delta\boldsymbol{r}^{\dagger} - \delta\boldsymbol{\alpha}^{\dagger}\left[\boldsymbol{S}\left(\boldsymbol{\alpha}^0\right)\right]^{\dagger}\right)\right\rangle \\ &= C_{rc}\left(\boldsymbol{\alpha}^0\right) - C_{r\alpha}\left[\boldsymbol{S}\left(\boldsymbol{\alpha}^0\right)\right]^{\dagger} - \left[\boldsymbol{S}\left(\boldsymbol{\alpha}^0\right)\right]C_{\alpha r} + C_m, \end{aligned} \tag{1.32}$$

$$C_{rc}\left(\boldsymbol{\alpha}^0\right) \triangleq \left\langle \delta\boldsymbol{r}\delta\boldsymbol{r}^{\dagger}\right\rangle = \left[\boldsymbol{S}\left(\boldsymbol{\alpha}^0\right)\right]\left\langle \delta\boldsymbol{\alpha}\delta\boldsymbol{\alpha}^{\dagger}\right\rangle\left[\boldsymbol{S}\left(\boldsymbol{\alpha}^0\right)\right]^{\dagger} = \left[\boldsymbol{S}\left(\boldsymbol{\alpha}^0\right)\right]C_{\alpha}\left[\boldsymbol{S}\left(\boldsymbol{\alpha}^0\right)\right]^{\dagger}. \tag{1.33}$$

$$C_{\alpha} = \begin{pmatrix} C_{\alpha}^{11} & C_{\alpha}^{12} & \cdots & \cdots \\ C_{\alpha}^{21} & C_{\alpha}^{22} & \cdots & \cdots \\ \cdots & \cdots & \cdots & \cdots \\ \cdots & \cdots & \cdots & C_{\alpha}^{N_tN_t} \end{pmatrix}, C_{\alpha r} = \begin{pmatrix} C_{\alpha r}^{11} & C_{\alpha r}^{12} & \cdots & \cdots \\ C_{\alpha r}^{21} & C_{\alpha r}^{22} & \cdots & \cdots \\ \cdots & \cdots & \cdots & \cdots \\ \cdots & \cdots & \cdots & C_{\alpha r}^{N_tN_t} \end{pmatrix},$$

$$C_{m} = \begin{pmatrix} C_{m}^{11} & C_{m}^{12} & \cdots & \cdots \\ C_{m}^{21} & C_{m}^{22} & \cdots & \cdots \\ \cdots & \cdots & \cdots & \cdots \\ \cdots & \cdots & \cdots & C_{m}^{N_tN_t} \end{pmatrix}. \tag{1.34}$$

The covariance matrix of the computed responses, C_{rc}, has the symmetric structure

$$C_{rc} \triangleq \begin{pmatrix} C_{rc}^{11} & \cdots & C_{rc}^{1N_t} \\ \vdots & \ddots & \vdots \\ C_{rc}^{N_t1} & \cdots & C_{rc}^{N_tN_t} \end{pmatrix},$$

with components defined as

$$C_{rc}^{\nu\mu} = \sum_{\eta=1}^{\nu}\sum_{\rho=1}^{\mu} S^{\nu\eta}C_{\alpha}^{\eta\rho}\left(S^{\mu\rho}\right)^{\dagger} = \left(C_{rc}^{\mu\nu}\right)^{\dagger}; \quad \nu,\mu = 1,\ldots,N_t. \tag{1.35}$$

In component form, the expression presented in Equation 1.27 for the calibrated best-estimate parameter values becomes

$$\left(\boldsymbol{\alpha}^{be}\right)^{\nu} = \left(\boldsymbol{\alpha}^{0}\right)^{\nu} + \sum_{\mu=1}^{N_t}\left\{\left[C_{\alpha r}^{\nu\mu} - \sum_{\rho=1}^{\mu} C_{\alpha}^{\nu\rho}\left(S^{\dagger}\right)^{\mu\rho}\right]\left[\sum_{\eta=1}^{N_t} K_{d}^{\mu\eta}d^{\eta}\right]\right\}, \quad \nu = 1,\ldots,N_t, \tag{1.36}$$

where $K_d^{\nu\eta}$ denotes the corresponding (ν,η)-element of the block-matrix C_d^{-1}.

2. *The expression given in Equation 1.37 for computing the best-estimate predicted nominal values, denoted as* r^{be}, *for the calibrated (adjusted) responses:*

$$r^{be} = r_m + \left(C_m - C_{r\alpha}\left[S(\alpha^0)\right]^{\dagger}\right)\left[C_d(\alpha^0)\right]^{-1} d. \tag{1.37}$$

At a specific time node ν, each component $(r^{be})^{\nu}$ of r^{be} has the explicit form

$$\left(r^{be}\right)^{\nu} = (r_m)^{\nu} + \sum_{\mu=1}^{N_t}\left\{\left[C_m^{\nu\mu} - \sum_{\rho=1}^{\mu} C_{r\alpha}^{\nu\rho}\left(S^{\dagger}\right)^{\mu\rho}\right]\left[\sum_{\eta=1}^{N_t} K_d^{\mu\eta} d^{\eta}\right]\right\}, \quad \nu = 1,\ldots,N_t. \tag{1.38}$$

3. *The expressions given in Equation 1.39 for computing the best-estimate predicted covariance matrix, denoted as* C_{α}^{be}, *corresponding to the best-estimate parameters*

$$C_{\alpha}^{be} \triangleq \left\langle\left(\alpha-\alpha^{be}\right)\left(\alpha-\alpha^{be}\right)^{\dagger}\right\rangle = C_{\alpha} - \left[C_{\alpha d}(\alpha^0)\right]\left[C_d(\alpha^0)\right]^{-1}\left[C_{\alpha d}(\alpha^0)\right]^{\dagger}, \tag{1.39}$$

where

$$C_{\alpha d}(\alpha^0) \triangleq \left\langle\left(\alpha-\alpha^0\right)d^{\dagger}\right\rangle = \left(C_{\alpha r} - C_{\alpha}\left[S(\alpha^0)\right]^{\dagger}\right). \tag{1.40}$$

In component form, the block-matrix expression presented in Equation 1.39 becomes

$$\left(C_{\alpha}^{be}\right)^{\nu\mu} = C_{\alpha}^{\nu\mu} - \sum_{\eta=1}^{N_t}\sum_{\rho=1}^{N_t}\left[C_{\alpha r}^{\nu\rho} - \sum_{\pi=1}^{\rho} C_{\alpha}^{\nu\pi}\left(S^{\dagger}\right)^{\rho\pi}\right] K_d^{\rho\eta}\left[C_{r\alpha}^{\eta\mu} - \sum_{\pi=1}^{\eta} S^{\eta\pi} C_{\alpha}^{\pi\mu}\right] \tag{1.41}$$

4. *The expressions given in Equation 1.42 for computing the best-estimate predicted covariance matrix, denoted as* C_r^{be}, *corresponding to the best-estimate responses* r^{be}:

$$C_r^{be} \triangleq \left\langle\left(r - r(\alpha^{be})\right)\left(r - r(\alpha^{be})\right)^{\dagger}\right\rangle = C_m - \left[C_{rd}(\alpha^0)\right]\left[C_d(\alpha^0)\right]^{-1}\left[C_{rd}(\alpha^0)\right]^{\dagger}, \tag{1.42}$$

where

$$C_{rd}(\alpha^0) \triangleq \left\langle (r - r_m) d^{\dagger}\right\rangle = \left(C_m - C_{r\alpha}\left[S(\alpha^0)\right]^{\dagger}\right). \tag{1.43}$$

In component form, the block-matrix expression presented in Equation 1.42 becomes

$$\left(C_r^{be}\right)^{\nu\mu} = C_m^{\nu\mu} - \sum_{\eta=1}^{N_t}\sum_{\rho=1}^{N_t}\left[C_m^{\nu\rho} - \sum_{\pi=1}^{\rho} C_{r\alpha}^{\nu\pi}\left(S^{+}\right)^{\rho\pi}\right]K_d^{\rho\eta}\left[C_m^{\eta\mu} - \sum_{\pi=1}^{\eta} S^{\eta\pi}C_{\alpha r}^{\pi\mu}\right]. \tag{1.44}$$

5. *The expressions given in Equation 1.45 for computing the best-estimate predicted parameter-response covariance matrix, denoted as* $\mathbf{C}_{\alpha r}^{be}$:

$$\mathbf{C}_{r\alpha}^{be} \triangleq \left\langle\left(\boldsymbol{\alpha}-\boldsymbol{\alpha}^{be}\right)\left(\boldsymbol{r}-\boldsymbol{r}\left(\boldsymbol{\alpha}^{be}\right)\right)^{+}\right\rangle = \mathbf{C}_{r\alpha} - \left[\mathbf{C}_{rd}\left(\boldsymbol{\alpha}^{0}\right)\right]\left[\mathbf{C}_{d}\left(\boldsymbol{\alpha}^{0}\right)\right]^{-1}\left[\mathbf{C}_{\alpha d}\left(\boldsymbol{\alpha}^{0}\right)\right]^{+}. \tag{1.45}$$

In component form, the block-matrix expression presented in Equation 1.45 becomes

$$\left(C_{r\alpha}^{be}\right)^{\nu\mu} = C_{r\alpha}^{\nu\mu} - \sum_{\eta=1}^{N_t}\sum_{\rho=1}^{N_t}\left[C_m^{\nu\rho} - \sum_{\pi=1}^{\rho} C_{r\alpha}^{\nu\pi}\left(S^{+}\right)^{\rho\pi}\right]K_d^{\rho\eta}\left[C_{\alpha r}^{\eta\mu} - \sum_{\pi=1}^{\eta} S^{\eta\pi}C_{\alpha}^{\pi\mu}\right]. \tag{1.46}$$

6. *The expressions given in Equation 1.47 for the "χ^2-information consistency indicator," defined as*

$$\chi^2 \triangleq \boldsymbol{d}^{+}\left[\mathbf{C}_d\left(\boldsymbol{\alpha}^{0}\right)\right]^{-1}\boldsymbol{d} \tag{1.47}$$

As Equation 1.47 indicates, the experimental responses can be considered as random variables approximately described by a multivariate Gaussian distribution with means located at the nominal values of the computed responses, and with a covariance matrix $\mathbf{C}_d(\boldsymbol{\alpha}^0)$. Furthermore, the quantity χ^2 represents the square of the length of the vector $\boldsymbol{d}$, measuring (in the corresponding metric) the deviations between the experimental and nominally computed responses. Note that χ^2 is independent of calibrating (or adjusting) the original data and can be evaluated directly from the given data (i.e., given parameters and responses, together with their original uncertainties) after having inverted the deviation-vector uncertainty matrix $\mathbf{C}_d(\boldsymbol{\alpha}^0)$. Recall that the χ^2 (chi-square) distribution with n degrees of freedom of the continuous variable $(0 \le x < \infty)$ is defined as

$$P\left(x<\chi^2<x+dx\right) \triangleq k_n(x)dx = \frac{1}{2^{n/2}\Gamma\left(n/2\right)}x^{n/2-1}e^{-x/2}dx,\ x>0,\ (n=1,2,\ldots). \tag{1.48}$$

The mean and variance of the random variable x in Equation 1.48 are $\langle x\rangle = n$ and $\text{var}(x) = 2n$. As the dimension of $\boldsymbol{d}$ indicates, the number of degrees of freedom characteristic of the calibration under consideration is equal to the number of experimental responses. In the extreme case of absence of experimental responses, no actual calibration takes place, since $\boldsymbol{d} = \boldsymbol{R}(\boldsymbol{\alpha}^0)$, so that the best-estimate parameter values are just the original nominal values, that is, $(\boldsymbol{\alpha}^{be})^k = (\boldsymbol{\alpha}^0)$; an actual adjustment occurs only when at least one experimental response is included.

For model calibration (adjustment), it is important to assess whether (1) the response and data measurements are free of gross errors (blunders such as wrong settings and mistaken readings) and (2) the measurements are consistent with the assumptions regarding the respective means, variances, and covariances. The numerical value of the consistency indicator "χ^2 per degrees of freedom" is close to 1.0 when the distance between any two nominal response values, $\left|R_i^0 - R_j^0\right|$, is smaller or at least not much larger than the sum of the corresponding uncertainties, say $\sigma_i + \sigma_j$. In such cases, the data are customarily considered to be "consistent within the respective error norms," which are usually considered "under quadratic loss." However, when the numerical value of the consistency indicator differs considerably from 1.0, the respective (measured or computed) data points may be inconsistent or discrepant. In particular, the consistency indicator is larger than 1.0 when the distance between the mean values of two (sets of) measurements or two (sets of) computations of the same quantity is larger than the sum of the two accompanying standard deviations, that is, when the distances $\left|R_i^0 - R_j^0\right|$ are larger than $(\sigma_j + \sigma_k)$. In such cases, the respective data are usually considered to be inconsistent or discrepant, and the causes underlying possible discrepancies must be investigated. Inconsistencies can be caused by unrecognized or ill-corrected experimental effects (e.g., background, dead time of the counting electronics, instrumental resolution, sample impurities, and calibration errors). Note that the probability that two equally precise measurements yield a separation greater than $\sigma_i + \sigma_j = 2\sigma$ is very small, namely $erfc(1) \simeq 0.157$ for Gaussian sampling distributions with standard deviation σ. This means that there is a small probability that two nondiscrepant (i.e., belonging to the same distribution) measurements that are separated by more than two standard deviations (thus giving the appearance of being discrepant!) could actually occur in practice. In other words, there is a nonzero probability that nondiscrepant data may give the appearance of being discrepant. Since the probability is rather small, it is much more likely that apparently discrepant data actually indicate the presence of unrecognized errors. Methods for treating unrecognized errors have been developed by Cacuci and Ionescu-Bujor (2010) by applying the maximum entropy principle under quadratic loss to the discrepant data. Once the inconsistent data, if any, is discarded, the predictive modeling methodology by Cacuci and Ionescu-Bujor (2010) predicts best-estimate values for parameters and predicted responses, as well as best-estimate reduced uncertainties (i.e., "smaller" values for the variance–covariance matrices) for the predicted best-estimate parameters and responses.

It is also important to note that indiscriminate calibration of model parameters by experimental responses that significantly deviate from their respective computed values would significantly modify the resulting adjusted parameters. On the contrary, calibrating a parameter library by using measured responses that are very close to their respective computed values would cause minimal parameter modifications and a nearly perfect reproduction of the given responses by the adjusted library (as would be expected). In such a case, the given responses would be considered as being consistent with the parameter library, in contradistinction to adjustment by inconsistent experimental information, in which case the adjustment could fail because of inconsistencies.

Note in Equation 1.39 that a symmetric positive matrix is subtracted from the initial parameter covariance matrix $\mathbf{C}_{\alpha}$. Therefore, the best-estimate predicted parameter uncertainty matrix $\mathbf{C}_{\alpha}^{be}$ will have smaller variances on its main diagonal, so that the corresponding standard deviations will be smaller than the original standard deviations in the parameter covariance matrix $\mathbf{C}_{\alpha}$. In this sense, therefore, the predicted uncertainties accompanying the best-estimate parameter nominal values are "reduced" (i.e., are smaller than the original parameter standard deviations) by the calibration (adjustment) procedure, through the introduction of new information from experiments. Similarly, a symmetric positive matrix is subtracted in Equation 1.42 from the initial covariance matrix $\mathbf{C}_m$ of the experimental responses. Hence, the predicted standard deviations accompanying the predicted best-estimate response values are smaller than the original measured or computed standard deviations due to the introduction of new experimental information. Furthermore, Equation 1.45 indicates that the calibration (adjustment) procedure will introduce correlations between the calibrated (adjusted) parameters and responses even if the parameters and response were initially uncorrelated, since $\mathbf{C}_{r\alpha}^{be} \neq 0$ even if $\mathbf{C}_{r\alpha} = 0$, that is,

$$\mathbf{C}_{r\alpha}^{be} = \mathbf{C}_m \left[\mathbf{C}_{rc}\left(\boldsymbol{\alpha}^0\right) + \mathbf{C}_m \right]^{-1} \left[\mathbf{S}\left(\boldsymbol{\alpha}^0\right) \right] \mathbf{C}_{\alpha}, \text{ when } \mathbf{C}_{r\alpha} = 0 \tag{1.49}$$

It is important to note that the computation of the best-estimate parameter and response values, together with their corresponding best-estimate uncertainties, namely, Equations 1.27, 1.37, 1.39, 1.42, and 1.45, require the inversion of a single matrix, namely the matrix $\mathbf{C}_d(\boldsymbol{\alpha}^0)$ defined in Equation 1.32. This is usually advantageous in practice, since the order of the matrix $\mathbf{C}_d(\boldsymbol{\alpha}^0)$ is given by the number of measured (or computed responses), which is most often considerably smaller that the number of model parameters under consideration. On the contrary, when the number of parameters exceeds the number of responses, it is possible to derive alternative expressions for the best-estimate calibrated parameters and their corresponding best-estimate covariances, by performing all derivations in the "parameter space" rather than in "response space." Cacuci and Ionescu-Bujor (2010) have shown that the Sherman–Morrison–Woodbury extension can be employed to obtain the alternative expression

$$C_d^{-1} \triangleq \left(C_{rc} - C_{r\alpha}S^{\dagger} - SC_{\alpha r} + C_m\right)^{-1} = A^{-1} - A^{-1}S\left(C_{\alpha}^{-1} + S^{\dagger}A^{-1}S\right)^{-1} S^{\dagger}A^{-1},$$
$$A \triangleq C_m - C_{r\alpha}S^{\dagger} - SC_{\alpha r}. \qquad (1.50)$$

The above expression provides the bridge between the "response space" and "parameter space" formulations. This expression also highlights the fact that the formulation of the predictive modeling methodology in the response space requires a single inversion of a square symmetric matrix (namely, the inversion of the matrix C_d) of the same dimensions as the number of responses. In contradistinction, the formulation of the predictive modeling methodology in the parameter space requires the inversion of three symmetric matrices, two of which have dimensions equal to the number of parameters and one with dimensions equal to the number of responses. Hence, from a computational standpoint, the response space formulations should be used whenever possible.

Since the inverse matrix, C_d^{-1}, incorporates simultaneously all of the available information about the system parameters and responses, at all time nodes (i.e., $\nu = 1, 2, \ldots, N_t$), it follows that *the predictive modeling methodology calibrates simultaneously all model parameters and responses, over all spatial locations and over the entire time interval under consideration*. Note that at any time instance, ν, C_d^{-1} incorporates information not only from time nodes prior in time to the instance ν (i.e., information regarding the "past" and "present" states of the system) but also from time instances posterior in time to ν (i.e., information about the "future" states of the system). Through the matrix C_d^{-1}, at any specified time node ν, the calibrated best-estimates parameters $(\boldsymbol{\alpha}^{be})^{\nu}$ and responses $\boldsymbol{r}^{be}$, together with the corresponding calibrated best-estimate covariance matrices $\left(C_{\alpha}^{be}\right)^{\nu\mu}$, $\left(C_{r}^{be}\right)^{\nu\mu}$, and $\left(C_{\alpha r}^{be}\right)^{\nu\mu}$ will also incorporate automatically all of the available information about the system parameters and responses at all time nodes (i.e., $\nu = 1, 2, \ldots, N_t$). In this respect, the predictive methodology of Cacuci and Ionescu-Bujor (2010) is conceptually related to the *foresight* aspects encountered in decision analysis. It is also important to note that, in practice, the application of the methodology developed in this section involves two distinct computational stages. A complete sensitivity database (i.e., response sensitivities of all responses with respect to all parameters, at all times nodes ν, $\mu = 1, \ldots, N_t$) needs to be generated *off-line* prior to performing the data assimilation and model calibration (or data adjustment) stages. The response sensitivities are subsequently combined within the predictive modeling formulas provided in Equations 1.27, 1.37, 1.39, 1.42, and 1.45 with appropriate covariance matrices to compute calibrated best-estimate responses, parameters, and best-estimate covariance matrices. For large-scale highly nonlinear problems involving many time nodes, the matrix C_d becomes very large, so its inversion would require large amounts of computational resources. These difficulties can be reduced at the expense of using less than the complete information available at any specific time node. For example, even in time-dependent problems in which the entire time history is known (e.g., transient behavior of reactor systems), one may nevertheless choose to use only information up to the current time index and disregard the information about future system states. On the contrary, in dynamical problems such as climate or weather prediction, in which the time variable advances continuously and states beyond the current time are not known, information about future states cannot be reliably accounted for anyway. Thus, the most common way of reducing the dimensionality of the data assimilation and model calibration problem is to disregard information about future states and limit the amount of information assimilated about past states.

Note that if the model is perfect (i.e., $\mathbf{C}_{\alpha\alpha} = 0$ and $\mathbf{C}_{\alpha r} = 0$), then Equations 1.27, 1.37, 1.39, 1.42, and 1.45 would yield $\boldsymbol{\alpha}^{pred} = \boldsymbol{\alpha}^0$ and $\mathbf{r}^{pred} = \mathbf{r}^c(\boldsymbol{\alpha}^0, \boldsymbol{\beta}^0)$, predicted "perfectly," without any accompanying uncertainties (i.e., $\mathbf{C}_{rr}^{pred} = \mathbf{0}$, $\mathbf{C}_{\alpha\alpha}^{pred} = \mathbf{0}$, $\mathbf{C}_{\alpha r}^{pred} = \mathbf{0}$). In other words, for a perfect model, the predictive modeling methodology predicts values for the responses and the parameters that coincide with the model's values (assumed to be perfect), and the experimental measurements would have no effect on the predictions (as would be expected, since imperfect measurements could not possibly improve the perfect model's predictions).

On the other hand, if the measurements were perfect (i.e., $C_{rr} = 0$ and $C_{\alpha r} = 0$), but the model were imperfect, then Equations 1.27, 1.37, 1.39, 1.42, and 1.45 would yield $\boldsymbol{\alpha}^{pred} = \boldsymbol{\alpha}^0 - \mathbf{C}_{\alpha\alpha}\mathbf{S}_{r\alpha}^{\dagger}\left[\mathbf{S}_{r\alpha}\mathbf{C}_{\alpha\alpha}\mathbf{S}_{r\alpha}^{\dagger}\right]^{-1}\mathbf{r}^d\left(\boldsymbol{\alpha}^0\right)$, $\mathbf{C}_{\alpha\alpha}^{pred} = \mathbf{C}_{\alpha\alpha} - \mathbf{C}_{\alpha\alpha}\mathbf{S}_{r\alpha}^{\dagger}\left[\mathbf{S}_{r\alpha}\mathbf{C}_{\alpha\alpha}\mathbf{S}_{r\alpha}^{\dagger}\right]^{-1}\mathbf{S}_{r\alpha}\mathbf{C}_{\alpha\alpha}$, $\mathbf{r}^{pred} = \mathbf{r}^m$, $\mathbf{C}_{rr}^{pred} = \mathbf{0}$, $\mathbf{C}_{\alpha r}^{pred} = \mathbf{0}$. In other words, in the case of perfect measurements, the predicted values for the responses would coincide with the measured values (assumed to be perfect), but the model's uncertain parameters would be calibrated by taking the measurements into account to yield improved nominal values and reduced parameters uncertainties.

In general, the response is a nonlinear function of the model parameters. The impact of parameter nonlinearities on the predicted quantities (i.e., best-estimate parameters, responses, and associated covariance matrices) can be quantified by using the best-estimated covariance matrix given by Equation 1.39 in conjunction with the "sandwich formula" represented by Equation 1.33 to obtain the following relation for the best-estimate computed (as opposed to "predicted") response covariance matrix:

$$
\begin{aligned}
C_{rc}^{be} &= \left[S\left(\alpha^{be}\right)\right] C_{\alpha}^{be} \left[S\left(\alpha^{be}\right)\right]^{\dagger} \\
&= \left[S\left(\alpha^{be}\right)\right]\left\{C_{\alpha} - \left[C_{\alpha d}\left(\alpha^{0}\right)\right]\left[C_{d}\left(\alpha^{0}\right)\right]^{-1}\left[C_{\alpha d}\left(\alpha^{0}\right)\right]^{\dagger}\right\}\left[S\left(\alpha^{be}\right)\right]^{\dagger}.
\end{aligned}
\tag{1.51}
$$

Comparing Equation 1.51 with Equation 1.42 reveals that, in general, $C_{rc}^{be} \neq C_{r}^{be}$ since $S(\boldsymbol{\alpha}^{be}) \neq S(\boldsymbol{\alpha}^{0})$. Nevertheless, *when the model is perfect (i.e., free of higher-order numerical errors) and the response is exactly a linear function of the model parameters, the sensitivity matrix* **S** *is independent of the parameter values* $\boldsymbol{\alpha}$, that is,

$$S(\boldsymbol{\alpha}^{be}) = S(\boldsymbol{\alpha}^{0}) = S, \textit{for "perfect" and linear models.} \tag{1.52}$$

Using Equation 1.52 in Equation 1.51 reduces the latter expression to

$$\begin{aligned} C_{rc}^{be} &= S\left[C_{\alpha} - \left(C_{\alpha r} - C_{\alpha}S^{\dagger}\right)C_{d}^{-1}\left(C_{\alpha r} - SC_{\alpha}\right)\right]S^{\dagger} \\ &= C_{rc} - \left(C_{rc} - SC_{\alpha r}\right)\left[C_{rc} + C_{m} - C_{r\alpha}S^{\dagger} - SC_{\alpha r}\right]^{-1}\left(C_{rc} - C_{\alpha r}S^{\dagger}\right) \\ &= C_{r}^{be}, \quad \text{for responses that are } \textit{linear} \text{ in the model parameters.} \end{aligned} \tag{1.53}$$

Nonlinear relations between computed responses and model parameters can be treated by applying Equations 1.27, 1.37, 1.39, 1.42, and 1.45 iteratively, starting with the known nominal values of the quantities involved. The first iteration (in such an iterative procedure) would yield all of the major explicit results derived in Equations 1.27, 1.37, 1.39, 1.42, and 1.45. The subsequent iteration would use the results of Equations 1.27, 1.37, 1.39, 1.42, and 1.45 as the "prior information" in a second application of these formulas and compute the new ("second-iteration") best-estimate quantities by using these formulas once again. This iterative procedure would be continued until the best-estimated values would converge within a small, user-specified, convergence criterion. Although the actual application of Equations 1.27, 1.37, 1.39, 1.42, and 1.45 to a physical system is straightforward in principle, the computations can become very demanding in terms of data handling and computational speed requirements.

As indicated by Equations 1.39, 1.42, and 1.45, the incorporation of additional (experimental) information in the adjustment (calibration) process reduces the variances of the adjusted parameters and responses while also modifying their correlations. This calibration process takes place through the couplings introduced by the sensitivities of the participating responses. In the predictive modeling procedure, the sensitivities play an essential role as weighting functions for propagating the initial parameter covariance matrices and the initial covariance matrices for the measured responses to obtain the adjusted best-estimate predicted quantities. This role is readily underscored by noting that the combination of experimental and computational information would not be possible if the sensitivities were not available. This crucial role played by the response sensitivities is further underscored by the "predictive modeling of coupled multiphysics systems (PMCMPS)" methodology developed by Cacuci (2014), which generalizes the predictive modeling methodology of Cacuci and Ionescu-Bujor (2010) from a single multiphysics system to two or more coupled multiphysics systems. Noteworthily, the mathematical methodology underlying the PMCMPS methodology is constructed such that the systems can be treated sequentially rather than simultaneously, while preserving exactly the same results as if the systems had been treated simultaneously. Consequently, very large coupled systems, which would exceed available computational resources if treated simultaneously, can be treated with the PMCMPS methodology sequentially, without any loss of generality or information, requiring only the resources that would be needed if the systems were treated simultaneously.

1.3 Advantages and Disadvantages of Statistical and Deterministic Methods for Computing Response Sensitivities

The results presented in Section 1.1 have demonstrated that, without sensitivities, it is not possible to quantify the uncertainties induced in system responses by the uncertainties in the model. Furthermore, the results presented in Section 1.2 have demonstrated that, without sensitivities, it is not possible to perform predictive modeling, that is, to combine experimental and computational information in order to obtain best-estimate predicted values for system responses and parameters while reducing the accompanying predicted uncertainties. These uses of response sensitivities to system parameters clearly underscore the importance of computing the response sensitivities as efficiently and as accurately as possible.

The simplest way of estimating the local sensitivity of a model response is to use the model to compute a perturbed response, denoted as $R\left(\alpha_1^0,\ldots,\alpha_i^0+\delta\alpha_i,\ldots,\alpha_{N\alpha}^0\right.$, where $\delta\alpha_i$ denotes a small deviation (e.g., of the order of 1%) from the nominal parameter value, α_i^0, and subsequently estimate the local sensitivity, $\partial R/\partial\alpha_i$, by using a finite difference approximation of the form $\partial R/\partial\alpha_i \cong \left[R\left(\alpha_1^0,\ldots,\alpha_i^0+\delta\alpha_i,\ldots,\alpha_{N\alpha}^0\right)-R\left(\alpha_1^0,\ldots,\alpha_i^0,\ldots,\alpha_{N\alpha}^0\right)\right]/\delta\alpha_i$. This procedure, usually called the "brute-force method," is conceptually simple to use and does not need any additional model development. However, this brute-force method is seldom accurate, particularly for nonlinear responses, and requires at least N_α model computations. Thus, even this simple brute-force method becomes very expensive for large-scale models with many (hundreds or more) parameters. Furthermore, the trial-and-error process needed to select the perturbations $\delta\alpha_i$ would produce erroneous sensitivities if $\delta\alpha_i$ was chosen to be either too small (in which case computational round-off errors could overwhelm the correct values) or too large (in which case nonlinear effects would cause the finite-difference result to yield a poor approximation of the actual first-order derivative). The brute-force method is a typical example of a *non-intrusive method,* since it does not require any modification of the original computational model. Other methods for computing sensitivities are *intrusive,* in that they modify the solvers underlying the original computational model.

Methods for computing sensitivities can also be categorized as *deterministic* or *statistical.* Statistical methods construct an approximate response distribution (often called "response surface") in the parameter's space and subsequently use scatter plots, regression, rank transformation, correlation, and partial correlation analyses to identify approximate expectation values, variances, and covariances for the responses. These statistical quantities are subsequently used to construct quantities that play the role of (approximate) first-order response sensitivities. Thus, statistical methods commence with "uncertainty analysis" and subsequently attempt an approximate "sensitivity analysis" of the computed model response under consideration. The currently popular statistical methods for uncertainty and sensitivity analysis are broadly categorized as sampling-based methods, variance-based methods, and spectral methods. The most popular statistical methods for uncertainty and sensitivity analysis were reviewed by Saltelli et al. (2000), and by Cacuci et al. (2005) with the following major conclusions:

1. The main advantage of using statistical methods for uncertainty and sensitivity analysis is that they are conceptually easy to implement.
2. On the other hand, statistical methods for uncertainty and sensitivity analysis have three major inherent practical drawbacks, as follows:
    a. Since many thousands of simulations are needed to obtain reliable results, statistical methods are at best expensive (for small systems) or, at worst, impracticable (e.g., for large time-dependent systems).
    b. Response sensitivities cannot be computed exactly.
    c. Since the response sensitivities and parameter uncertainties are amalgamated, inherently and inseparably, within the results produced by statistical methods, improvements in parameter uncertainties cannot be directly propagated to improve response uncertainties; rather, the entire set of simulations and statistical postprocessing must be repeated anew. In particular, a "foolproof" statistical method for analyzing correctly models involving highly correlated parameters does not seem to exist currently, so that particular care must be used when interpreting regression results for such models.

In contradistinction with statistical methods, the deterministic methods for computing response sensitivities rely on analytical or numerical differentiation of the computational model under investigation to compute local response sensitivities exactly. The commonly used deterministic methods (namely, the *direct method*, the *Green's function method*, and the *forward sensitivity analysis procedure*) require at least N_α computations to obtain the exact values of the first-order local response sensitivities and require at least $N_\alpha(N_\alpha+1)/2$ computations to obtain the second-order response sensitivities for N_α model parameters. Thus, these deterministic methods also become impractical for large systems with many parameters, because of prohibitively large computational requirements.

For large-scale systems, in which the number of system parameters and/or parameter variations to be considered exceeds the number of responses of interest, the *adjoint sensitivity analysis procedure* (ASAP) *for nonlinear systems*, introduced by Cacuci (1981a, 1981b), is by far the most efficient deterministic method for computing exactly and exhaustively the local first-order response sensitivities. The ASAP *requires a single large-scale computation for computing all N_α first-order derivatives of a scalar-valued response with respect to all of the model's parameters*. However, the ASAP requires development of an appropriate adjoint sensitivity model. Very few additional resources are required to develop the adjoint sensitivity model if this adjoint model is developed simultaneously with the original model. However, if the adjoint model is developed *a posteriori*, after the original model is already operational, considerable skills may be required for the successful development and implementation of the adjoint sensitivity model. The remarkable efficiency of the ASAP stems from the fact that the adjoint sensitivity system is linear in the adjoint function and is independent of any parameter variations. Hence, the adjoint sensitivity system needs to be solved only once, for each response, in order to obtain the adjoint functions. In particular, if the original model is linear in the state (i.e., dependent) variables, the adjoint sensitivity equations can be solved independently of the original model. In turn, the adjoint functions are used in simple quadrature of definite integrals to obtain the sensitivities to all system parameters, without needing to solve repeatedly large systems of differential and/or integral equations.

Building upon the ASAP (Cacuci 1981a, 1981b), this book presents the *second-order adjoint sensitivity analysis methodology* (2nd-ASAM) for computing exactly, comprehensively, and efficiently both the first-order and the second-order response sensitivities with respect to the system's parameters. The 2nd-ASAM is especially useful for the computation of response sensitivities for systems that are characterized by a very large number of parameters. For realistic systems, the first-order and, especially, the second-order sensitivities of responses of systems characterized by many (e.g., over 100) parameters can be computed comprehensively and exactly only by using the 2nd-ASAM. In practice, the larger the number of system parameters, the more useful the 2nd-ASAM becomes.

2

Illustrative Application of the 2nd-ASAM to a Linear Evolution Problem

Chapter 1 presented several motivations underlying the development of the 2nd-ASAM. To facilitate the understanding of the general mathematical derivation of the 2nd-ASAM, this chapter will illustrate the main steps for implementing the 2nd-ASAM by using a simple evolution model that admits a closed-form, easily tractable, analytical solution. Furthermore, this paradigm illustrative model comprises sufficiently many parameters (over a thousand), just as typically found in realistic physical systems, to preclude the use of statistical and or other deterministic methods for computing exactly all of the first- and second-order response sensitivities. This paradigm model will also apply the formulas introduced in Chapter 1 to illustrate the main applications of the first- and second-order response sensitivities for quantifying the uncertainties in model response, as well as for performing predictive modeling by combining computational with experimental results to reduce the uncertainties in predicted results.

The paradigm model considered in this chapter simulates a typical evolution or attenuation of a quantity that will be denoted as $\rho(t)$, satisfying the following linear conservation equation:

$$\frac{d\rho(t)}{dt} + \rho(t)\sum_{i=1}^{N} w_i\beta_i = q\sum_{i=1}^{N} w_i\beta_i,\ 0 < t \le t_f, \tag{2.1}$$

$$\rho(0) = \rho_{in},\ at\ t = 0. \tag{2.2}$$

The dependent variable ρ (t), could model the volumetric mass concentration of nitric acid of the liquid phase in a spent fuel dissolver (Peltz and Cacuci 2016), which evolves from an initial-time value, $t_0 = 0$, to a final-time value, t_f. The quantity ρ_{in} represents an initial condition, the quantity q represents a source, the quantities w_i and β_i, $i = 1,\ldots, N$ represent various material properties. Alternatively, as will be discussed in Chapter 4, $\rho(t)$ could represent the propagation (attenuation) of uncollided photons through a multimaterial homogenized one-dimensional slab of thickness t_f, containing photon sources and extending in the spatial direction t from one face located at $t_0 = 0$ to the other face located at $t = t_f$.

The total number of model parameters for this problem is $N_\alpha = 2N + 2$. All of these model parameters are considered to be time-independent scalars. For example, the spent fuel dissolver model (Peltz and Cacuci 2016) comprises

1292 parameters. The response of interest for the above model would be a measurement, denoted as $\rho(t_1)$, of $\rho(t)$ at some time instance $t = t_1$, $0 < t_1 < t_f$. The following functional can represent mathematically such a measurement:

$$\rho(t_1) = \int_0^{t_f} \rho(t)\delta(t - t_1)\, dt, \tag{2.3}$$

where $\delta(t - t_1)$ denotes the well-known Dirac-delta (impulse) functional.

Equations 2.1 and 2.2 can be readily solved by using the "integrating factor" method to obtain

$$\rho(t) = q + (\rho_{in} - q)\exp\left(-t\sum_{i=1}^{N} w_i\beta_i\right), \tag{2.4}$$

so the expression of the model's response is

$$\rho(t_1) = q + (\rho_{in} - q)\exp\left(-t_1\sum_{i=1}^{N} w_i\beta_i\right). \tag{2.5}$$

In practice, neither the analytical expression for the dependent variable $\rho(t)$ provided in Equation 2.4 nor the analytical expression for the response $\rho(t_1)$ provided in Equation 2.5 is available. Only the nominal (mean) parameter values, denoted as ρ_{in}^0, q^0, w_i^0, and β_i^0, and uncertainties accompanying these nominal values would be available. The sensitivities of the response $\rho(t_1)$ to the $2N_\alpha + 2$ model parameters ρ_{in}, q, w_i, and β_i would thus need to be computed by using the available information regarding the model parameters (nominal values, standard deviations, and correlations) in conjunction with some intrusive or nonintrusive methods for sensitivity and/or uncertainty analysis, which would be applied to Equations 2.1 through 2.3. The large number, $N_\alpha = O(10^3)$, of model parameters precludes the use of any statistical method for performing sensitivity and uncertainty analysis of the response $\rho(t_1)$ using the model, since the computational costs would be prohibitive. For such models, involving many parameters but relatively few responses, the 2nd-ASAM, introduced by Cacuci (2015a, 2015b, 2016a, 2016b), is best suited for computing exactly and efficiently the first- and second-order response sensitivities. The general theory for the 2nd-ASAM for linear systems (Cacuci 2015a, 2015b) will be presented in Chapter 3, and the general theory for the 2nd-ASAM for nonlinear systems (Cacuci 2016a, 2016b) will be presented in Chapter 7. However, the principles underlying the 2nd-ASAM for linear systems will be illustrated in the remainder of this chapter by applying it to compute the first and second-order sensitivities of the response $\rho(t_1)$ defined in Equation 2.3 for the paradigm model defined by Equations 2.1 and 2.2. The advantages of using the 2nd-ASAM will thus become apparent, providing the essential reason for using this methodology whenever possible.

2.1 Exact Computation of the First-Order Response Sensitivities

As has been demonstrated within the adjoint sensitivity analysis methodology developed by Cacuci (1981a, 1981b, 2003), the most general definition of the first-order sensitivities of the response $\rho(t_1)$ to the model parameters is provided by the Gateaux differential (G-differential), $\delta\rho(t_1)$, of the response $\rho(t_1)$. Applying the definition of the G-differential to Equation 2.3 yields the following expression for $\delta\rho(t_1)$:

$$\delta\rho(t_1) \triangleq \left\{\frac{d}{d\varepsilon}\int_0^{t_f}\left[\rho(t)+\varepsilon\delta\rho(t)\right]\delta(t-t_1)dt\right\}_{\varepsilon=0} = \int_0^{t_f}\delta\rho(t)\delta(t-t_1)dt. \quad (2.6)$$

As Equation 2.6 indicates, the response $\rho(t_1)$ depends *implicitly* on all of the model parameters. The G-differential, $\delta\rho(t)$, of the state function $\rho(t)$ is the solution of the system of equations obtained by G-differentiating Equations 2.1 and 2.2 around the nominal parameter values. Thus, applying the definition of the G-differential to Equations 2.1 and 2.2 yields

$$\frac{d}{d\varepsilon}\left\{\frac{d\left[\rho^0(t)+\varepsilon\delta\rho(t)\right]}{dt}+\left[\rho^0(t)+\varepsilon\delta\rho(t)\right]\sum_{i=1}^{N}\left(w_i^0+\varepsilon\delta w_i\right)\left(\beta_i^0+\varepsilon\delta\beta_i\right)\right\}_{\varepsilon=0}$$
$$=\frac{d}{d\varepsilon}\left\{\left[q^0+\varepsilon\delta q\right]\sum_{i=1}^{N}\left(w_i^0+\varepsilon\delta w_i\right)\left(\beta_i^0+\varepsilon\delta\beta_i\right)\right\}_{\varepsilon=0}, \quad 0<t\le t_f, \quad (2.7)$$

$$\frac{d}{d\varepsilon}\left\{\rho^0(t)+\varepsilon\delta\rho(t)\right\}_{\varepsilon=0}=\frac{d}{d\varepsilon}\left\{\rho_{in}^0+\varepsilon\delta\rho_{in}\right\}_{\varepsilon=0}, \quad at\ t=0. \quad (2.8)$$

Performing the differentiation with respect to ε in Equations 2.7 and 2.8 and setting $\varepsilon = 0$ in the resulting expressions yields the following system of equations to be satisfied by the G-differential $\delta\rho(t)$:

$$\frac{d\left[\delta\rho(t)\right]}{dt}+\delta\rho(t)\sum_{i=1}^{N}w_i^0\beta_i^0=$$
$$\left[q^0-\rho^0(t)\right]\sum_{i=1}^{N}\left(w_i^0\delta\beta_i+\beta_i^0\delta w_i\right)+\delta q\sum_{i=1}^{N}w_i^0\beta_i^0, \quad 0<t\le t_f, \quad (2.9)$$

$$\delta\rho(0)=\delta\rho_{in}, \quad at\ t=0. \quad (2.10)$$

The system comprised in Equations 2.9 and 2.10 is called the *first-level forward sensitivity system* (1st-LFSS) (Cacuci 2015a, 2015b, 2016a, 2016b) since it could in principle be solved $N_\alpha = O(10^3)$ times to obtain, albeit approximately, the sensitivities of $\rho(t_1)$ to every model parameter. In practice, however, such a procedure would be prohibitively expensive computationally. The alternative methodology, which avoids the need for solving repeatedly the 1st-LFSS, is the adjoint sensitivity analysis methodology (ASAM), originally introduced by Cacuci (1981a, 1981b). The ASAM proceeds by considering $\delta\rho(t)$ to be an element of a Hilbert space $\mathcal{H}^{(1)}$ endowed with the following inner product, denoted as $\langle \rho_1(t), \rho_2(t)\rangle$, between two (square-integrable) functions $\rho_1(t)$ and $\rho_2(t)$:

$$\langle \rho_1(t), \rho_2(t)\rangle \triangleq \int_0^{t_f} \rho_1(t)\rho_2(t)dt. \tag{2.11}$$

Next, introduce a square-integrable function $\psi^{(1)}(t) \in \mathcal{H}^{(1)}$ to construct the following inner product with Equation 2.9:

$$\begin{aligned}&\int_0^{t_f} \psi^{(1)}(t)\left[\frac{d[\delta\rho(t)]}{dt} + \delta\rho(t)\sum_{i=1}^{N} w_i\beta_i^0\right]dt \\ &= \int_0^{t_f} \psi^{(1)}(t)\left\{\left[q^0 - \rho^0(t)\right]\sum_{i=1}^{N}\left(w_i^0\delta\beta_i + \beta_i^0\delta w_i\right) + \delta q\sum_{i=1}^{N} w_i^0\beta_i^0\right\}dt.\end{aligned} \tag{2.12}$$

Integrate the left side of Equation 2.12 by parts once, so as to transfer the differential operation from $\delta\rho(t)$ onto $\psi^{(1)}(t)$, to obtain

$$\begin{aligned}\int_0^{t_f} \psi^{(1)}(t)\left[\frac{d[\delta\rho(t)]}{dt} + \delta\rho(t)\sum_{i=1}^{N_\alpha} w_i\beta_i^0\right]dt &= \int_0^{t_f} \delta\rho(t)\left[-\frac{d\psi^{(1)}(t)}{dt} + \psi^{(1)}(t)\sum_{i=1}^{N_\alpha} w_i\beta_i^0\right]dt \\ &+ \psi^{(1)}(t_f)\delta\rho(t_f) - \psi^{(1)}(0)\delta\rho(0).\end{aligned} \tag{2.13}$$

Since the function $\psi^{(1)}(t)$ is still an arbitrary (other than being square-integrable) function at this time, the first term on the right side of Equation 2.13 can be required to represent the same functional as the right side of Equation 2.6, by imposing the following equality:

$$-\frac{d\psi^{(1)}(t)}{dt} + \psi^{(1)}(t)\sum_{i=1}^{N} w_i^0\beta_i^0 = \delta(t - t_1),\ \ 0 < t \le t_f. \tag{2.14}$$

It, therefore, follows from Equations 2.6 and 2.12 through 2.14, that

$$\int_0^{t_f} \psi^{(1)}(t)\left\{\left[q^0-\rho^0(t)\right]\sum_{i=1}^{N}\left(w_i^0\delta\beta_i+\beta_i^0\delta w_i\right)+\delta q\sum_{i=1}^{N} w_i^0\beta_i^0\right\}dt$$
$$-\psi^{(1)}\left(t_f\right)\delta\rho\left(t_f\right)+\psi^{(1)}(0)\delta\rho(0)=\int_0^{t_f}\delta\rho(t)\delta(t-t_1)dt=\delta\rho(t_1). \quad (2.15)$$

The appearance in Equation 2.15 of the unknown quantity $\delta\rho(t_f)$ can now be eliminated by requiring that the function $\psi^{(1)}(t)$ satisfy the condition

$$\psi^{(1)}\left(t_f\right)=0. \quad (2.16)$$

It follows from Equations 2.15 and 2.16 that the response sensitivity $\delta\rho(t_1)$ is given by the following expression in terms of the function $\psi^{(1)}(t)$:

$$\delta\rho(t_1)=\int_0^{t_f}\psi^{(1)}(t)\left\{\left[q^0-\rho^0(t)\right]\sum_{i=1}^{N}\left(w_i^0\delta\beta_i+\beta_i^0\delta w_i\right)+\delta q\sum_{i=1}^{N}w_i^0\beta_i^0\right\}dt+\psi^{(1)}(0)\delta\rho_{in}. \quad (2.17)$$

The system comprising Equations 2.14 and 2.16 is called the *first-level adjoint sensitivity system* (1st-LASS) (Cacuci 2015a, 2016b), and the function $\psi^{(1)}(t)$ is called the *first-level adjoint function*; the superscript 1 indicates *first level*. In terms of $\psi^{(1)}(t)$, the partial sensitivities of $\rho(t_1)$ with respect to the variations in the model parameters are readily obtained from Equation 2.17; their respective expressions are as follows:

$$\frac{\partial\rho(t_1)}{\partial\rho_{in}}=\psi^{(1)}(0), \quad (2.18)$$

$$\frac{\partial\rho(t_1)}{\partial q}=\left(\sum_{i=1}^{N}w_i^0\beta_i^0\right)\int_0^{t_f}\psi^{(1)}(t)dt, \quad (2.19)$$

$$\frac{\partial\rho(t_1)}{\partial\beta_i}=w_i^0\int_0^{t_f}\psi^{(1)}(t)\left[q^0-\rho^0(t)\right]dt,\ \ i=1,\dots,N, \quad (2.20)$$

$$\frac{\partial\rho(t_1)}{\partial w_i}=\beta_i^0\int_0^{t_f}\psi^{(1)}(t)\left[q^0-\rho^0(t)\right]dt,\ \ i=1,\dots,N. \quad (2.21)$$

It is evident from Equations 2.18 through 2.21 that the sensitivities of the response $\rho(t_1)$ can be computed by fast quadratures once the first-level adjoint function $\psi^{(1)}(t)$ has been obtained by solving the 1st-LASS, that is, Equations 2.14 and 2.16. Notably, the 1st-LASS needs to be solved once only since it does not depend on any variations in the model parameters or state functions. This is a general characteristic of the 1st-LASS, as will be seen repeatedly throughout this book. By using the integrating factor method for Equations 2.14 and 2.16, the explicit solution of the 1st-LASS is readily obtained in the form

$$\psi^{(1)}(t)=\left[1-H(t-t_1)\right]\exp\left[(t-t_1)\sum_{i=1}^{N_\alpha} w_i^0\beta_i^0\right], \tag{2.22}$$

where $H(t_1-t)$ is the customary Heaviside unit-step functional, defined as

$$H(t-t_1)\triangleq\begin{cases}1, & t\geq t_1;\\ 0, & t<t_1.\end{cases} \tag{2.23}$$

Inserting the result from Equation 2.22 into Equations 2.18 through 2.22, respectively, yields the following explicit expressions for the first-order sensitivities of the response $\rho(t_1)$:

$$\frac{\partial\rho(t_1)}{\partial\rho_{in}}=\exp\left(-t_1\sum_{i=1}^{N} w_i^0\beta_i^0\right), \tag{2.24}$$

$$\frac{\partial\rho(t_1)}{\partial q}=1-\exp\left(-t_1\sum_{i=1}^{N} w_i^0\beta_i^0\right), \tag{2.25}$$

$$\frac{\partial\rho(t_1)}{\partial\beta_i}=\left(q^0-\rho_{in}^0\right)w_i^0 t_1\exp\left(-t_1\sum_{i=1}^{N} w_i^0\beta_i^0\right),\quad i=1,\ldots,N_\alpha, \tag{2.26}$$

$$\frac{\partial\rho(t_1)}{\partial w_i}=\left(q^0-\rho_{in}^0\right)\beta_i^0 t_1\exp\left(-t_1\sum_{i=1}^{N} w_i^0\beta_i^0\right),\quad i=1,\ldots,N_\alpha. \tag{2.27}$$

The correctness of the above expressions can be readily verified by computing the first-order sensitivities directly from the expression of the model response presented in Equation 2.5. For large-scale systems, however, the closed-form expression of the model response would not be available analytically (such as given in Equation 2.5). "Blunders" in computing the first-order response sensitivities using the first-level adjoint function $\psi^{(1)}(t)$ can be identified, however, by using forward computations with altered parameter values $\left(\beta_i^0+\delta\beta_i\right)$ in conjunction with first-order difference formulas of the form

$$\frac{\partial\rho\left(t_1;\beta_i^0\right)}{\partial\beta_i}\cong\frac{\rho\left(t_1;\beta_i^0+\delta\beta_i\right)-\rho\left(t_1;\beta_i^0\right)}{\delta\beta_i}. \tag{2.28}$$

Particularly important are the sensitivities to initial conditions, such as exemplified by Equation 2.18, which indicates that the value of the first-level adjoint function $\psi^{(1)}(t)$ at the initial time value $t = 0$ is proportional to the response sensitivity to initial conditions. Since the value of the first-level adjoint function at $t = 0$ can be obtained only after computing the entire evolution of $\psi^{(1)}(t)$, from the final time $t = t_f$ up to the initial-time $t = 0$, it becomes apparent that the sensitivities to initial conditions provide a stringent verification procedure for assessing the accuracy of the solution of the 1st-LASS. As will be generally shown in Chapters 3 and 7, the sensitivities to initial and boundary conditions indeed provide stringent accuracy metrics for assessing the accuracy of the solution of the 1st-LASS.

2.2 Exact Computation of the Second-Order Response Sensitivities

Anticipating the general theoretical consideration underlying the 2nd-ASAM, it is noted now that the second-order sensitivities of $\rho(t_1)$ with respect to the model parameters should be obtainable by considering each first-order sensitivity as a "system response" and applying to it the procedure that has been outlined thus far in this chapter. In other words, the expressions of the second-order response sensitivities are provided by the G-differentials of the expressions shown in Equations 2.18 through 2.21.

2.2.1 Computing the Second-Order Response Sensitivities Corresponding to the First-Order Sensitivities $\partial\rho(t_1)/\partial\beta_i$

Consider the expression of the response sensitivities with respect to the model parameters β_i provided in Equation 2.20. Taking the G-differential of Equation 2.20 yields

$$\delta\left[\frac{\partial\rho(t_1)}{\partial\beta_i}\right] \triangleq \frac{d}{d\varepsilon}\left\{(w_i+\varepsilon\delta w_i)\int_0^{t_1}\left[\psi^{(1)}(t)+\varepsilon\delta\psi^{(1)}(t)\right]\left[q+\varepsilon\delta q-\rho(t)-\varepsilon\delta\rho(t)\right]dt\right\}_{\varepsilon=0}$$

$$= w_i\int_0^{t_1}\delta\psi^{(1)}(t)\left[q-\rho(t)\right]dt - w_i\int_0^{t_1}\psi^{(1)}(t)\delta\rho(t)dt$$

$$+(\delta w_i)\int_0^{t_1}\psi^{(1)}(t)\left[q-\rho(t)\right]dt+(\delta q)w_i\int_0^{t_1}\psi^{(1)}(t)dt,\ i=1,\dots,N_\alpha. \tag{2.29}$$

The last two terms on the right side of Equation 2.29 can be computed immediately, since the adjoint function $\psi^{(1)}(t)$ and the forward function $\rho(t)$ are known. However, the first and second terms on the right side of Equation 2.29 contain the variation $\delta\psi^{(1)}(t)$ in the adjoint function and, respectively, the variation $\delta\rho(t)$ in the forward function, neither of which are available. The variation $\delta\psi^{(1)}(t)$ of the first-level adjoint function $\psi^{(1)}(t)$ is related to the parameter variations through the G-differential of the 1st-LASS, which is computed by applying the definition of the G-differential to Equations 2.14 and 2.16. Thus, taking the G-differential of Equation 2.14 yields

$$\frac{d}{d\varepsilon}\left\{-\frac{d\left[\psi^{(1)}(t)+\varepsilon\delta\psi^{(1)}(t)\right]}{dt}+\left[\psi^{(1)}(t)+\varepsilon\delta\psi^{(1)}(t)\right]\sum_{i=1}^{N}(w_i+\varepsilon\delta w_i)(\beta_i+\varepsilon\delta\beta_i)\right\}_{\varepsilon=0}=0,$$

or

$$-\frac{d\left[\delta\psi^{(1)}(t)\right]}{dt}+\delta\psi^{(1)}(t)\sum_{i=1}^{N}w_i\beta_i=-\psi^{(1)}(t)\sum_{i=1}^{N}(w_i\delta\beta_i+\beta_i\delta w_i)\quad 0<t\le t_f. \tag{2.30}$$

Furthermore, taking the G-differential of Equation 2.16 yields

$$\delta\psi^{(1)}\left(t_f\right)=0. \tag{2.31}$$

Altogether, Equations 2.9, 2.10, 2.30, and 2.31 constitute a well-posed system of equations that could, in principle, be solved to obtain the variations $\delta\psi^{(1)}(t)$ and $\delta\rho(t)$ in terms of the parameter variations. However, such a procedure would be just as impractical computationally as solving the 1st-LFSS. Therefore, the need for solving these equations, which depend on parameter variations, will be circumvented within the 2nd-ASAM by expressing the first two terms on the right side of Equation 2.29 in an alternative way so as to eliminate the appearance of $\delta\psi^{(1)}(t)$ and $\delta\rho(t)$. For this purpose, we introduce another Hilbert space, denoted as $\mathcal{H}^{(2)}$, which comprises, as elements, two-component vectors of the form $\psi_i^{(2)}(t)\triangleq\left[\psi_{i1}^{(2)}(t),\psi_{i2}^{(2)}(t)\right]$, with square-integrable functions $\psi_{i1}^{(2)}(t)$. The Hilbert space $\mathcal{H}^{(2)}$ will be endowed with the following inner product between two elements $\psi_i^{(2)}(t) \triangleq \left[\psi_{i1}^{(2)}(t), \psi_{i2}^{(2)}(t)\right] \in \mathcal{H}^{(2)}$ and $\varphi_i^{(2)}(t) \triangleq \left[\varphi_{i1}^{(2)}(t), \varphi_{i2}^{(2)}(t)\right] \in \mathcal{H}^{(2)}$:

$$\left\langle \psi_i^{(2)}(t), \varphi_i^{(2)}(t) \right\rangle \triangleq \sum_{j=1}^{2} \int_0^{t_f} \psi_{ij}^{(2)}(t) \varphi_{ij}^{(2)}(t) dt. \tag{2.32}$$

Omitting henceforth, for simplicity, the superscript 0, which denoted nominal values, Equations 2.30 and 2.9 are written in matrix form as follows:

$$\begin{pmatrix} -\dfrac{d}{dt} + \sum_{i=1}^{N} w_i \beta_i & 0 \\ 0 & \dfrac{d}{dt} + \sum_{i=1}^{N} w_i \beta_i \end{pmatrix} \begin{pmatrix} \delta\psi^{(1)}(t) \\ \delta\rho(t) \end{pmatrix} = \begin{pmatrix} -\psi^{(1)}(t) \sum_{i=1}^{N} (w_i \delta\beta_i + \beta_i \delta w_i) \\ [q - \rho(t)] \sum_{i=1}^{N} (w_i \delta\beta_i + \beta_i \delta w_i) + \delta q \sum_{i=1}^{N} w_i \beta_i \end{pmatrix}, \; 0 < t \le t_f. \tag{2.33}$$

Using the definition in Equation 2.32, construct the inner product of Equation 2.33 with an arbitrary element $\psi_1^{(2)}(t) \triangleq \left[\psi_{11}^{(2)}(t), \psi_{12}^{(2)}(t)\right] \in \mathcal{H}^{(2)}$ to obtain the following relation:

$$\int_0^{t_f} \left[\psi_{11}^{(2)}(t), \psi_{12}^{(2)}(t)\right] \begin{pmatrix} -\dfrac{d}{dt} + \sum_{i=1}^{N} w_i \beta_i & 0 \\ 0 & \dfrac{d}{dt} + \sum_{i=1}^{N} w_i \beta_i \end{pmatrix} \begin{pmatrix} \delta\psi^{(1)}(t) \\ \delta\rho(t) \end{pmatrix} dt$$
$$= \int_0^{t_f} \left[\psi_{11}^{(2)}(t), \psi_{12}^{(2)}(t)\right] \begin{pmatrix} -\psi^{(1)}(t) \sum_{i=1}^{N} (w_i \delta\beta_i + \beta_i \delta w_i) \\ [q - \rho(t)] \sum_{i=1}^{N} (w_i \delta\beta_i + \beta_i \delta w_i) + \delta q \sum_{i=1}^{N} w_i \beta_i \end{pmatrix} dt, \; 0 < t \le t_1. \tag{2.34}$$

Integrate by parts the left side of Equation 2.34 to transfer the differential operations on $\delta\psi^{(1)}(t)$ and $\delta\rho(t)$ to differential operations on $\psi_{11}^{(2)}(t)$ and $\psi_{12}^{(2)}(t)$ to obtain the following result:

$$\int_0^{t_f} \psi_{11}^{(2)}(t)\left[-\frac{d}{dt}\delta\psi^{(1)}(t)+\delta\psi^{(1)}(t)\sum_{i=1}^{N} w_i\beta_i\right]+\psi_{12}^{(2)}(t)\left[\frac{d}{dt}\delta\rho(t)+\delta\rho(t)\sum_{i=1}^{N} w_i\beta_i\right]dt$$
$$=-\psi_{11}^{(2)}(t_f)\delta\psi^{(1)}(t_f)+\psi_{11}^{(2)}(0)\delta\psi^{(1)}(0)+\psi_{12}^{(2)}(t_f)\delta\rho(t_f)-\psi_{12}^{(2)}(0)\delta\rho(0)$$
$$+\int_0^{t_f}\delta\psi^{(1)}(t)\left[\frac{d\psi_{11}^{(2)}(t)}{dt}+\psi_{11}^{(2)}(t)\sum_{i=1}^{N} w_i\beta_i\right]dt+\int_0^{t_1}\delta\rho(t)\left[-\frac{d\psi_{12}^{(2)}(t)}{dt}+\psi_{12}^{(2)}(t)\sum_{i=1}^{N} w_i\beta_i\right]dt. \tag{2.35}$$

Replacing the left side of Equation 2.35 with the right side of Equation 2.34 yields the relation

$$\int_0^{t_f}\left[\psi_{11}^{(2)}(t),\psi_{12}^{(2)}(t)\right]\begin{pmatrix} -\psi^{(1)}(t)\sum_{i=1}^{N}(w_i\delta\beta_i+\beta_i\delta w_i) \\ [q-\rho(t)]\sum_{i=1}^{N}(w_i\delta\beta_i+\beta_i\delta w_i)+\delta q\sum_{i=1}^{N} w_i\beta_i \end{pmatrix}dt$$
$$=-\psi_{11}^{(2)}(t_f)\delta\psi^{(1)}(t_f)+\psi_{11}^{(2)}(0)\delta\psi^{(1)}(0)+\psi_{12}^{(2)}(t_f)\delta\rho(t_f)-\psi_{12}^{(2)}(0)\delta\rho(0)$$
$$+\int_0^{t_f}\delta\psi^{(1)}(t)\left[\frac{d\psi_{11}^{(2)}(t)}{dt}+\psi_{11}^{(2)}(t)\sum_{i=1}^{N} w_i\beta_i\right]dt+\int_0^{t_1}\delta\rho(t)\left[-\frac{d\psi_{12}^{(2)}(t)}{dt}+\psi_{12}^{(2)}(t)\sum_{i=1}^{N} w_i\beta_i\right]dt. \tag{2.36}$$

The last two terms on the right side of Equation 2.36 can now be identified with the first two terms on the right side of Equation 2.29 by requiring that

$$\frac{d\psi_{11}^{(2)}(t)}{dt}+\psi_{11}^{(2)}(t)\sum_{i=1}^{N} w_i\beta_i=w_i\left[q-\rho(t)\right]=w_i\left(q-\rho_{in}\right)\exp\left(-t\sum_{i=1}^{N} w_i\beta_i\right), \tag{2.37}$$

$$-\frac{d\psi_{12}^{(2)}(t)}{dt}+\psi_{12}^{(2)}(t)\sum_{i=1}^{N} w_i\beta_i=-w_i\psi^{(1)}(t)=-w_i\left[1-H(t-t_1)\right]\exp\left[(t-t_1)\sum_{i=1}^{N} w_i\beta_i\right]. \tag{2.38}$$

Using the results obtained in Equations 2.37 and 2.38 together with Equation 2.36 in Equation 2.29 transforms the latter into the form

$$\delta\left[\frac{\partial\rho(t_1)}{\partial\beta_i}\right]=\int_0^{t_f}\left[\psi_{11}^{(2)}(t),\psi_{12}^{(2)}(t)\right]\begin{pmatrix}-\psi^{(1)}(t)\sum_{i=1}^{N}(w_i\delta\beta_i+\beta_i\delta w_i)\\ [q-\rho(t)]\sum_{i=1}^{N}(w_i\delta\beta_i+\beta_i\delta w_i)+\delta q\sum_{i=1}^{N}w_i\beta_i\end{pmatrix}dt$$
$$+\psi_{11}^{(2)}(t_f)\delta\psi^{(1)}(t_f)-\psi_{11}^{(2)}(0)\delta\psi^{(1)}(0)-\psi_{12}^{(2)}(t_f)\delta\rho(t_f)+\psi_{12}^{(2)}(0)\delta\rho(0)$$
$$+(\delta w_i)\int_0^{t_1}\psi^{(1)}(t)[q-\rho(t)]dt+(\delta q)w_i\int_0^{t_1}\psi^{(1)}(t)dt,\quad i=1,\ldots,N_\alpha. \tag{2.39}$$

Since Equation 2.31 stipulates that $\delta\psi^{(1)}(t_f)=0$ and since Equation 2.10 stipulates that $\delta\rho(0)=\delta\rho_{in}$, it follows that only the quantities $\delta\psi^{(1)}(0)$ and $\delta\rho(t_f)$ remain unknown on the right side of Equation 2.39. These unknown quantities are now eliminated from Equation 2.39 by requiring that

$$\psi_{11}^{(2)}(0)=0, \tag{2.40}$$

$$\psi_{12}^{(2)}(t_f)=0 \tag{2.41}$$

Thus, Equation 2.39 reduces to

$$\delta\left[\frac{\partial\rho(t_1)}{\partial\beta_i}\right]=\int_0^{t_f}\left[\psi_{11}^{(2)}(t),\psi_{12}^{(2)}(t)\right]\begin{pmatrix}-\psi^{(1)}(t)\sum_{i=1}^{N}(w_i\delta\beta_i+\beta_i\delta w_i)\\ [q-\rho(t)]\sum_{i=1}^{N}(w_i\delta\beta_i+\beta_i\delta w_i)+\delta q\sum_{i=1}^{N}w_i\beta_i\end{pmatrix}dt$$
$$+\psi_{12}^{(2)}(0)\delta\rho_{in}+(\delta w_i)\int_0^{t_f}\psi^{(1)}(t)[q-\rho(t)]dt+(\delta q)w_i\int_0^{t_f}\psi^{(1)}(t)dt,\quad i=1,\ldots,N_\alpha. \tag{2.42}$$

Identifying in Equation 2.42 the coefficients of $\delta\rho_{in}$, δq, $\delta\beta_i$, and δw_i yields the following expression for the respective second-order sensitivities of the response $\rho(t_1)$:

$$\frac{\partial^2\rho(t_1)}{\partial\beta_i\,\partial\rho_{in}}=\psi_{12}^{(2)}(0),\quad i=1,\ldots,N_\alpha, \tag{2.43}$$

$$\frac{\partial^2 \rho(t_1)}{\partial \beta_i \,\partial q} = w_i \int_0^{t_f} \psi^{(1)}(t)\,dt + \left(\sum_{i=1}^{N} w_i \beta_i\right) \int_0^{t_f} \psi_{12}^{(2)}(t)\,dt, \; i = 1,\dots,N_\alpha, \tag{2.44}$$

$$\frac{\partial^2 \rho(t_1)}{\partial \beta_i \,\partial \beta_j} = w_j \int_0^{t_f} \psi_{12}^{(2)}(t)\left[q - \rho(t)\right]dt - w_j \int_0^{t_f} \psi_{11}^{(2)}(t)\,\psi^{(1)}(t)\,dt; \; i,j = 1,\dots,N_\alpha, \tag{2.45}$$

$$\begin{aligned} \frac{\partial^2 \rho(t_1)}{\partial \beta_i \,\partial w_j} &= \beta_j \int_0^{t_f} \psi_{12}^{(2)}(t)\left[q - \rho(t)\right]dt - \beta_j \int_0^{t_f} \psi_{11}^{(2)}(t)\,\psi^{(1)}(t)\,dt \\ &+ \delta_{ij} \int_0^{t_f} \psi^{(1)}(t)\left[q - \rho(t)\right]dt; \; i,j = 1,\dots,N_\alpha; \; \delta_{ij} = \begin{cases} 1, & i = j; \\ 0, & i \neq j. \end{cases} \end{aligned} \tag{2.46}$$

The function $\psi_{11}^{(2)}$ is the solution of Equation 2.37, subject to the initial-time condition given in Equation 2.40, while the function $\psi_{12}^{(2)}$ is the solution of Equation 2.38, subject to the final-time condition given in Equation 2.41. The vector-valued function $\psi_1^{(2)}(t) \triangleq \left[\psi_{11}^{(2)}(t), \psi_{12}^{(2)}(t)\right] \in \mathcal{H}^{(2)}$ is called the *second-level adjoint function,* and Equations 2.37, 2.38, 2.40, and 2.41 constitute the second-level adjoint sensitivity system (2nd-LASS). Note that the 2nd-LASS is independent of parameter variations and therefore needs to be solved only once to obtain $\psi_1^{(2)}(t) \triangleq \left[\psi_{11}^{(2)}(t), \psi_{12}^{(2)}(t)\right] \in \mathcal{H}^{(2)}$. Subsequently, the second-level adjoint function $\psi_1^{(2)}(t)$ is used in Equations 2.43 through 2.46 to compute efficiently and exactly the respective second-order response sensitivities.

Using the integrating factor method yields the following solution for Equations 2.37 and 2.40:

$$\psi_{11}^{(2)}(t) = w_i\left(q - \rho_{in}\right) t \exp\left(-t \sum_{i=1}^{N} w_i \beta_i\right). \tag{2.47}$$

Similarly, using the integrating factor method yields the following solution for Equations 2.38 and 2.41:

$$\psi_{12}^{(2)}(t) = \begin{cases} w_i (t - t_1) \exp\left[(t - t_1) \sum_{i=1}^{N} w_i \beta_i\right], & t < t_1; \\ 0, & t > t_1. \end{cases} \tag{2.48}$$

Evaluating the function $\psi_{12}^{(2)}(t)$ at $t = 0$ in Equation 2.48 and introducing the resulting expression into Equation 2.43 yields

$$\frac{\partial^2 \rho(t_1)}{\partial \beta_i \, \partial \rho_{in}} = -w_i t_1 \exp\left(-t_1 \sum_{i=1}^{N} w_i \beta_i\right), \quad i = 1, \ldots, N. \tag{2.49}$$

Replacing the results given in Equations 2.22 and 2.48 in Equation 2.44 and carrying out the respective integrations yields

$$\frac{\partial^2 \rho(t_1)}{\partial \beta_i \, \partial q} = t_1 w_i \exp\left(-t_1 \sum_{i=1}^{N} w_i \beta_i\right), \; i = 1, \ldots, N. \tag{2.50}$$

Replacing the results given in Equations 2.22, 2.47, and 2.48 in Equation 2.45 and carrying out the respective integrations yields

$$\frac{\partial^2 \rho(t_1)}{\partial \beta_i \, \partial \beta_j} = (\rho_{in} - q) w_i w_j t_1^2 \exp\left(-t_1 \sum_{i=1}^{N} w_i \beta_i\right); \; i, j = 1, \ldots, N. \tag{2.51}$$

Replacing the results given in Equations 2.22, 2.47, and 2.48 in Equation 2.46 and carrying out the respective integrations yields

$$\frac{\partial^2 \rho(t_1)}{\partial \beta_i \, \partial w_j} = (q - \rho_{in})(t_1 \delta_{ij} - t_1^2 w_i \beta_j) \exp\left(-t_1 \sum_{i=1}^{N} w_i \beta_i\right); \; i, j = 1, \ldots, N. \tag{2.52}$$

Note that right sides of Equations 2.37 and 2.38 depend on the index $i = 1, \ldots, N$, since the model parameters w_i depend on this index. Strictly speaking, therefore, the second-level adjoint sensitivity function $\psi_1^{(2)}(t) \triangleq \left[\psi_{11}^{(2)}(t), \psi_{12}^{(2)}(t)\right]$ is a function of the index $i = 1, \ldots, N$. Hence, in the most unfavorable situation, the 2nd-LASS, comprising Equations 2.37, 2.38, 2.40, and 2.41, would need to be solved numerically for each distinct value w_i, for a total of N times. Even in this worst-case scenario, however, only the right sides (i.e., "sources") of Equations 2.37 and 2.38 would need to be modified, which is relatively easy to implement computationally. The left sides of these equations, which contain the differential operators and which would therefore involve "solvers" that would be much more difficult to modify computationally, remain unchanged, in any case. In many practical situations, however, it is possible to reduce drastically the number of computations involving the 2nd-LASS. For example, in the case of the 2nd-LASS, comprising Equations 2.37, 2.38, 2.40, and 2.41, a simple change of the dependent variable to $\psi_1^{(2)}(t) \triangleq \left[\psi_{11}^{(2)}(t), \psi_{12}^{(2)}(t)\right] = \left[w_i\varphi_{11}^{(2)}(t), w_i\varphi_{12}^{(2)}(t)\right]$ would transform this 2nd-LASS into the following form:

$$\frac{d\varphi_{11}^{(2)}(t)}{dt} + \varphi_{11}^{(2)}(t)\sum_{i=1}^{N} w_i\beta_i = \left[q - \rho(t)\right], \tag{2.37a}$$

$$-\frac{d\varphi_{12}^{(2)}(t)}{dt} + \varphi_{12}^{(2)}(t)\sum_{i=1}^{N} w_i\beta_i = -\psi^{(1)}(t), \tag{2.38a}$$

$$\varphi_{11}^{(2)}(0) = 0, \tag{2.40a}$$

$$\varphi_{12}^{(2)}\left(t_f\right) = 0. \tag{2.41a}$$

The above (alternative) 2nd-LASS, comprising Equations 2.37a, 2.38a, 2.40a, and 2.41a, is *independent* of the index $i = 1, \ldots, N$ and *would need to be solved* (numerically or analytically) *only once*, to obtain the functions $\varphi_{11}^{(2)}(t)$ and $\varphi_{12}^{(2)}(t)$. The components of the second-level adjoint function $\psi_1^{(2)}(t)$ would subsequently be obtained by simply multiplying the functions $\varphi_{11}^{(2)}(t)$ and $\varphi_{12}^{(2)}(t)$ by the respective model parameters w_i.

2.2.2 Computing the Second-Order Response Sensitivities Corresponding to the First-Order Sensitivities $\partial\rho(t_1)/\partial w_i$

The procedure used in Section 2.2.1 will now be applied to the first-order response sensitivities given by Equation 2.21. Thus, taking the G-differential of Equation 2.21 yields

$$\begin{aligned}
\delta\left[\frac{\partial\rho(t_1)}{\partial w_i}\right] &\triangleq \frac{d}{d\varepsilon}\left\{(\beta_i + \varepsilon\delta\beta_i)\int_0^{t_1}\left[\psi^{(1)}(t) + \varepsilon\delta\psi^{(1)}(t)\right]\left[q + \varepsilon\delta q - \rho(t) - \varepsilon\delta\rho(t)\right]dt\right\}_{\varepsilon=0} \\
&= \beta_i\int_0^{t_1}\delta\psi^{(1)}(t)\left[q - \rho(t)\right]dt - \beta_i\int_0^{t_1}\psi^{(1)}(t)\delta\rho(t)dt \\
&\quad + (\delta\beta_i)\int_0^{t_1}\psi^{(1)}(t)\left[q - \rho(t)\right]dt + (\delta q)\beta_i\int_0^{t_1}\psi^{(1)}(t)dt, \; i = 1, \ldots, N_\alpha.
\end{aligned} \tag{2.53}$$

As expected from a comparison of Equations 2.20 and 2.21, comparing Equation 2.53 to Equation 2.29 readily indicates that the right sides of the latter equations differ only in that the model parameter β_i plays in Equation 2.53 the same role as the model parameter w_i played in Equation 2.29. This observation will greatly simplify the following endeavors: (i) the construction of the 2nd-LASS needed for computing the second-order response sensitivities implicitly comprised in Equation 2.53; (ii) the subsequent solution of this 2nd-LASS to obtain the corresponding second-level adjoint function; and (iii) the use of this adjoint function to compute the respective second-order sensitivities. Thus, analogous to the considerations used to construct Equations 2.32 through 2.36, use an arbitrary function $\psi_2^{(2)}(t) \triangleq \left[\psi_{21}^{(2)}(t), \psi_{22}^{(2)}(t)\right] \in \mathcal{H}^{(2)}$ to obtain the following relation, which is formally identical to Equation 2.36, except that the role previously played by $\psi_1^{(2)}(t)$ is now played by the yet-undetermined function $\psi_2^{(2)}(t)$:

$$\int_0^{t_f} \left[\psi_{21}^{(2)}(t), \psi_{22}^{(2)}(t)\right] \begin{pmatrix} -\psi^{(1)}(t)\sum_{i=1}^{N}(w_i\delta\beta_i + \beta_i\delta w_i) \\ [q-\rho(t)]\sum_{i=1}^{N}(w_i\delta\beta_i + \beta_i\delta w_i) + \delta q\sum_{i=1}^{N} w_i\beta_i \end{pmatrix} dt$$

$$= -\psi_{21}^{(2)}(t_f)\delta\psi^{(1)}(t_f) + \psi_{21}^{(2)}(0)\delta\psi^{(1)}(0) + \psi_{22}^{(2)}(t_f)\delta\rho(t_f) - \psi_{22}^{(2)}(0)\delta\rho(0) \tag{2.54}$$

$$+\int_0^{t_f} \delta\psi^{(1)}(t)\left[\frac{d\psi_{21}^{(2)}(t)}{dt} + \psi_{21}^{(2)}(t)\sum_{i=1}^{N} w_i\beta_i\right]dt + \int_0^{t_1} \delta\rho(t)\left[-\frac{d\psi_{22}^{(2)}(t)}{dt} + \psi_{22}^{(2)}(t)\sum_{i=1}^{N} w_i\beta_i\right]dt.$$

The last two terms on the right side of Equation 2.54 are now identified with the first two terms on the right side of Equation 2.53 by requiring that

$$\frac{d\psi_{21}^{(2)}(t)}{dt} + \psi_{21}^{(2)}(t)\sum_{i=1}^{N} w_i\beta_i = \beta_i\left[q-\rho(t)\right] = \beta_i\left(q-\rho_{in}\right)\exp\left(-t\sum_{i=1}^{N} w_i\beta_i\right), \tag{2.55}$$

$$-\frac{d\psi_{22}^{(2)}(t)}{dt} + \psi_{22}^{(2)}(t)\sum_{i=1}^{N} w_i\beta_i = -\beta_i\psi^{(1)}(t) = -\beta_i\left[1-H(t-t_1)\right]\exp\left[(t-t_1)\sum_{i=1}^{N} w_i\beta_i\right], \tag{2.56}$$

$$\psi_{21}^{(2)}(0) = 0, \tag{2.57}$$

$$\psi_{22}^{(2)}(t_f) = 0. \tag{2.58}$$

Using the results given in Equations 2.54 through 2.58 in Equation 2.53 reduces the latter to

$$\delta\left[\frac{\partial\rho(t_1)}{\partial w_i}\right]=\int_0^{t_f}\left[\psi_{21}^{(2)}(t),\psi_{22}^{(2)}(t)\right]\begin{pmatrix}-\psi^{(1)}(t)\sum_{i=1}^{N}(w_i\delta\beta_i+\beta_i\delta w_i)\\ \left[q-\rho(t)\right]\sum_{i=1}^{N}(w_i\delta\beta_i+\beta_i\delta w_i)+\delta q\sum_{i=1}^{N}w_i\beta_i\end{pmatrix}dt$$

$$+\psi_{22}^{(2)}(0)\delta\rho_{in}+(\delta\beta_i)\int_0^{t_f}\psi^{(1)}(t)\left[q-\rho(t)\right]dt+(\delta q)\beta_i\int_0^{t_f}\psi^{(1)}(t)dt,\quad i=1,\ldots,N. \tag{2.59}$$

Identifying in Equation 2.59, the coefficients of $\delta\rho_{in}$, (δq), $\delta\beta_i$, and δw_i yield the following expression for the respective second-order sensitivities of the response $\rho(t_1)$:

$$\frac{\partial^2\rho(t_1)}{\partial w_i\,\partial\rho_{in}}=\psi_{22}^{(2)}(0),\quad i=1,\ldots,N, \tag{2.60}$$

$$\frac{\partial^2\rho(t_1)}{\partial w_i\,\partial q}=\beta_i\int_0^{t_f}\psi^{(1)}(t)dt+\left(\sum_{i=1}^{N}w_i\beta_i\right)\int_0^{t_f}\psi_{22}^{(2)}(t)dt,\ i=1,\ldots,N, \tag{2.61}$$

$$\frac{\partial^2\rho(t_1)}{\partial w_i\,\partial\beta_j}=w_j\int_0^{t_f}\psi_{22}^{(2)}(t)\left[q-\rho(t)\right]dt-w_j\int_0^{t_f}\psi_{21}^{(2)}(t)\psi^{(1)}(t)dt$$
$$+\delta_{ij}\int_0^{t_f}\psi^{(1)}(t)\left[q-\rho(t)\right]dt;\quad i,j=1,\ldots,N, \tag{2.62}$$

$$\frac{\partial^2\rho(t_1)}{\partial w_i\,\partial w_j}=\beta_j\int_0^{t_f}\psi_{22}^{(2)}(t)\left[q-\rho(t)\right]dt-\beta_j\int_0^{t_f}\psi_{21}^{(2)}(t)\psi^{(1)}(t)dt;\ i,j=1,\ldots,N. \tag{2.63}$$

Using the integrating factor method to solve the 2nd-LASS defined by Equations 2.55 through 2.58 yields the following expressions for the components of the second-level adjoint function $\psi_2^{(2)}(t)\triangleq\left[\psi_{21}^{(2)}(t),\psi_{22}^{(2)}(t)\right]\in\mathcal{H}^{(2)}$:

$$\psi_{21}^{(2)}(t)=\beta_i\left(q-\rho_{in}\right)t\exp\left(-t\sum_{i=1}^{N}w_i\beta_i\right), \tag{2.64}$$

$$\psi_{22}^{(2)}(t) = \begin{cases} \beta_i (t - t_1)\exp\left[(t - t_1)\sum_{i=1}^{N} w_i \beta_i\right], & \text{for } t < t_1; \\ 0, & \text{for } t > t_1. \end{cases} \quad (2.65)$$

Evaluating the function $\psi_{22}^{(2)}(t)$ at $t = 0$ and replacing the resulting expression in Equation 2.60 yields

$$\frac{\partial^2 \rho(t_1)}{\partial w_i \, \partial \rho_{in}} = -\beta_i t_1 \exp\left(-t_1 \sum_{i=1}^{N} w_i \beta_i\right), \quad i = 1, \ldots, N. \quad (2.66)$$

Replacing the results given in Equations 2.22 and 2.65 in Equation 2.61 and carrying out the respective integrations yields

$$\frac{\partial^2 \rho(t_1)}{\partial w_i \, \partial q} = t_1 \beta_i \exp\left(-t_1 \sum_{i=1}^{N} w_i \beta_i\right), \; i = 1, \ldots, N. \quad (2.67)$$

Replacing the results given in Equations 2.22, 2.64, and 2.65 in Equation 2.62 and carrying out the respective integrations yields

$$\frac{\partial^2 \rho(t_1)}{\partial w_i \, \partial \beta_j} = (q - \rho_{in})(t_1 \delta_{ij} - t_1^2 \beta_i w_j)\exp\left(-t_1 \sum_{i=1}^{N} w_i \beta_i\right); \quad i, j = 1, \ldots, N. \quad (2.68)$$

Replacing the results given in Equations 2.22, 2.64, and 2.65 in Equation 2.63 and carrying out the respective integrations yields

$$\frac{\partial^2 \rho(t_1)}{\partial w_i \, \partial w_j} = (\rho_{in} - q) t_1^2 \beta_i \beta_j \exp\left(-t_1 \sum_{i=1}^{N} w_i \beta_i\right); \quad i, j = 1, \ldots, N. \quad (2.69)$$

It is important to note that the left sides of the 2nd-LASS, cf. Equations 2.55 and 2.56, involve the same differential operators as the left sides of the 2nd-LASS, cf. Equations 2.37 and 2.38, which were obtained in the previous section, Section 2.2.1. Hence, the same "solvers" that are used for solving the 2nd-LASS in Section 2.2.1, that is, Equations 2.37 and 2.38, are also to be used for solving the 2nd-LASS in this section, that is, Equations 2.55 and 2.56.

Just as for the 2nd-LASS obtained in the previous Section 2.2.1, the right sides of Equations 2.55 and 2.56 depend on the index $i = 1,\ldots,N$, since the model parameters β_i depend on this index. Hence, the second-level adjoint sensitivity function $\psi_2^{(2)}(t) \triangleq \left[\psi_{21}^{(2)}(t), \psi_{22}^{(2)}(t)\right]$ is actually a function of the index $i = 1,\ldots,N$; thus, in the most unfavorable situation, the 2nd-LASS, comprising Equations 2.55 through 2.58, would need to be solved anew for each distinct value β_i, for a total of N times, as was discussed in Section 2.2.1. Even in this worst-case scenario, however, only the right sides (i.e., "sources") of Equations 2.55 and 2.56 would need to be modified, which is relatively easy to do. The left sides of these equations (which would involve solvers that would be much more difficult to modify computationally) remain unchanged, in any case. However, just as was the case in Section 2.2.1, it is also possible to reduce drastically the number of computations involving the 2nd-LASS comprising Equations 2.55 through 2.58, by using a similar change of the dependent variable, namely $\psi_2^{(2)}(t) \triangleq \left[\psi_{21}^{(2)}(t), \psi_{22}^{(2)}(t)\right] = \left[\beta_i \varphi_{21}^{(2)}(t), \beta_i \varphi_{22}^{(2)}(t)\right]$. This change of the dependent variables transforms the 2nd-LASS defined by Equations 2.55 through 2.58 into the following form:

$$\frac{d\varphi_{21}^{(2)}(t)}{dt} + \varphi_{21}^{(2)}(t) \sum_{i=1}^{N} w_i \beta_i = \left[q - \rho(t)\right], \tag{2.55a}$$

$$-\frac{d\varphi_{22}^{(2)}(t)}{dt} + \varphi_{22}^{(2)}(t) \sum_{i=1}^{N} w_i \beta_i = -\psi^{(1)}(t), \tag{2.56a}$$

$$\varphi_{21}^{(2)}(0) = 0, \tag{2.57a}$$

$$\varphi_{22}^{(2)}(t_f) = 0. \tag{2.58a}$$

The above (alternative) 2nd-LASS, comprising Equations 2.55a through 2.58a, is *independent* of the index $i = 1, \ldots, N$ and *would need to be solved* (numerically or analytically) *only once*, to obtain the functions $\varphi_{21}^{(2)}(t)$ and $\varphi_{22}^{(2)}(t)$. The components of the second-level adjoint function $\psi_2^{(2)}(t)$ would subsequently be obtained by simply multiplying the functions $\varphi_{21}^{(2)}(t)$ and $\varphi_{22}^{(2)}(t)$ by the respective model parameters β_i.

2.2.3 Computing the Second-Order Response Sensitivities Corresponding to the First-Order Sensitivities $\partial\rho(t_1)/\partial q$

The computation of the 2nd-LASS and second-order sensitivities corresponding to the first-order sensitivities provided by Equation 2.19 will be performed next. Thus, taking the G-differential of Equation 2.19 yields

$$\begin{aligned} \delta\left[\frac{\partial\rho(t_1)}{\partial q}\right] &= \frac{d}{d\varepsilon}\left\{\sum_{i=1}^{N}(w_i + \varepsilon\delta w_i)(\beta_i + \varepsilon\delta\beta_i)\int_0^{t_1}\left[\psi^{(1)}(t) + \varepsilon\delta\psi^{(1)}(t)\right]dt\right\}_{\varepsilon=0} \\ &= \left(\sum_{i=1}^{N_\alpha} w_i\beta_i\right)\int_0^{t_1}\delta\psi^{(1)}(t)dt + \sum_{i=1}^{N}(\beta_i\delta w_i + w_i\delta\beta_i)\int_0^{t_1}\psi^{(1)}(t)dt. \end{aligned} \tag{2.70}$$

Analogous to the previous considerations used to construct Equations 2.36 and 2.54, respectively, use an arbitrary function $\psi_3^{(2)}(t) \triangleq \left[\psi_{31}^{(2)}(t), \psi_{32}^{(2)}(t)\right] \in \mathcal{H}^{(2)}$ to obtain the following relation, which is formally identical to Equation 2.36 and 2.54, except that the roles previously played by $\psi_1^{(2)}(t)$ and $\psi_2^{(2)}(t)$, respectively, are now played by the yet-undetermined function $\psi_3^{(2)}(t)$:

$$\int_0^{t_f} \left[\psi_{31}^{(2)}(t), \psi_{32}^{(2)}(t)\right] \begin{pmatrix} -\psi^{(1)}(t)\sum_{i=1}^{N}(w_i\delta\beta_i + \beta_i\delta w_i) \\ [q-\rho(t)]\sum_{i=1}^{N}(w_i\delta\beta_i + \beta_i\delta w_i) + \delta q\sum_{i=1}^{N} w_i\beta_i \end{pmatrix} dt$$

$$= -\psi_{31}^{(2)}(t_f)\delta\psi^{(1)}(t_f) + \psi_{31}^{(2)}(0)\delta\psi^{(1)}(0) + \psi_{32}^{(2)}(t_f)\delta\rho(t_f) - \psi_{32}^{(2)}(0)\delta\rho(0) \tag{2.71}$$

$$+\int_0^{t_f} \delta\psi^{(1)}(t)\left[\frac{d\psi_{31}^{(2)}(t)}{dt} + \psi_{31}^{(2)}(t)\sum_{i=1}^{N} w_i\beta_i\right]dt + \int_0^{t_1} \delta\rho(t)\left[-\frac{d\psi_{32}^{(2)}(t)}{dt} + \psi_{32}^{(2)}(t)\sum_{i=1}^{N} w_i\beta_i\right]dt.$$

The last two terms on the right side of Equation 2.71 are now identified with the first term on the right side of Equation 2.70 by requiring that

$$\frac{d\psi_{31}^{(2)}(t)}{dt} + \psi_{31}^{(2)}(t)\sum_{i=1}^{N} w_i\beta_i = \sum_{i=1}^{N} w_i\beta_i, \tag{2.72}$$

$$-\frac{d\psi_{32}^{(2)}(t)}{dt} + \psi_{32}^{(2)}(t)\sum_{i=1}^{N} w_i\beta_i = 0, \tag{2.73}$$

$$\psi_{31}^{(2)}(0) = 0, \tag{2.74}$$

$$\psi_{32}^{(2)}(t_f) = 0. \tag{2.75}$$

Note again that the left sides of the above 2nd-LASS, that is, Equations 2.72 and 2.73, involve the same differential operators as the left sides of the previous 2nd-LASS, namely Equations 2.37 and 2.38, and, respectively, Equations 2.55 and 2.56, which were previously obtained in Sections 2.2.1 and 2.2.2. Hence, the same solvers would also be used for solving the above 2nd-LASS, comprising Equations 2.72 through 2.75; only the source terms on the right sides of these differential equations are distinct. Note also that, in contradistinction to the 2nd-LASS derived in Sections 2.2.1 and 2.2.2, the right sides of Equations 2.72 and 2.73 are independent of the index $i = 1,\ldots, N$. Hence, the 2nd-LASS comprising Equations 2.72 through 2.75 would need to be solved once only, to obtain the second-level adjoint function $\psi_3^{(2)}(t) \triangleq \left[\psi_{31}^{(2)}(t), \psi_{32}^{(2)}(t)\right]$.

Using the results given in Equations 2.71 through 2.75 in Equation 2.70 reduces the latter to

$$\delta\left[\frac{\partial\rho(t_1)}{\partial q}\right]=\int_0^{t_f}\left[\psi_{31}^{(2)}(t),\psi_{32}^{(2)}(t)\right]\begin{pmatrix}-\psi^{(1)}(t)\sum_{i=1}^{N}(w_i\delta\beta_i+\beta_i\delta w_i)\\ \left[q-\rho(t)\right]\sum_{i=1}^{N}(w_i\delta\beta_i+\beta_i\delta w_i)+\delta q\sum_{i=1}^{N}w_i\beta_i\end{pmatrix}dt$$
$$+\psi_{32}^{(2)}(0)\delta\rho_{in}+\sum_{i=1}^{N}(\beta_i\delta w_i+w_i\delta\beta_i)\int_0^{t_1}\psi^{(1)}(t)dt,\quad i=1,\ldots,N_\alpha. \tag{2.76}$$

Identifying in Equation 2.76 the respective coefficients of the parameter variations yields the following expressions:

$$\frac{\partial^2\rho(t_1)}{\partial q\,\partial\rho_{in}}=\psi_{32}^{(2)}(0), \tag{2.77}$$

$$\frac{\partial^2\rho(t_1)}{\partial q^2}=\left(\sum_{i=1}^{N}w_i\beta_i\right)\int_0^{t_f}\psi_{32}^{(2)}(t)dt, \tag{2.78}$$

$$\frac{\partial^2\rho(t_1)}{\partial q\,\partial\beta_j}=-w_j\int_0^{t_f}\psi_{31}^{(2)}(t)\psi^{(1)}(t)dt+w_j\int_0^{t_f}\psi_{32}^{(2)}(t)\left[q-\rho(t)\right]dt$$
$$+w_j\int_0^{t_f}\psi^{(1)}(t)dt,\qquad j=1,\ldots,N, \tag{2.79}$$

$$\frac{\partial^2\rho(t_1)}{\partial q\,\partial w_j}=-\beta_j\int_0^{t_f}\psi_{31}^{(2)}(t)\psi^{(1)}(t)dt+\beta_j\int_0^{t_f}\psi_{32}^{(2)}(t)\left[q-\rho(t)\right]dt$$
$$+\beta_j\int_0^{t_f}\psi^{(1)}(t)dt,\qquad j=1,\ldots,N. \tag{2.80}$$

Applying the integrating factor method to solve Equations 2.72 through 2.75 yields the following solutions:

$$\psi_{31}^{(2)}(t)=1-\exp\left(-t\sum_{i=1}^{N}w_i\beta_i\right), \tag{2.81}$$

$$\psi_{32}^{(2)}(t) \equiv 0. \tag{2.82}$$

Now replacing Equations 2.81 and 2.82 in Equations 2.77 through 2.80 and carrying out the respective integrations yields

$$\frac{\partial^2 \rho(t_1)}{\partial q \, \partial \rho_{in}} \equiv 0, \tag{2.83}$$

$$\frac{\partial^2 \rho(t_1)}{\partial q^2} \equiv 0, \tag{2.84}$$

$$\frac{\partial^2 \rho(t_1)}{\partial q \, \partial \beta_j} = w_j t_1 \exp\left(-t_1 \sum_{i=1}^{N} w_i \beta_i\right), \quad j = 1,\ldots,N, \tag{2.85}$$

$$\frac{\partial^2 \rho(t_1)}{\partial q \, \partial w_j} = \beta_j t_1 \exp\left(-t_1 \sum_{i=1}^{N} w_i \beta_i\right), \quad j = 1,\ldots,N. \tag{2.86}$$

2.2.4 Computing the Second-Order Response Sensitivities Corresponding to the First-Order Sensitivities $\partial\rho(t_1)/\partial\rho_{in}$

The last set of second-order response sensitivities will be calculated in this section by taking the G-differential of Equation 2.18. Since the model responses need to be written in the form of an inner product in order to apply the adjoint sensitivity analysis methodology, Equation 2.18 is first rewritten in the form

$$\frac{\partial \rho(t_1)}{\partial \rho_{in}} = \psi^{(1)}(0) = \int_0^{t_f} \psi^{(1)}(t)\delta(t)\,dt. \tag{2.87}$$

Taking the G-differential of Equation 2.87 yields

$$\delta\left[\frac{\partial \rho(t_1)}{\partial \rho_{in}}\right] = \frac{d}{d\varepsilon}\left\{\int_0^{t_f} \left[\psi^{(1)}(t) + \varepsilon \delta\psi^{(1)}(t)\right]\delta(t)\,dt\right\}_{\varepsilon=0} = \int_0^{t_f} \delta\psi^{(1)}(t)\delta(t)\,dt. \tag{2.88}$$

Analogous to the previous considerations, use an arbitrary function $\psi_4^{(2)}(t) \triangleq \left[\psi_{41}^{(2)}(t), \psi_{42}^{(2)}(t)\right] \in \mathcal{H}^{(2)}$ to obtain the following relation, which is formally identical to Equations 2.36, 2.54, and 2.71, except that the roles previously played by $\psi_1^{(2)}(t)$, $\psi_2^{(2)}(t)$, and $\psi_3^{(2)}(t)$, respectively, are now played by the yet-undetermined function $\psi_4^{(2)}(t)$:

$$\int_0^{t_f} \left[\psi_{41}^{(2)}(t), \psi_{42}^{(2)}(t)\right] \begin{pmatrix} -\psi^{(1)}(t)\sum_{i=1}^{N}(w_i\delta\beta_i + \beta_i\delta w_i) \\ [q-\rho(t)]\sum_{i=1}^{N}(w_i\delta\beta_i + \beta_i\delta w_i) + \delta q\sum_{i=1}^{N} w_i\beta_i \end{pmatrix} dt$$

$$= -\psi_{41}^{(2)}(t_f)\delta\psi^{(1)}(t_f) + \psi_{41}^{(2)}(0)\delta\psi^{(1)}(0) + \psi_{42}^{(2)}(t_f)\delta\rho(t_f) - \psi_{42}^{(2)}(0)\delta\rho(0)$$
$$+\int_0^{t_f} \delta\psi^{(1)}(t)\left[\frac{d\psi_{41}^{(2)}(t)}{dt} + \psi_{41}^{(2)}(t)\sum_{i=1}^{N} w_i\beta_i\right]dt$$
$$+\int_0^{t_1} \delta\rho(t)\left[-\frac{d\psi_{42}^{(2)}(t)}{dt} + \psi_{42}^{(2)}(t)\sum_{i=1}^{N} w_i\beta_i\right]dt. \quad (2.89)$$

The last two terms on the right side of Equation 2.89 are now identified with the right side of Equation 2.88 by requiring that

$$\frac{d\psi_{41}^{(2)}(t)}{dt} + \psi_{41}^{(2)}(t)\sum_{i=1}^{N} w_i\beta_i = \delta(t), \quad (2.90)$$

$$-\frac{d\psi_{42}^{(2)}(t)}{dt} + \psi_{42}^{(2)}(t)\sum_{i=1}^{N} w_i\beta_i = 0, \quad (2.91)$$

$$\psi_{41}^{(2)}(0) = 0, \quad (2.92)$$

$$\psi_{42}^{(2)}(t_f) = 0. \quad (2.93)$$

The above system of equations, comprising Equations 2.90 through 2.93, constitutes the 2nd-LASS for computing the second-order sensitivities stemming from $\partial\rho(t_1)/\partial\rho_{in}$. Note again that the left sides of Equations 2.90 and 2.91 involve the same differential operators as the left sides of the 2nd-LASS that were previously obtained in Sections 2.2.1 through 2.2.3. Hence, the same solvers would also be used for solving Equations 2.90 through 2.93; only the source terms on the right sides of these differential equations are distinct.

Since the right sides of Equations 2.90 through 2.93 are independent of the index $i = 1,\ldots, N$, these equations would need to be solved once only, to obtain the second-level adjoint function $\psi_4^{(2)}(t) \triangleq \left[\psi_{41}^{(2)}(t), \psi_{42}^{(2)}(t)\right]$.

Using the results given in Equations 2.90 through 2.93 in Equation 2.89 reduces the latter to

$$\delta\left[\frac{\partial \rho(t_1)}{\partial \rho_{in}}\right] = \int_0^{t_f} \left[\psi_{41}^{(2)}(t), \psi_{42}^{(2)}(t)\right] \begin{pmatrix} -\psi^{(1)}(t) \sum_{i=1}^{N} (w_i \delta\beta_i + \beta_i \delta w_i) \\ \left[q - \rho(t)\right] \sum_{i=1}^{N} (w_i \delta\beta_i + \beta_i \delta w_i) + \delta q \sum_{i=1}^{N} w_i \beta_i \end{pmatrix} dt$$

$$+ \psi_{42}^{(2)}(0)\delta\rho_{in}, \quad i = 1,\ldots, N. \tag{2.94}$$

Identifying in Equation 2.94 the respective coefficients of the parameter variations yields the following expressions:

$$\frac{\partial^2 \rho(t_1)}{\partial \rho_{in}^2} = \psi_{42}^{(2)}(0), \tag{2.95}$$

$$\frac{\partial^2 \rho(t_1)}{\partial \rho_{in}\, \partial q} = \left(\sum_{i=1}^{N} w_i \beta_i\right) \int_0^{t_f} \psi_{42}^{(2)}(t)\, dt, \tag{2.96}$$

$$\frac{\partial^2 \rho(t_1)}{\partial \rho_{in}\, \partial \beta_j} = -w_j \int_0^{t_f} \psi_{41}^{(2)}(t) \psi^{(1)}(t)\, dt + w_j \int_0^{t_f} \psi_{42}^{(2)}(t)\left[q - \rho(t)\right] dt; \; j = 1,\ldots, N, \tag{2.97}$$

$$\frac{\partial^2 \rho(t_1)}{\partial \rho_{in}\, \partial w_j} = -\beta_j \int_0^{t_f} \psi_{41}^{(2)}(t) \psi^{(1)}(t)\, dt + \beta_j \int_0^{t_f} \psi_{42}^{(2)}(t)\left[q - \rho(t)\right] dt; \; j = 1,\ldots, N. \tag{2.98}$$

Applying the integrating factor method to solve Equations 2.90 through 2.93 yields the following solutions:

$$\psi_{41}^{(2)}(t) = H(t) \exp\left(-t \sum_{i=1}^{N} w_i \beta_i\right), \tag{2.99}$$

$$\psi_{42}^{(2)}(t) \equiv 0. \tag{2.100}$$

Replacing now Equations 2.99 and 2.100 in Equations 2.77 and 2.80 and carrying out the respective integrations yields

$$\frac{\partial^2 \rho(t_1)}{\partial \rho_{in}^2} \equiv 0, \tag{2.101}$$

$$\frac{\partial^2 \rho(t_1)}{\partial \rho_{in}\, \partial q} \equiv 0, \tag{2.102}$$

$$\frac{\partial^2 \rho(t_1)}{\partial \rho_{in}\, \partial \beta_j} = -w_j t_1 \exp\left(-t_1 \sum_{i=1}^{N} w_i \beta_i\right), \qquad j = 1, \ldots, N, \tag{2.103}$$

$$\frac{\partial^2 \rho(t_1)}{\partial \rho_{in}\, \partial w_j} = -\beta_j t_1 \exp\left(-t_1 \sum_{i=1}^{N} w_i \beta_i\right), \qquad j = 1, \ldots, N. \tag{2.104}$$

2.2.5 Discussion of the Essential Features of the 2nd-ASAM

The results obtained by applying the 2nd-ASAM to the paradigm evolution problem analyzed in Sections 2.1 and 2.2 indicate the following major characteristics of this powerful methodology for computing exactly and efficiently the first- and second-order sensitivities of model responses with respect to model parameters:

1. It is evident from Equations 2.18 through 2.21 that the first-order sensitivities of the response $\rho(t_1)$ can be computed by fast quadratures once the first-level adjoint function $\psi^{(1)}(t)$ has been obtained by solving the 1st-LASS, that is, Equations 2.14 and 2.16. The 1st-LASS needs to be solved once only, since it does not depend on any variations in the model parameters or state functions. This is a general characteristic of the 1st-LASS, as will be seen repeatedly throughout this book. In particular, if the original forward problem is linear, the 1st-LASS is independent of the forward state functions and can therefore be solved independently of the original forward system.
2. The magnitudes of the first-order *relative* sensitivities are used to rank the importance of the respective parameters in affecting the response (e.g., the importance of the respective parameter's uncertainty in contributing to the overall uncertainty in the response). For the paradigm illustrative evolution problem considered in this chapter, Equations 2.20 and 2.21 indicate that the relative sensitivities of the response to the parameters β_i, $[\partial\rho(t_1)/\partial\beta_i][\beta_i/\rho(t_1)]$, and the

relative sensitivities of the response to the parameters w_i, $[\partial\rho(t_1)/\partial w_i]$ $[w_i/\rho(t_1)]$, respectively, happen to be *identical*, for all of these $2N$ model parameters, since

$$\frac{\partial\rho(t_1)}{\partial\beta_i}\frac{\beta_i}{\rho(t_1)}=\frac{\beta_i w_i}{\rho(t_1)}\int_0^{t_f}\psi^{(1)}(t)\left[q-\rho(t)\right]dt\equiv\frac{\partial\rho(t_1)}{\partial w_i}\frac{w_i}{\rho(t_1)},\quad i=1,\ldots,N. \qquad (2.105)$$

It is known that statistical methods that use *a priori* screening techniques to reduce the number of model parameters that are actually considered in the respective statistical uncertainty/sensitivity analysis are very likely to *fail* to achieve their goal for problems that have many parameters with identical (or nearly identical) relative sensitivities. This illustrative paradigm problem, which has many (N_α = 1292) model parameters (Cacuci et al. 2016; Peltz and Cacuci 2016) that have identical (or nearly identical) relative sensitivities, would be a prime candidate for illustrating the failure of statistical methods for sensitivity and uncertainty analysis.

3. The 2nd-ASAM considers each first-order sensitivity as a response and develops a corresponding 2nd-LASS for computing the respective second-order sensitivities. For each first-order sensitivity, the solution of each of the 2nd-LASS is a two-component (vector) *second-level adjoint sensitivity function* of the form $\psi_i^{(2)}(t)\triangleq\left[\psi_{i1}^{(2)}(t),\psi_{i2}^{(2)}(t)\right]$, $i=1,\ldots,N_\alpha$, where N_α denotes the total number of distinct model parameters. Hence, although these second-level adjoint sensitivity functions are independent of any parameter variations, the 2nd-LASS would need to be solved with N_α distinct right sides (i.e., source terms), in the most unfavorable situation. Even in this worst-case scenario, only the right sides of the 2nd-LASS would need to be modified, which is relatively easy to program computationally. The left sides of these equations (which contain the differential operators and which would therefore involve solvers that would be much more difficult to modify computationally) remain unchanged, in any case.
4. In many practical situations, it is possible to reduce drastically the number of computations involving the 2nd-LASS. Occasionally, the solutions of some of the 2nd-LASS can be written down by inspection, without actually solving the corresponding 2nd-LASS. *For the paradigm evolution problem analyzed in this chapter, for example, the 2nd-LASS would need to be solved only four times, to compute the second-level adjoint functions* $\psi_j^{(2)}(t)\triangleq\left[\psi_{j1}^{(2)}(t),\psi_{j2}^{(2)}(t)\right]$, $j=1,\ldots,4$, *which would suffice for obtaining all of the* $N_\alpha(N_\alpha+1)/2$ = 835278 *distinct second-order sensitivities of the scalar response* $\rho(t_1)$ to all of the N_α = 1292 *model parameters.*

Of course, such a very large reduction in the number of large-scale computations cannot be expected in every practical problem, but in most cases the number of computations required for computing the complete set of second-order response sensitivities is far less than the number, N_α, of model parameters.

5. It is important to note that many of the *mixed* second-order sensitivities are obtained *twice*, stemming from distinct 2nd-LASS. This fact enables the 2nd-ASAM to provide an *inherent independent verification of the correctness and accuracy of the second-level adjoint sensitivity functions that are used to compute the respective mixed second-order sensitivities.* This fact becomes readily apparent by considering, for example, the second-order sensitivities involving the model parameters β_i. Thus, on the one hand, the sensitivities $[\partial^2\rho(t_1)/\partial\beta_i\partial\rho_{in}]$ are obtained in Equation 2.43 in terms of the components of the second-level adjoint sensitivity function $\psi_1^{(2)}(t)$, but, on the other hand, the same sensitivities are also obtained in Equation 2.97 in terms of the components of the second-level adjoint sensitivity function $\psi_4^{(2)}(t)$. Similarly, the sensitivities $[\partial^2\rho(t_1)/\partial\beta_i\partial\rho]$ are obtained in Equation 2.44 in terms of the components of the second-level adjoint sensitivity function $\psi_1^{(2)}(t)$, on the one hand, but the same sensitivities are also obtained in Equation 2.85 in terms of the components of the second-level adjoint sensitivity function $\psi_3^{(2)}(t)$, on the other hand. Likewise, the sensitivities $[\partial^2\rho(t_1)/\partial\beta_i\partial w]$ are obtained in Equation 2.46 in terms of the components of the second-level adjoint sensitivity function $\psi_1^{(2)}(t)$, but the same sensitivities are also obtained in Equation 2.62 in terms of the second-level adjoint sensitivity function $\psi_2^{(2)}(t)$.
6. The specific second-order sensitivities to be computed can be selected *a priori*, based on the magnitude/importance of the first-order sensitivities.
7. The *un-mixed second-order sensitivities* of the form $\partial^2\rho(t_1)/\partial\alpha_i^2$ are obtained only once, so they can be independently verified only by recomputations using finite difference formulas of the form

$$\frac{\partial^2\rho(t_1)}{\partial\alpha_i^2} \cong \frac{\rho\left(t_1;\alpha_i^0+\delta\alpha_i\right)-2\rho\left(t_1;\alpha_i^0\right)+\rho\left(t_1;\alpha_i^0-\delta\alpha_i\right)}{\left|\delta\alpha_i\right|^2}. \tag{2.106}$$

Of course, using such finite difference formulas yields *approximate*, rather than exact, values for the respective second-order sensitivities.

2.2.6 Illustrative Use of Response Sensitivities for Predictive Modeling

The role of the first-order sensitivities for predictive modeling will be illustrated in this section by assimilating a single measurement of the response

$\rho(t_1)$ for the paradigm dissolver model analyzed in this chapter. The methodology of Cacuci and Ionescu-Bujor (2010), which was summarized in Chapter 1, will be used along with the following *a priori* known information:

1. The nominal (mean) value of the measurement, denoted as $\rho^{meas}(t_1)$, and the measurement's variance, denoted as $Var[\rho^{meas}(t_1)]$.
2. The parameter mean values, denoted as α_i^0, and the parameter covariance, denoted as $\mathrm{cov}(\alpha_i, \alpha_j)$, of the N_α imprecisely known system (model) parameters.
3. The response sensitivities, $\partial\rho(t_1)/\partial\alpha_i$, computed using the 1st-LASS in Section 2.1.
4. The second-order response sensitivities will not be used.
5. The expected value, $E[\rho^{comp}(t_1)]$, of the computed response, given by Equation 1.18 in Chapter 1.
6. The variance, $Var[\rho^{comp}(t_1)]$, of the computed response $\rho^{comp}(t_1)$, given by Equation 1.19 in Chapter 1, which can be written in the form $Var\left[\rho^{comp}(t_1)\right] = \mathbf{S}_{r\alpha}\mathbf{C}_{\alpha\alpha}\mathbf{S}_{r\alpha}^{\dagger}$, where $\mathbf{C}_{\alpha\alpha}$ represents the parameter covariance matrix and $\mathbf{S}_{r\alpha}$ represents the row vector of sensitivities $[\partial\rho(t_1)/\partial\alpha_1, \ldots, \partial\rho(t_1)/\partial\alpha_{N_\alpha+1}]$.
7. The response $\rho(t)$ is assumed to be uncorrelated with the model parameters α_i.

Application of Equation 1.27 from Chapter 1 yields the following result for the optimally predicted best-estimate nominal values, $\boldsymbol{\alpha}^{pred}$, for the "calibrated" model parameters:

$$\alpha^{pred} = \alpha^0 - \mathbf{C}_{\alpha\alpha}\mathbf{S}_{r\alpha}^{\dagger}\left\{Var\left[\rho^{comp}(t_1)\right] + Var\left[\rho^{meas}(t_1)\right]\right\}^{-1}\left[\rho^{comp}(t_1) - \rho^{meas}(t_1)\right],$$

which can be written for each component, for each $i = 1,\ldots,N_\alpha$, as follows:

$$\alpha_i^{pred} = \alpha_i^0 - \frac{\left[\rho^{comp}(t_1) - \rho^{meas}(t_1)\right]}{Var\left[\rho^{comp}(t_1)\right] + Var\left[\rho^{meas}(t_1)\right]}\sum_{j=1}^{N_\alpha+1}\frac{\partial\rho(t_1)}{\partial\alpha_j}\mathrm{cov}(\alpha_i, \alpha_j). \tag{2.107}$$

As Equation 2.107 indicates, the calibrated parameter value α_i^{pred} may become larger or smaller than the original mean (nominal) parameter value α_i^0, depending on the sign of the second term on the right side of this equation. If the parameters are uncorrelated, Equation 2.107 takes on the simpler form

$$\alpha_i^{pred} = \alpha_i^0 - \frac{\left[\rho^{comp}(t_1) - \rho^{meas}(t_1)\right]}{Var\left[\rho^{comp}(t_1)\right] + Var\left[\rho^{meas}(t_1)\right]}\frac{\partial\rho(t_1)}{\partial\alpha_i}Var(\alpha_i). \tag{2.108}$$

Application of Equation 1.39 from Chapter 1 yields the following result for the predicted covariance matrix, denoted as $\mathbf{C}_{\alpha\alpha}^{pred}$, for the predicted nominal parameter values:

$$\mathbf{C}_{\alpha\alpha}^{pred} = \mathbf{C}_{\alpha\alpha} - \frac{\left(\mathbf{C}_{\alpha\alpha}\mathbf{S}_{r\alpha}^{\dagger}\right)\left(\mathbf{C}_{\alpha\alpha}\mathbf{S}_{r\alpha}^{\dagger}\right)^{\dagger}}{Var\left[\rho^{comp}(t_1)\right] + Var\left[\rho^{meas}(t_1)\right]},$$

which implies that

$$\mathbf{C}_{\alpha\alpha}^{pred} - \mathbf{C}_{\alpha\alpha} = -\frac{\sum_{j=1}^{N_\alpha+1}\left[\partial\rho(t_1)/\partial\alpha_i\right]^2\left(\mathbf{C}_{\alpha\alpha}\right)^2}{Var\left[\rho^{comp}(t_1)\right] + Var\left[\rho^{meas}(t_1)\right]}. \tag{2.109}$$

Since the right side of Equation 2.109 represents a negative-definite matrix, it follows that the diagonal elements in this equation obey the inequality

$$Var\left(\alpha_i^{pred}\right) - Var(\alpha_i) < 0. \tag{2.110}$$

The above inequality is equivalent to the inequality

$$Std.\ Dev.\left(\alpha_i^{pred}\right) < Std.Dev(\alpha_i), \tag{2.111}$$

which indicates that *the predicted uncertainties in the predicted (calibrated) parameters are smaller (i.e., are "reduced") than the original uncertainties in the model parameters.*

Application of Equation 1.37 from Chapter 1 yields the following result for the optimally predicted best-estimate nominal values, denoted as $\rho^{pred}(t_1)$, for the model responses:

$$\rho^{pred}(t_1) = \rho^{meas}(t_1) + \frac{Var\left[\rho^{meas}(t_1)\right]\left[\rho^{comp}(t_1) - \rho^{meas}(t_1)\right]}{Var\left[\rho^{comp}(t_1)\right] + Var\left[\rho^{meas}(t_1)\right]}. \tag{2.112}$$

The above expression can be recast in the form

$$\rho^{pred}(t_1) - \rho^{meas}(t_1) = x\left[\rho^{comp}(t_1) - \rho^{meas}(t_1)\right], \tag{2.113}$$

with

$$0 \le x \triangleq \frac{Var\left[\rho^{meas}(t_1)\right]}{Var\left[\rho^{comp}(t_1)\right] + Var\left[\rho^{meas}(t_1)\right]} \le 1. \tag{2.114}$$

As Equations 2.113 and 2.114 indicate, if the measurement is performed with perfect accuracy, then $Var[\rho^{meas}(t_1)] = 0$ and, consequently, $\rho^{pred}(t_1) = \rho^{meas}(t_1)$. In other words, the predicted values for the responses coincide with the measured values when the latter are perfectly well measured. On the other hand, if the computation is perfect, then the predicted response values would coincide with the computed values, that is, $\rho^{pred}(t_1) = \rho^{meas}(t_1)$, since $Var[\rho^{comp}(t_1)] = 0$. In such an improbable case, the experimental measurements would have no effect on the predictions, since imperfect measurements could not possibly improve a "perfect" model's predictions. Finally, if both the computation and the experiment are affected by uncertainties, then Equations 2.113 and 2.114 indicate that the predicted response value falls, as would be expected, between the computed and measured response values, that is,

$$\rho^{meas}(t_1) \le \rho^{pred}(t_1) \le \rho^{comp}(t_1). \tag{2.115}$$

The above argument has tacitly assumed that both the computed and measured responses have positive nominal values. For responses that have negative nominal values, the above argument is repeated using the responses' absolute values, leading to the same general conclusion.

Application of Equation 1.42 from Chapter 1 yields the following result for the predicted variance for the predicted response:

$$Var\left[\rho^{pred}(t_1)\right] = Var\left[\rho^{meas}(t_1)\right] - \frac{\left\{Var\left[\rho^{meas}(t_1)\right]\right\}^2}{Var\left[\rho^{comp}(t_1)\right] + Var\left[\rho^{meas}(t_1)\right]}. \tag{2.116}$$

The above expression implies that

$$Var\left[\rho^{pred}(t_1)\right] - Var\left[\rho^{meas}(t_1)\right] < 0, \quad Var\left[\rho^{pred}(t_1)\right] - Var\left[\rho^{comp}(t_1)\right] < 0. \tag{2.117}$$

The inequalities shown in Equation 2.117 indicate that the predicted standard deviation will be smaller that both the measured and the computed response standard deviations, respectively. In this sense, the uncertainties in the predicted response will be reduced by comparison to the uncertainties for both the computed and the measured response.

Application of Equation 1.45 from Chapter 1 yields the following result for the predicted correlation matrix, $\mathbf{C}_{\alpha r}^{pred}$, between the predicted model parameters and responses:

$$\mathbf{C}_{\alpha r}^{pred} = \left(\mathbf{C}_{\alpha\alpha}\mathbf{S}_{r\alpha}^{\dagger}\right) \frac{Var\left[\rho^{meas}(t_1)\right]}{Var\left[\rho^{comp}(t_1)\right] + Var\left[\rho^{meas}(t_1)\right]}, \tag{2.118}$$

or, in component form, for each $i = 1,\ldots,N_\alpha$,

$$Corr\left[\alpha_i^{pred},\rho^{pred}(t_1)\right]=\frac{Var\left[\rho^{meas}(t_1)\right]}{Var\left[\rho^{comp}(t_1)\right]+Var\left[\rho^{meas}(t_1)\right]}\sum_{j=1}^{N_\alpha+1}\frac{\partial\rho(t_1)}{\partial\alpha_j}\operatorname{cov}\left(\alpha_i,\alpha_j\right). \tag{2.119}$$

As indicated by the expressions in Equations 2.118 or 2.119, even if the response is *a priori* uncorrelated with the model parameters (i.e., $Corr[\alpha_i, \rho(t_1)] = 0$), an imperfect measurement (i.e., $Var[\rho^{meas}(t_1)] \neq 0$) will introduce nonzero posterior correlations (i.e., $Corr[\alpha_i^{pred}, \rho^{pred}(t_1)] \neq 0$) between the predicted calibrated parameter values and the predicted best-estimate response. Only if the measurement(s) were perfect would initially uncorrelated response parameters remain uncorrelated after the measurements, in which case the predicted responses would also coincide with the measured ones as was discussed in the foregoing, following Equation 2.114.

Application of Equation 1.47 from Chapter 1 yields the following result for the consistency indicator

$$\chi^2 \triangleq \frac{\left[\rho^{comp}(t_1)-\rho^{meas}(t_1)\right]^2}{Var\left[\rho^{comp}(t_1)\right]+Var\left[\rho^{meas}(t_1)\right]}, \tag{2.120}$$

For a single measurement, the above consistency indicator is helpful to identify possibly inconsistent data, which would be flagged, for example, when the "distance" between the measurement and computation is larger than the sum of the respective standard deviations of the computed and measured responses, that is, when $\rho^{comp}(t_1) - \rho^{meas}(t_1)| > Std.Dev.[\rho^{meas}(t_1)] + Std.Dev.[\rho^{comp}(t_1)]$ due to unrecognized errors.

The predictive modeling methodology of Cacuci and Ionescu-Bujor (2010) can also be applied for "inverse predictive modeling" by using measurements in order to identify unknown model parameters α_i. For example, for the paradigm dissolver model illustratively analyzed in this chapter, the "inverse predictive mode" would employ Equation 2.108 to start the iteration by using an "initial estimate," $(\alpha_i^0)^{estimated}$, instead of the unknown initially nominal value α_i^0. In other words, in the inverse predictive mode, Equation 2.108 would be replaced, for the initial iteration, by the equation

$$\alpha_i^{pred} = \left(\alpha_i^0\right)^{estimated} - \frac{\left[\rho^{comp}(t_1)-\rho^{meas}(t_1)\right]}{Var\left[\rho^{comp}(t_1)\right]+Var\left[\rho^{meas}(t_1)\right]}\sum_{j=1}^{N_\alpha+1}\frac{\partial\rho(t_1)}{\partial\alpha_j}\operatorname{cov}\left(\alpha_i,\alpha_j\right). \tag{2.121}$$

Subsequently, the predicted response and parameter values, as well as the respective predicted covariance matrices, would be computed repeatedly by using the newly computed "predicted values" to replace the corresponding values obtained in the previous iteration, assimilating new experimental measurements, etc., until the computed values agree with the corresponding measured values within an *a priori* user-selected error criterion. Cacuci (2017) illustrated such an "inverse predictive modeling" application in the context of inverse predictive modeling of radiation transport through optically thick media in the presence of counting uncertainties.

3

The 2nd-ASAM for Linear Systems

The field of nuclear science and engineering provided the earliest use (Wigner 1945) of adjoint operators for performing sensitivity analysis of a large-scale linear system comprising many parameters. The large system analyzed by Wigner (1945) was a model of a nuclear reactor, in which the model parameters were the thousands of neutron cross sections influencing the reactor's behavior, and the model response was the reactor's multiplication factor. The field of nuclear science and engineering also provided pioneering works, during the 1970s, for computing second-order response sensitivities of responses associated with the forward and adjoint neutron and radiation transport and/or diffusion equations (Mitani 1973; Seki 1973; Gandini 1978; Greenspan et al. 1978; Cacuci et al. 1980). These works generally indicated that the second-order sensitivities of responses (e.g., reaction rates, the system's effective multiplication factor) to model parameters (e.g., cross sections and densities) were computationally intensive to obtain, requiring $O\left(N_\alpha^2\right)$ large-scale computations per response, for a system comprising N_α model parameters, so only the second-order sensitivities that were deemed *a priori* to be "important" were actually computed. Furthermore, it turned out that the second-order sensitivities thus computed were considerably smaller than the corresponding first-order sensitivities. These results likely gave rise to the generally held opinion that second-order sensitivities are generally insignificant in reactor physics, which may, in turn, have led to diminishing interest in developing efficient methods for computing second-order sensitivities for nuclear engineering systems.

While the interest in computing second-order response sensitivities practically vanished in the nuclear engineering field in the 1990s, interest in this topic became increasingly evident in other fields, driven mostly by the knowledge that second-order (Hessian) sensitivity information accelerates the convergence of optimization algorithms. In structural mechanics, for example, the interest has been focused primarily on developing specific adjoint methods for computing the second-order sensitivity of structural responses to variations of structural stiffness parameters (see, e.g., Haug 1981; Haug and Ehle 1982; Haftka 1982; Dems and Mroz 1985; Haftka and Mroz 1986; Haug, et al. 1986). In the context of parametric circuit analysis and optimization (see, e.g., Ye et al. 2010; Negm et al. 2014), time-domain second-order adjoint sensitivities for linear circuits were also computed, albeit approximately. In the atmospheric and earth sciences (see, e.g., Wang et al. 1992, 1995; Le Dimet et al. 2002; Cacuci, Navon, and Ionescu-Bujor 2005, 2014; Sandu and Zhang 2008; Daescu and Navon 2014; Práger and Kelemen 2014), so-called second-order adjoint models have been used for the following purposes:

1. Compute products between the Hessian of the cost functional and a vector (representing a perturbation in sensitivity analysis, a search direction in optimization, an eigenvector, etc.)
2. Perform sensitivity analysis of the cost function with respect to distributed observations
3. Study the evolution of the condition number (the ratio of the largest to smallest eigenvalues) of the Hessian during minimization
4. Perform specific sensitivity studies in three-dimensional atmospheric chemical transport models

The methods mentioned in the foregoing were all developed for specific, rather than general, applications for which they usually estimated, rather than computed exactly and inclusively, second-order response sensitivities to the model's parameters. Cacuci (2015a) conceived the generally applicable 2nd-ASAM for linear systems for computing exactly and most efficiently *all* of the second-order functional derivatives of model responses to parameters in models of linear systems (physical, engineering, biological, etc.). While the main steps in applying the 2nd-ASAM to a paradigm linear system have already been illustrated in Chapter 1, this chapter will present the general formulation and theoretical underpinnings of the 2nd-ASAM, following the material originally developed by Cacuci (2015a).

3.1 Mathematical Modeling of a General Linear System

A *linear* physical system is represented mathematically by means of N_u coupled operator equations of the form $\sum_{j=1}^{N_u} L_{ij}(\boldsymbol{\alpha})u_j(\mathbf{x})=Q_i(\boldsymbol{\alpha}),\ i=1,\ldots,N_u,\quad \mathbf{x}\in\Omega_x$, in which the operators $L_{ij}(\boldsymbol{\alpha})$ act *linearly on the state functions* $u_j(\mathbf{x})$. The above system of equations can be written in matrix form as

$$\mathbf{L}\left[\boldsymbol{\alpha}(\mathbf{x})\right]\mathbf{u}(\mathbf{x})=\mathbf{Q}\left[\boldsymbol{\alpha}(\mathbf{x})\right],\quad \mathbf{x}\in\Omega_x. \tag{3.1}$$

For notational convenience, all vectors in this book are considered to be column vectors. Transposition will be indicated by a dagger (†) superscript. The quantities appearing in Equation 3.1 are defined as follows:

1. $\mathbf{x} \triangleq (x_1, \ldots, x_{J_x})$ denotes the J_x-dimensional phase-space position vector and the symbol "$\triangleq$" will be used to denote "is defined as" or "is by definition"; $\mathbf{x} \in \Omega_x \subset \mathbb{R}^{J_x}$, where Ω_x is a subset of the J_x-dimensional real vector space $\mathbb{R}^{J_x}$.
2. $\mathbf{u}(\mathbf{x}) \triangleq [u_1(\mathbf{x}), \ldots, u_{N_u}(\mathbf{x})]$ denotes a N_u-dimensional column vector whose components represent the system's dependent variables (also called "state functions"); $\mathbf{u}(\mathbf{x}) \in \mathscr{E}_u$, where $\mathscr{E}_u$ denotes a normed linear space over the scalar field $\mathscr{F}$ of real numbers.
3. $\boldsymbol{\alpha}(\mathbf{x}) \triangleq [\alpha_1(\mathbf{x}), \ldots, \alpha_{N_\alpha}(\mathbf{x})]$ denotes a N_α-dimensional column vector whose components are the system's parameters; $\boldsymbol{\alpha} \in \mathscr{E}_\alpha$, where $\mathscr{E}_\alpha$ also denotes a normed linear space.
4. $\mathbf{Q}[\boldsymbol{\alpha}(\mathbf{x})] \triangleq [Q_1(\boldsymbol{\alpha}), \ldots, Q_{N_u}(\boldsymbol{\alpha})]$ denotes a N_u-dimensional column vector whose elements represent inhomogeneous source terms that depend either linearly or nonlinearly on $\boldsymbol{\alpha}$; $\mathbf{Q} \in \mathscr{E}_Q$, where $\mathscr{E}_Q$ is also a normed linear space; the components of $\mathbf{Q}$ may be operators, rather than just functions, acting on $\boldsymbol{\alpha}(\mathbf{x})$ and $\mathbf{x}$.
5. $\mathbf{L}(\boldsymbol{\alpha}) \triangleq [L_1(\boldsymbol{\alpha}), \ldots, L_{N_u}(\boldsymbol{\alpha})]$ denotes a N_u-component column vector whose components are *operators* (including differential, difference, integral, distributions, and infinite matrices) acting *linearly* on $\mathbf{u}$ and *nonlinearly* on $\boldsymbol{\alpha}$.
6. All of the equalities in this work are considered to hold in the weak ("distributional") sense, since the right sides ("sources") of Equation 3.1 and of other various equations to be derived in this chapter may contain distributions ("generalized functions/functionals"), particularly Dirac distributions and derivatives and/or integrals thereof.

In view of the definitions given above, $\mathbf{L}(\boldsymbol{\alpha})$ represents the mapping $\mathbf{L}: \mathscr{D} \subset \mathscr{E} \to \mathscr{E}_Q$, where $\mathscr{D} = \mathscr{D}_u \oplus \mathscr{D}_\alpha$, $\mathscr{D}_u \subset \mathscr{E}_u$, $\mathscr{D}_\alpha \subset \mathscr{E}_\alpha$, and $\mathscr{E} = \mathscr{E}_u \oplus \mathscr{E}_\alpha$. Note that an arbitrary element $\mathbf{e} \in \mathscr{E}$ is of the form $\mathbf{e} = (\mathbf{u}, \boldsymbol{\alpha})$. If differential operators appear in Equation 3.1, then a corresponding set of boundary and/or initial conditions (which are essential to define $\mathscr{D}$) must also be given. Since we consider here only systems that are linear in the state function $\mathbf{u}(\mathbf{x}) \in \mathscr{E}_u$, the accompanying boundary and/or initial conditions must also be linear in $\mathbf{u}(\mathbf{x})$, so they can therefore be represented in operator form as

$$\left[\mathbf{B}(\boldsymbol{\alpha})\mathbf{u} - \mathbf{C}(\boldsymbol{\alpha})\right]_{\partial\Omega_x} = \mathbf{0}, \quad \mathbf{x} \in \partial\Omega_x, \tag{3.2}$$

where $\partial\Omega_x$ denotes the boundary of Ω_x, and where $\mathbf{B}(\boldsymbol{\alpha})\mathbf{u}$ acts *linearly* on $\mathbf{u}$ and *nonlinearly* on $\boldsymbol{\alpha}$, while $\mathbf{C}(\boldsymbol{\alpha})$ denotes an operator that acts *nonlinearly* on $\boldsymbol{\alpha}$.

The vector-valued function $\mathbf{u}(\mathbf{x})$ is considered to be the unique nontrivial solution of the physical problem described by Equations 3.1 and 3.2. The system response (i.e., result of interest), associated with the problem modeled by Equations 3.1 and 3.2 is considered to be a real-valued nonlinear *functional* of $(\mathbf{u},\boldsymbol{\alpha})$, denoted as $R(\mathbf{u},\boldsymbol{\alpha})$, and defined on a domain D_R; such a model response can be generally represented in operator form as

$$R(\mathbf{u},\boldsymbol{\alpha}):\mathcal{D}_R \subset \mathcal{E} \to \mathcal{F}, \tag{3.3}$$

where $\mathcal{F}$ denotes the field of real scalars.

The nominal solution of Equations 3.1 and 3.2 is denoted as $\mathbf{u}^0(\mathbf{x})$ and is obtained by solving these equations at the nominal parameter values $\boldsymbol{\alpha}^0(\mathbf{x})$. In other words, $\mathbf{u}^0(\mathbf{x})$ and $\boldsymbol{\alpha}^0(\mathbf{x})$ satisfy the following equations:

$$\mathbf{L}\left(\boldsymbol{\alpha}^0\right)\mathbf{u}^0 = \mathbf{Q}\left(\boldsymbol{\alpha}^0\right), \quad \mathbf{x} \in \Omega_x, \tag{3.4}$$

$$\mathbf{B}\left(\boldsymbol{\alpha}^0\right)\mathbf{u}^0 = \mathbf{C}\left(\boldsymbol{\alpha}^0\right), \quad \mathbf{x} \in \partial\Omega_x. \tag{3.5}$$

Equations 3.4 and 3.5 represent the *base-case* or *nominal* state of the physical system. After solving Equations 3.4 and 3.5, the *nominal solution*, $\mathbf{u}^0(\mathbf{x})$, thus obtained is used to obtain the *nominal value* $R(\mathbf{e}^0)$ of the response $R(\mathbf{e})$, using the nominal values $\mathbf{e}^0 \triangleq (\mathbf{u}^0,\boldsymbol{\alpha}^0) \in \mathcal{E}$ of the model's state function and parameters. Throughout this chapter, a superscript 0 will be used to denote nominal values.

The model parameters are considered to be imprecisely known quantities, so their actual values may differ from their nominal values by quantities denoted as $\delta\alpha_i(\mathbf{x}) \triangleq \alpha_i(\mathbf{x}) - \alpha_i^0(\mathbf{x})$, $i = 1,\ldots,N_\alpha$. Since the model parameters $\boldsymbol{\alpha}(\mathbf{x})$ and the state functions $\mathbf{u}(\mathbf{x})$ are related to each other through Equations 3.1 and 3.2, it follows that the vector of "parameter variations" $\delta\boldsymbol{\alpha} \triangleq \left(\delta\alpha_1,\ldots,\delta\alpha_{N_\alpha}\right) \in \mathcal{E}_\alpha$ will cause corresponding variations $\delta\mathbf{u} \triangleq \left(\delta u_1,\ldots,\delta u_{N_u}\right) \in \mathcal{E}_u$ in the state functions around the nominal solution $\mathbf{u}^0(\mathbf{x})$. All of these variations will cause variations in the response around the nominal response value $R(\mathbf{e}^0)$. *Sensitivity analysis* aims at quantifying the response variations that are induced by variations $\mathbf{h} \triangleq (\delta\boldsymbol{\alpha}, \delta\mathbf{u}) \in \mathcal{E} = \mathcal{E}_\alpha \oplus \mathcal{E}_u$ in a neighborhood around $\mathbf{e}^0 \triangleq (\boldsymbol{\alpha}^0, \mathbf{u}^0) \in \mathcal{E}$.

3.2 The 1st-LASS for Computing Exactly and Efficiently First-Order Sensitivities of Scalar-Valued Responses for Linear Systems

The general forward and adjoint sensitivity analysis methodology for computing exactly and efficiently the *first-order sensitivities of nonlinear responses associated with general nonlinear systems* was formulated and developed by Cacuci (1981a, 1981b). The linear version of this methodology will be presented in this section since it provides the basis for the subsequent generalizations needed for computing the second-order response sensitivities. Cacuci (1981a, 1981b) has shown that the most general definition of the first-order total sensitivity of an operator-valued model response $\mathbf{R}(\mathbf{e})$ to variations $\mathbf{h} \triangleq (\delta\boldsymbol{\alpha}, \delta\mathbf{u})$ in the model parameters and state functions in a neighborhood around $\mathbf{e}^0 = (\boldsymbol{\alpha}^0, \mathbf{u}^0) \in \mathscr{E}$ is given by the first-order G-differential (also called the *Gateaux variation*) of $\mathbf{R}(\mathbf{e})$. This G-differential will be denoted as $\delta\mathbf{R}(\mathbf{e}^0; \mathbf{h})$ and is defined as

$$\delta\mathbf{R}\left(\mathbf{e}^0;\mathbf{h}\right) \triangleq \left\{\frac{d}{d\varepsilon}\left[\mathbf{R}\left(\mathbf{e}^0+\varepsilon\mathbf{h}\right)\right]\right\}_{\varepsilon=0} \triangleq \lim_{\varepsilon\to 0}\frac{\mathbf{R}\left(\mathbf{e}^0+\varepsilon\mathbf{h}\right)-\mathbf{R}\left(\mathbf{e}^0\right)}{\varepsilon} \tag{3.6}$$

for $\varepsilon \in \mathscr{F}$ and all (i.e., arbitrary) vectors $\mathbf{h} \in \mathscr{E} = \mathscr{E}_{\boldsymbol{\alpha}} \times \mathscr{E}_u$ in a neighborhood $(\mathbf{e}^0 + \varepsilon\mathbf{h})$ around $\mathbf{e}^0 = (\boldsymbol{\alpha}^0, \mathbf{u}^0) \in \mathscr{E}$. The G-differential $\delta\mathbf{R}(\mathbf{e}^0; \mathbf{h})$ is an operator defined on the same domain as $R(\mathbf{e})$, has the same range as $\mathbf{R}(\mathbf{e})$, and provides the total first-order sensitivity of $\mathbf{R}(\mathbf{e})$ to variations in the model's parameters and state functions. The G-differential $\delta\mathbf{R}(\mathbf{e}^0; \mathbf{h})$ satisfies the relation $\mathbf{R}(\mathbf{e}^0 + \varepsilon\mathbf{h}) - \mathrm{R}(\mathbf{e}^0) = \delta\mathbf{R}(\mathbf{e}^0; \mathbf{h}) + \Delta(\mathbf{h})$, with $\lim_{\varepsilon\to 0}\left[\Delta(\varepsilon\mathbf{h})\right]/\varepsilon = 0$.

The existence of the G-differential $\delta\mathbf{R}(\mathbf{e}^0; \mathbf{h})$ does not guarantee its numerical computability. Numerical methods require that $\delta\mathbf{R}(\mathbf{e}^0; \mathbf{h})$ be linear in $\mathbf{h} \triangleq (\delta\boldsymbol{\alpha}, \delta\mathbf{u})$ in a neighborhood $(\mathbf{e}^0 + \varepsilon\mathbf{h})$ around $\mathbf{e}^0 = (\mathbf{u}^0, \boldsymbol{\alpha}^0) \in \mathscr{E}$. The necessary and sufficient conditions for the G-differential $\delta\mathbf{R}(\mathbf{e}^0; \mathbf{h})$ of a nonlinear operator $\mathbf{R}(\mathbf{e})$ to be linear in $\mathbf{h}$ in a neighborhood $(\mathbf{e}^0 + \varepsilon\mathbf{h})$ around $\mathbf{e}^0 = (\boldsymbol{\alpha}^0, \mathbf{u}^0) \in \mathscr{E}$, and thus admit G-derivatives, are as follows:

1. $\mathbf{R}(\mathbf{e})$ satisfies a weak Lipschitz condition at $\mathbf{e}^0$; (3.7)
2. $\mathbf{R}(\mathbf{e})$ satisfies the following condition:

$$\mathbf{R}\left(\mathbf{e}^0+\varepsilon\mathbf{h}_1+\varepsilon\mathbf{h}_2\right)-\mathbf{R}\left(\mathbf{e}^0+\varepsilon\mathbf{h}_1\right)-\mathbf{R}\left(\mathbf{e}^0+\varepsilon\mathbf{h}_2\right)+\mathbf{R}\left(\mathbf{e}^0\right)=o(\varepsilon);\ \mathbf{h}_1,\mathbf{h}_2\in\mathscr{E};\quad \varepsilon\in\mathscr{F}. \tag{3.8}$$

If the conditions described in Equations 3.7 and 3.8 are not satisfied, the first-order derivatives of the response and operators in Equations 3.1 through 3.3 are not guaranteed to exist, so the computation of higher-order response sensitivities (G-derivatives) would be moot. Therefore, the conditions given in Equations 3.7 and 3.8 are henceforth considered to be satisfied by the operators underlying the physical system modeled by Equations 3.1 and 3.2 as well as by the system's response defined by Equation 3.3.

When the conditions given in Equations 3.7 and 3.8 are satisfied, the G-differential $\delta\mathbf{R}(\mathbf{e}^0; \mathbf{h})$ can be written in the form

$$\delta\mathbf{R}\left(\mathbf{e}^0;\mathbf{h}\right)=\left\{\frac{\partial\mathbf{R}(\boldsymbol{\alpha},\mathbf{u})}{\partial\mathbf{u}}\right\}_{\left(\boldsymbol{\alpha}^0,\mathbf{u}^0\right)}\delta\mathbf{u}+\left\{\frac{\partial\mathbf{R}(\boldsymbol{\alpha},\mathbf{u})}{\partial\boldsymbol{\alpha}}\right\}_{\left(\boldsymbol{\alpha}^0,\mathbf{u}^0\right)}\delta\boldsymbol{\alpha},\quad \mathbf{x}\in\partial\Omega_x, \tag{3.9}$$

where $\partial\mathbf{R}(\boldsymbol{\alpha},\mathbf{u})/\partial\mathbf{u}$ and $\partial\mathbf{R}(\boldsymbol{\alpha},\mathbf{u})/\partial\boldsymbol{\alpha}$ denote the *partial G-derivatives* of $\mathbf{R}(\mathbf{e})$ with respect to $\mathbf{u}$ and $\boldsymbol{\alpha}$.

Applying the definition given in Equation 3.6 to the scalar-valued *nonlinear functional* response $R(\boldsymbol{\alpha},\mathbf{u})$ defined in Equation 3.3, and assuming that this functional also satisfies the conditions expressed in Equations 3.7 and 3.8, yields an expression formally similar to Equation 3.9, namely:

$$\delta R\left(\boldsymbol{\alpha}^0,\mathbf{u}^0;\mathbf{h}\right)=\left\{\frac{\partial R(\boldsymbol{\alpha},\mathbf{u})}{\partial \mathbf{u}}\right\}_{\left(\alpha^0,\mathbf{u}^0\right)}\delta\mathbf{u}+\left\{\frac{\partial R(\boldsymbol{\alpha},\mathbf{u})}{\partial \boldsymbol{\alpha}}\right\}_{\left(\alpha^0,\mathbf{u}^0\right)}\delta\boldsymbol{\alpha}, \tag{3.10}$$

where the column vectors $\partial R(\boldsymbol{\alpha},\mathbf{u})/\partial\mathbf{u}\triangleq\left[\partial R(\boldsymbol{\alpha},\mathbf{u})/\partial u_1,\ldots,\partial R(\boldsymbol{\alpha},\mathbf{u})/\partial u_{N_u}\right]$ and $\partial R(\boldsymbol{\alpha},\mathbf{u})/\partial\boldsymbol{\alpha}\triangleq\left[\partial R(\boldsymbol{\alpha},\mathbf{u})/\partial \alpha_1,\ldots,\partial R(\boldsymbol{\alpha},\mathbf{u})/\partial \alpha_{N_\alpha}\right]$, respectively, denote the partial gradients at $\mathbf{e}^0$ of R($\mathbf{e}$) with respect to $\mathbf{u}$ and $\boldsymbol{\alpha}$. It is convenient to refer to the quantities $[\partial R(\boldsymbol{\alpha},\mathbf{u})/\partial\boldsymbol{\alpha}]\delta\boldsymbol{\alpha}$ and $[\partial R(\boldsymbol{\alpha},\mathbf{u})/\partial\mathbf{u}]\delta\mathbf{u}$ as the *direct effect term* and the *indirect effect term*, respectively, because the direct effect term $[\partial R(\boldsymbol{\alpha},\mathbf{u})/\partial\boldsymbol{\alpha}]\delta\boldsymbol{\alpha}$ can be evaluated immediately while the indirect effect term $[\partial R(\boldsymbol{\alpha},\mathbf{u})/\partial\mathbf{u}]\delta\mathbf{u}$ can be quantified only after determining $\delta\mathbf{u}$ as function of $\delta\boldsymbol{\alpha}$. Since the system's state vector $\mathbf{u}$ and parameters $\boldsymbol{\alpha}$ are related to each other through Equations 3.1 and 3.2, it follows that $\delta\mathbf{u}$ and $\delta\boldsymbol{\alpha}$ must also be related to each other. Therefore, the total sensitivity $\delta R(\mathbf{e}^0;\mathbf{h})$ of $R(\mathbf{e})$ at $\mathbf{e}^0$ can only be evaluated after determining the vector of variations $\delta\mathbf{u}$ in terms of the vector of parameter variations $\delta\boldsymbol{\alpha}$.

The first-order relationship between $\delta\mathbf{u}$ and $\delta\boldsymbol{\alpha}$ is determined by taking the G-differentials of Equations 3.1 and 3.2. Thus, applying the definition of the G-differential to Equations 3.1 and 3.2 yields, respectively,

$$\mathbf{L}\left(\boldsymbol{\alpha}^0\right)\delta\mathbf{u}=\mathbf{Q}^{(1)}\left(\boldsymbol{\alpha}^0,\mathbf{u}^0;\delta\boldsymbol{\alpha}\right),\quad \mathbf{x}\in\Omega_x, \tag{3.11}$$

$$\mathbf{B}_F^{(1)}\left(\boldsymbol{\alpha}^0,\mathbf{u}^0;\,\delta\boldsymbol{\alpha},\delta\mathbf{u}\right)=\mathbf{0},\quad \mathbf{x}\in\partial\Omega_x, \tag{3.12}$$

where the superscript *(1)* indicates first level, the letter *F* (which will be used as "operator" and "subscript") indicates *forward equations* (in contradistinction to the adjoint equations that will be derived later in this section), the letter *B* indicates *boundary and/or initial conditions*, and where the following definitions were used:

$$\mathbf{Q}^{(1)}(\boldsymbol{\alpha},\mathbf{u};\delta\boldsymbol{\alpha})\triangleq\frac{\partial\left[\mathbf{Q}(\boldsymbol{\alpha})-\left[\mathbf{L}(\boldsymbol{\alpha})\mathbf{u}\right]\right]}{\partial\boldsymbol{\alpha}}\delta\boldsymbol{\alpha};\quad \mathbf{x}\in\Omega_x, \tag{3.13}$$

$$\mathbf{B}_F^{(1)}(\boldsymbol{\alpha},\mathbf{u};\delta\boldsymbol{\alpha},\delta\mathbf{u})\triangleq\mathbf{B}(\boldsymbol{\alpha})\delta\mathbf{u}+\frac{\partial\left[\mathbf{B}(\boldsymbol{\alpha})\mathbf{u}-\mathbf{C}(\boldsymbol{\alpha})\right]}{\partial\boldsymbol{\alpha}}\delta\boldsymbol{\alpha},\quad \mathbf{x}\in\partial\Omega_x. \tag{3.14}$$

The partial G-derivatives appearing in Equation 3.13 are defined as follows:

$$\frac{\partial \mathbf{Q}(\boldsymbol{\alpha})}{\partial \boldsymbol{\alpha}} \triangleq \begin{pmatrix} \frac{\partial Q_1}{\partial \alpha_1} & \cdots & \frac{\partial Q_1}{\partial \alpha_{N_\alpha}} \\ \vdots & \ddots & \vdots \\ \frac{\partial Q_{N_u}}{\partial \alpha_1} & \cdots & \frac{\partial Q_{N_u}}{\partial \alpha_{N_\alpha}} \end{pmatrix};$$

$$\frac{\partial \left[\mathbf{L}(\boldsymbol{\alpha})\mathbf{u}\right]}{\partial \boldsymbol{\alpha}} \triangleq \begin{pmatrix} \frac{\partial \left[\sum_{j=1}^{N_u} L_{1,j}(\boldsymbol{\alpha}) u_j\right]}{\partial \alpha_1} & \cdots & \frac{\partial \left[\sum_{j=1}^{N_u} L_{1,j}(\boldsymbol{\alpha}) u_j\right]}{\partial \alpha_{N_\alpha}} \\ \vdots & \ddots & \vdots \\ \frac{\partial \left[\sum_{j=1}^{N_u} L_{N_u,j}(\boldsymbol{\alpha}) u_j\right]}{\partial \alpha_1} & \cdots & \frac{\partial \left[\sum_{j=1}^{N_u} L_{N_u,j}(\boldsymbol{\alpha}) u_j\right]}{\partial \alpha_{N_\alpha}} \end{pmatrix}. \tag{3.15}$$

The partial G-derivatives that appear in Equation 3.14 are also matrices, having components similar to those defined in Equation 3.15. It is convenient to refer to the system of equations comprising Equations 3.11 and 3.12 as the *First-Level Forward Sensitivity System (1st-LFSS)*. For a given vector of parameter variations $\delta\boldsymbol{\alpha}$ in a neighborhood around $\boldsymbol{\alpha}^0$, the 1st-LFSS needs to be solved to obtain $\delta\mathbf{u}$ as a function of $\delta\boldsymbol{\alpha}$. In turn, the solution $\delta\mathbf{u}$ thus obtained could be used in Equation 3.10 to compute the total response sensitivity $\delta R(\mathbf{e}^0;\mathbf{h})$. *Computing the (total) response sensitivity* $\delta R(\mathbf{e}^0;\mathbf{h})$ *by using the ($\delta\boldsymbol{\alpha}$-dependent) solution* $\delta\mathbf{u}$ *of the 1st-LFSS is called the* forward sensitivity analysis procedure *(FSAP)*. From the standpoint of computational costs and effort, the FSAP requires $O(N_\alpha)$ large-scale forward computations. Therefore, the FSAP is advantageous to employ *only if*, in the problem under consideration, the number N_α of model parameters is considerably less than N_r, so that the number of responses of interest would exceed the number of system parameters and/or parameter variations of interest. This is rarely the case in practice, however, since most problems of practical interest are characterized by many parameters (i.e., $\boldsymbol{\alpha}$ has many components) and comparatively few responses. In such situations, it is not economical to employ the FSAP since it becomes prohibitively expensive to repeatedly solve the $\delta\boldsymbol{\alpha}$-dependent 1st-LFSS in order to determine the variations $\delta\mathbf{u}$ for all possible variations $\delta\boldsymbol{\alpha}$ in the model parameters $\boldsymbol{\alpha}$.

In most practical situations, the number of model parameters significantly exceeds the number of functional responses of interest, that is, $N_r << N_\alpha$.

In such cases, *the adjoint sensitivity analysis methodology* developed by Cacuci (1981a, 1981b) *is the most efficient method for computing exactly the total response first-order sensitivity* $\delta R(\mathbf{e}^0;\mathbf{h})$, *since its computation requires only* $O(N_r)$ *large-scale computations.* As shown by Cacuci (1981a, 1981b), the development of the adjoint sensitivity analysis methodology for computing the first-order G-differential $\delta R(\mathbf{e}^0;\mathbf{h})$ requires the introduction of adjoint operators, which can be conveniently accomplished by considering that the spaces $\mathcal{E}_u$ and $\mathcal{E}_Q$ are Hilbert spaces. Therefore, the Hilbert spaces corresponding to $\mathcal{E}_u$ and $\mathcal{E}_Q$ will be denoted as $\mathcal{H}_u(\Omega_x)$ and $\mathcal{H}_Q(\Omega_x)$, respectively. The elements of $\mathcal{H}_u(\Omega_x)$ and $\mathcal{H}_Q(\Omega_x)$ are, as before, vector-valued functions defined on the open set $\Omega_x \subset \mathbb{R}^{Jx}$, with smooth boundary $\partial\Omega_x$. On $\mathcal{H}_u(\Omega_x)$, the inner product of two vectors $\mathbf{u}^{(a)} \in \mathcal{H}_u$ and $\mathbf{u}^{(b)} \in \mathcal{H}_u$ will be denoted as $\langle \mathbf{u}^{(a)},\mathbf{u}^{(b)}\rangle_u$, while the inner product on $\mathcal{H}_Q(\Omega_x)$ of two vectors $\mathbf{Q}^{(a)} \in \mathcal{H}_Q$ and $\mathbf{Q}^{(b)} \in \mathcal{H}_Q$ will be denoted as $\langle \mathbf{Q}^{(a)},\mathbf{Q}^{(b)}\rangle_Q$.

As Equation 3.10 indicates, the total sensitivity $\delta R(\mathbf{e}^0;\mathbf{h})$ is separately linear in $\delta\mathbf{u}$ and in $\delta\boldsymbol{\alpha}$, respectively, which implies that the following relations hold uniquely in view of the Riesz representation theorem:

$$\left\{\frac{\partial R(\boldsymbol{\alpha},\mathbf{u})}{\partial \mathbf{u}}\right\}_{(\boldsymbol{\alpha}^0,\mathbf{u}^0)} \delta\mathbf{u} = \left\langle \frac{\partial R(\boldsymbol{\alpha},\mathbf{u})}{\partial \mathbf{u}},\, \delta\mathbf{u}\right\rangle_u, \quad \delta\mathbf{u} \in \mathcal{H}_u. \tag{3.16}$$

$$\left\{\frac{\partial R(\boldsymbol{\alpha},\mathbf{u})}{\partial \boldsymbol{\alpha}}\right\}_{(\boldsymbol{\alpha}^0,\mathbf{u}^0)} \delta\boldsymbol{\alpha} = \left\langle \frac{\partial R(\boldsymbol{\alpha},\mathbf{u})}{\partial \boldsymbol{\alpha}},\, \delta\boldsymbol{\alpha}\right\rangle_{\alpha}, \quad \delta\boldsymbol{\alpha} \in \mathcal{H}_{\alpha}, \tag{3.17}$$

where the inner product in the Hilbert space $\mathcal{H}_{\alpha}$ is denoted as $\langle \bullet,\bullet\rangle_{\alpha}$.

The system that is adjoint to the 1st-LFSS is now constructed by recalling from the geometry of Hilbert spaces that the following relationship holds for an arbitrary vector $\boldsymbol{\psi}^{(1)} \in \mathcal{H}_Q$:

$$\left\langle \boldsymbol{\psi}^{(1)}, \mathbf{L}(\boldsymbol{\alpha}^0)\delta\mathbf{u}\right\rangle_Q = \left\langle \mathbf{L}^*(\boldsymbol{\alpha}^0)\boldsymbol{\psi}^{(1)},\delta\mathbf{u}\right\rangle_u + \left\{P^{(1)}\left(\boldsymbol{\alpha}^0,\mathbf{u}^0;\boldsymbol{\psi}^{(1)};\delta\mathbf{u}\right)\right\}_{\partial\Omega_x}. \tag{3.18}$$

In the above equation, the formal adjoint operator $\mathbf{L}^*(\boldsymbol{\alpha}^0)$ is the N_u x N_u matrix

$$\mathbf{L}^*(\boldsymbol{\alpha}^0) \equiv \left[L^*_{ji}(\boldsymbol{\alpha}^0)\right],\ (i,j=1,\ldots,N_u), \tag{3.19}$$

comprising elements $L^*_{ji}(\boldsymbol{\alpha}^0)$ obtained by transposing the formal adjoints of the operators $L_{ij}(\boldsymbol{\alpha}^0)$, while $\{P^{(1)}(\boldsymbol{\alpha},\mathbf{u};\boldsymbol{\psi}^{(1)};\delta\mathbf{u})\}_{\partial\Omega x}$ denotes the associated bilinear form evaluated on $\partial\Omega_x$. The domain of $\mathbf{L}^*(\boldsymbol{\alpha}^0)$ is determined by selecting appropriate adjoint boundary and/or initial conditions, represented here in operator form as

$$\mathbf{B}^*(\boldsymbol{\alpha}^0)\boldsymbol{\psi}^{(1)} - \mathbf{C}^*(\boldsymbol{\alpha}^0) = \mathbf{0},\ \mathbf{x} \in \partial\Omega_x. \tag{3.20}$$

The above boundary conditions for $\mathbf{L}^*(\boldsymbol{\alpha}^0)$ are obtained by requiring that

1. They must be independent of *unknown* values of $\delta\mathbf{u}$ and $\delta\boldsymbol{\alpha}$;
2. The substitution of the boundary and/or initial conditions represented by Equations 3.12 and 3.20 into the expression of $\{P^{(1)}(\boldsymbol{\alpha}, \mathbf{u}; \boldsymbol{\psi}^{(1)}; \delta\mathbf{u})\}_{\partial\Omega x}$ must cause all terms containing unknown values of $\delta\mathbf{u}$ to vanish.

This selection of the boundary conditions for $\mathbf{L}^*(\boldsymbol{\alpha}^0)$ reduces the bilinear concomitant $\{P^{(1)}(\boldsymbol{\alpha}, \mathbf{u}; \boldsymbol{\psi}^{(1)}; \delta\mathbf{u})\}_{\partial\Omega x}$ to a quantity that contains boundary terms involving only known values of $\delta\boldsymbol{\alpha}$, $\boldsymbol{\psi}$, and, possibly, $\boldsymbol{\alpha}^0$; this quantity will be denoted by $\hat{P}^{(1)}\left(\boldsymbol{\alpha},\mathbf{u};\boldsymbol{\psi}^{(1)};\delta\boldsymbol{\alpha}\right)$. In general, $\hat{P}^{(1)}\left(\boldsymbol{\alpha},\mathbf{u};\boldsymbol{\psi}^{(1)};\delta\boldsymbol{\alpha}\right)$ does not automatically vanish as a result of these manipulations, although it may do so in particular instances. In principle, $\hat{P}^{(1)}\left(\boldsymbol{\alpha},\mathbf{u};\boldsymbol{\psi}^{(1)};\delta\boldsymbol{\alpha}\right)$ could be forced to vanish by considering extensions of $\mathbf{L}(\boldsymbol{\alpha}^0)$, in the operator sense, but such extensions seldom need to be used in practice.

Using now the 1st-LFSS, cf. Equations 3.11 and 3.12, together with the adjoint boundary conditions represented by Equation 3.20, in Equation 3.18 and rearranging the resulting equation yields

$$\left\langle \mathbf{L}^*\left(\boldsymbol{\alpha}^0\right)\boldsymbol{\psi}^{(1)},\, \delta\mathbf{u}\right\rangle_u = \left\langle \boldsymbol{\psi}^{(1)},\, \mathbf{Q}^{(1)}\left(\boldsymbol{\alpha}^0,\mathbf{u}^0;\delta\boldsymbol{\alpha}\right)\right\rangle_Q - \hat{P}^{(1)}\left(\boldsymbol{\alpha}^0,\mathbf{u}^0;\boldsymbol{\psi}^{(1)};\delta\boldsymbol{\alpha}\right) \quad (3.21)$$

Since $\boldsymbol{\psi}$ is not completely defined yet, we now complete its definition by requiring that the left side of Equation 3.21 and the right side of Equation 3.16 represent the same functional, which is accomplished by requiring that the following equality be satisfied in the weak (distributional) sense:

$$\mathbf{L}^*\left(\boldsymbol{\alpha}^0\right)\boldsymbol{\psi}^{(1)} = \frac{\partial R\left(\boldsymbol{\alpha}^0,\mathbf{u}^0\right)}{\partial \mathbf{u}}, \quad (3.22)$$

where the superscript 0 emphasizes the fact that the function $\boldsymbol{\psi}^{(1)}$ satisfies Equations 3.22 and 3.20 for the nominal parameter values $\boldsymbol{\alpha}^0$.

The system of Equations 3.22 and 3.20 will be called the *First-Level Adjoint Sensitivity System* (*1st-LASS*) and its solution, that is, the function $\boldsymbol{\psi}^{(1)}$, will be called the *first-level adjoint sensitivity function*. Furthermore, Equations 3.11 through 3.22 can now be used in Equation 3.10 to obtain the following expression for the total sensitivity $\delta R(\mathbf{e}^0;\mathbf{h})$ of $R(\mathbf{e})$ at

$$\delta R\left(\mathbf{e}^0;\mathbf{h}\right) = \left\{\frac{\partial R(\boldsymbol{\alpha},\mathbf{u})}{\partial \boldsymbol{\alpha}}\right\}_{\left(\boldsymbol{\alpha}^0,\mathbf{u}^0\right)} \delta\boldsymbol{\alpha} + \left\langle \boldsymbol{\psi}^{(1)},\mathbf{Q}^{(1)}\left(\boldsymbol{\alpha}^0,\mathbf{u}^0;\delta\boldsymbol{\alpha}\right)\right\rangle_Q - \hat{P}^{(1)}\left(\boldsymbol{\alpha}^0,\mathbf{u}^0;\boldsymbol{\psi}^{(1)};\delta\boldsymbol{\alpha}\right)$$

$$\triangleq DR\left(\boldsymbol{\alpha}^0,\mathbf{u}^0;\boldsymbol{\psi}^{(1)};\delta\boldsymbol{\alpha}\right). \quad (3.23)$$

As the right side of Equation 3.23 indicates, the desired elimination of all unknown values of $\delta\mathbf{u}$ from the expression of total first-order response sensitivity, $\delta R(\mathbf{e}^0;\mathbf{h})$, of $R(\mathbf{e})$ at $\mathbf{e}^0$ has been accomplished. Instead of depending on $\delta\mathbf{u}$, the first-order response sensitivity $\delta R(\mathbf{e}^0;\mathbf{h})$ now depends on the adjoint function $\boldsymbol{\psi}^{(1)} \in \mathcal{H}_Q$, a fact that is emphasized in the above expression by noting explicitly the identity $\delta R(\mathbf{e}^0;\mathbf{h}) \equiv DR(\boldsymbol{\alpha}^0,\mathbf{u}^0;\boldsymbol{\psi}^{(1)};\delta\boldsymbol{\alpha})$.

To emphasize that $\mathcal{H}_Q(\Omega_x)$ becomes the Hilbert space on which the adjoint function $\boldsymbol{\psi}^{(1)} \in \mathcal{H}_Q$ is defined, *the Hilbert space $\mathcal{H}_Q$ will henceforth be denoted as $\mathcal{H}_\psi^{(1)}(\Omega_x)$, and the inner product $\langle \bullet,\bullet \rangle_Q$ will be denoted as $\langle \bullet,\bullet \rangle_\psi^{(1)}$, so that* $\boldsymbol{\psi}^{(1)} \in \mathcal{H}_\psi^{(1)}(\Omega_x)$. This notation will be convenient for subsequent developments, to be presented in Section 3.3, where the second-level adjoint sensitivity systems will involve adjoint functions that will be designated as $\boldsymbol{\psi}^{(2)} \in \mathcal{H}_\psi^{(2)}(\Omega_x)$, and where the inner products in $\mathcal{H}_\psi^{(2)}(\Omega_x)$ will be denoted as $\langle \bullet,\bullet \rangle_\psi^{(2)}$.

As indicated by Equation 3.23, *the total first-order response sensitivity $DR(\boldsymbol{\alpha}^0,\mathbf{u}^0;\boldsymbol{\psi}^{(1)};\delta\boldsymbol{\alpha})$ can be computed after solving Equations 3.20 and 3.22* only once *to obtain the adjoint function $\boldsymbol{\psi}^{(1)} \in \mathcal{H}_\psi^{(1)}(\Omega_x)$. Equations 3.20 and 3.22 will be called the* 1st-LASS, *and its solution, $\boldsymbol{\psi}^{(1)} \in \mathcal{H}_\psi^{(1)}(\Omega_x)$, will be called the first-level adjoint function.* It is very important to note that the 1st-LASS is independent of the variation $\delta\mathbf{u}$ in the original state functions $\mathbf{u}$ and is also independent of variations $\delta\boldsymbol{\alpha}$ in the model parameters $\boldsymbol{\alpha}$.

By successively setting ($\delta\alpha_i = 1$, $\delta\alpha_j = 0$ *for* $j \neq i$, $i = 1,\ldots, N_\alpha$) in the expression of $DR(\boldsymbol{\alpha}^0,\mathbf{u}^0;\boldsymbol{\psi}^{(1)};\delta\boldsymbol{\alpha})$ given in Equation 3.23, we obtain by means of simple integrations the individual sensitivities of the response $R(\mathbf{e})$ with respect to each of the model parameters α_i, evaluated at the nominal values $\mathbf{e}^0 = (\boldsymbol{\alpha}^0,\mathbf{u}^0)$.

In the vast majority of practical situations, the G-differential $\delta\mathbf{Q}(\boldsymbol{\alpha}^0;\delta\boldsymbol{\alpha})$ of the source term $\mathbf{Q}(\boldsymbol{\alpha}^0)$ is linear in $\delta\boldsymbol{\alpha}$, which implies that

$$\mathbf{Q}^{(1)}\left(\boldsymbol{\alpha}^0,\mathbf{u}^0;\delta\boldsymbol{\alpha}\right) = \sum_{i_1=1}^{N_\alpha} \frac{\partial\left[\mathbf{Q}\left(\boldsymbol{\alpha}^0\right) - \mathbf{L}\left(\boldsymbol{\alpha}^0\right)\mathbf{u}^0\right]}{\partial\alpha_{i_1}}\delta\alpha_{i_1}. \tag{3.24}$$

Since $\hat{P}\left(\boldsymbol{\alpha}^0,\mathbf{u}^0;\boldsymbol{\psi}^{(1)};\delta\boldsymbol{\alpha}\right)$ is also linear in $\delta\boldsymbol{\alpha}$, it follows that it can be expressed as

$$\hat{P}^{(1)}\left(\boldsymbol{\alpha}^0,\mathbf{u}^0;\boldsymbol{\psi}^{(1)};\delta\boldsymbol{\alpha}\right) = \sum_{i_1=1}^{N_\alpha} \frac{\partial\hat{P}^{(1)}\left(\boldsymbol{\alpha}^0,\mathbf{u}^0;\boldsymbol{\psi}^{(1)}\right)}{\partial\alpha_{i_1}}\delta\alpha_{i_1}. \tag{3.25}$$

Therefore, Equation 3.23 can be expressed in the form

$$DR\left(\boldsymbol{\alpha}^0,\mathbf{u}^0;\boldsymbol{\psi}^{(1)};\delta\boldsymbol{\alpha}\right) = \sum_{i_1=1}^{N_\alpha} R_{i_1}^{(1)}\left(\boldsymbol{\alpha}^0,\mathbf{u}^0;\boldsymbol{\psi}^{(1)}\right)\delta\alpha_{i_1}, \tag{3.26}$$

where the quantities $R_{i_1}^{(1)}\left(\boldsymbol{\alpha}^0;\mathbf{u}^0;\boldsymbol{\psi}^{(1)}\right)$ denote the *first-order partial sensitivities (G-derivatives)* of the response $R(\mathbf{e})$ with respect to each of the model parameters α_i, evaluated at $\boldsymbol{\alpha}^0$, $\mathbf{u}^0$, $\boldsymbol{\psi}^{(1)}$, and are given by the following expression:

$$R_{i_1}^{(1)}\left(\boldsymbol{\alpha},\mathbf{u};\boldsymbol{\psi}^{(1)}\right) \triangleq \frac{\partial R(\boldsymbol{\alpha},\mathbf{u})}{\partial \alpha_{i_1}} - \frac{\partial \hat{P}^{(1)}\left(\boldsymbol{\alpha},\mathbf{u};\boldsymbol{\psi}^{(1)}\right)}{\partial \alpha_{i_1}} \tag{3.27}$$

$$+ \int_{\Omega_x} \boldsymbol{\psi}^{(1)}(\mathbf{x}) \bullet \left\{ \frac{\partial\left[\mathbf{Q}(\boldsymbol{\alpha}) - \mathbf{L}(\boldsymbol{\alpha})\mathbf{u}\right]}{\partial \alpha_{i_1}} \right\} dx, \quad i_1 = 1,\ldots,N_\alpha.$$

For subsequent generalizations to higher levels and orders, the standardized notation for the 1st-LASS is summarized below, keeping in mind that the equalities in respective operator equations hold in the weak, distributional sense:

$$\mathbf{A}_{11}^{(1)}(\boldsymbol{\alpha})\boldsymbol{\psi}^{(1)}(\mathbf{x}) = \boldsymbol{\Theta}^{(1)}(\boldsymbol{\alpha},\mathbf{u}), \ \mathbf{x} \in \Omega_x, \ \boldsymbol{\psi}^{(1)} \in \mathcal{H}_\psi^{(1)}(\Omega_x), \tag{3.28}$$

$$\mathbf{A}_{11}^{(1)}(\boldsymbol{\alpha}) \triangleq \mathbf{L}^*(\boldsymbol{\alpha}), \quad \boldsymbol{\Theta}^{(1)}(\boldsymbol{\alpha},\mathbf{u}) \triangleq \frac{\partial R(\boldsymbol{\alpha},\mathbf{u})}{\partial \mathbf{u}},$$

$$\mathbf{B}_A^{(1)}\left(\boldsymbol{\alpha};\boldsymbol{\psi}^{(1)}\right) \triangleq \mathbf{B}^*\left(\boldsymbol{\alpha}^0\right)\boldsymbol{\psi}^{(1)} - \mathbf{C}^*\left(\boldsymbol{\alpha}^0\right) = \mathbf{0}, \ \mathbf{x} \in \partial\Omega_x, \tag{3.29}$$

where the superscript *(1)* indicates first level, the letter *A* (used as "operator" and "subscript") indicates *adjoint*, the letter *B* indicates *boundary and/or initial conditions*. It is very important to note that this adjoint system is independent not only of the variations $\delta\mathbf{u}$ and $\delta\boldsymbol{\alpha}$ but is also independent of the nominal values $\mathbf{u}^0$ of $\mathbf{u}$. *It is important to note that this advantageous situation arises if and only if the original equations that model the physical problem, that is, Equations 3.1 and 3.2, are linear in the state variable* $\mathbf{u}$. For linear problems, therefore, the 1st-LASS, namely Equations 3.28 and 3.29, can be solved independently of the solution $\mathbf{u}^0$ of the original equations, a fact that simplifies considerably the choice of numerical methods for solving it.

3.3 The 2nd-LASS for Computing Exactly and Efficiently First-Order Sensitivities of Scalar-Valued Responses for Linear Systems

The *second-order total sensitivity* of the model response $R(\mathbf{e})$ to variations $\mathbf{h} \triangleq (\delta\boldsymbol{\alpha}, \delta\mathbf{u})$ in the model parameters and state functions around $\mathbf{e}^0 = (\boldsymbol{\alpha}^0,\mathbf{u}^0)$ is given by the *second-order G-differential* of $R(\mathbf{e})$ at $\mathbf{e}^0 = (\boldsymbol{\alpha}^0,\mathbf{u}^0)$ in arbitrary directions $\mathbf{h} \triangleq (\delta\boldsymbol{\alpha}, \delta\mathbf{u})$. This *second-order G-differential* of $R(\mathbf{e})$ is assumed to exist and will be denoted as $\delta^{(2)}R(\mathbf{e}^0;\mathbf{h}_1,\mathbf{h}_2)$. By definition, $\delta^{(2)}R(\mathbf{e}^0;\mathbf{h}_1,\mathbf{h}_2)$ is defined inductively as "the total first-variation (or first-order differential) of the first-order total response variation (or first-order differential)," as follows:

$$\begin{aligned}\delta^{(2)}R\left(\mathbf{e}^0;\mathbf{h}_1,\mathbf{h}_2\right) &\triangleq \left\{\frac{d}{d\varepsilon_1}\left[\delta R\left(\mathbf{e}^0+\varepsilon_1\mathbf{h}_1;\mathbf{h}_2\right)\right]\right\}_{\varepsilon_1=0} \\ &= \left\{\frac{\partial^2}{\partial\varepsilon_1\,\partial\varepsilon_2}\left[R\left(\mathbf{e}^0+\varepsilon_1\mathbf{h}_1+\varepsilon_2\mathbf{h}_2\right)\right]\right\}_{\varepsilon_1=\varepsilon_2=0},\end{aligned} \tag{3.30}$$

for ε_1, $\varepsilon_2 \in \mathcal{F}$, and all (i.e., arbitrary) vectors of variations $\mathbf{h}_i \in \mathcal{E}$; $i = 1,2$, around $\mathbf{e}^0=(\boldsymbol{\alpha}^0,\mathbf{u}^0) \in \mathcal{E}$. In order for the second-order response sensitivities (i.e., second-order partial G-derivatives) to model parameters to exist, the first-order sensitivities must satisfy the conditions generally expressed in Equations 3.7 and 3.8. Under these conditions, which will henceforth be assumed to be satisfied, the total second-order response differential $\delta^{(2)}R(\mathbf{e}^0;\mathbf{h}_1,\mathbf{h}_2)$ is bilinear in the vectors of variations $\mathbf{h}_i \in \mathcal{E}$; $i = 1,2$. Ultimately, the total second-order response differential $\delta^{(2)}R(\mathbf{e}^0;\mathbf{h})$ will be shown to be bilinear in the parameter variations $\delta\alpha_i$, so that Equation 3.30 will ultimately be expressed in the following form:

$$\delta^{(2)}R\left(\mathbf{e}^0;\mathbf{h}_1,\mathbf{h}_2\right)=\sum_{i_2,i_1=1}^{N_\alpha} R^{(2)}_{i_2 i_1}\,\delta\alpha_{i_2}\,\delta\alpha_{i_1}. \tag{3.31}$$

In Equation 3.31, the quantities $R^{(2)}_{i_2 i_1}$ do *not* depend on the parameter variations and represent the *mixed second-order partial sensitivities (G-derivatives)* of the model response $R(\mathbf{e})$ at $\mathbf{e}^0$. The exact and efficient computation of the second-order partial sensitivities (G-derivatives) $R^{(2)}_{i_2 i_1}$ will be the objective of the developments to be presented in this section.

As Equation 3.27 indicates, the first-order partial sensitivities are functionals of the form $R^{(1)}_{i_1}\left(\boldsymbol{\alpha},\mathbf{u};\boldsymbol{\psi}^{(1)}\right)\in\mathcal{H}_\alpha\oplus\mathcal{H}_u(\Omega_x)\oplus\mathcal{H}^{(1)}_\psi(\Omega_x)\to\mathcal{F}$. The product Hilbert space $\mathcal{H}^{(1)}_\psi(\Omega_x)\oplus\mathcal{H}_u(\Omega_x)$ will play a prominent role in the derivations to follow. For notational convenience, therefore, it will be denoted as $\mathcal{H}^{(2)}_\psi(\Omega_x)\triangleq\mathcal{H}^{(1)}_\psi(\Omega_x)\oplus\mathcal{H}_u(\Omega_x)$. An arbitrary vector in $\mathcal{H}^{(2)}_\psi(\Omega_x)$ will be denoted as $\boldsymbol{\varphi}^{(2)}\triangleq\left(\mathbf{u};\boldsymbol{\psi}^{(1)}\right)\in\mathcal{H}_u(\Omega_x)\oplus\mathcal{H}^{(1)}_\psi(\Omega_x)$, with the respective "nominal value" denoted as $\boldsymbol{\varphi}^{(2,0)}\triangleq\left(\mathbf{u}^0;\boldsymbol{\psi}^{(1,0)}\right)\in\mathcal{H}_u(\Omega_x)\oplus\mathcal{H}^{(1)}_\psi(\Omega_x)$. A vector of arbitrary variations around $\boldsymbol{\varphi}^{(2,0)}\triangleq(\mathbf{u}^0;\boldsymbol{\psi}^{(1,0)})$ will be denoted as $\delta\boldsymbol{\varphi}^{(2)}\triangleq(\delta\mathbf{u};\delta\boldsymbol{\psi}^{(1)})$. The arguments of $\boldsymbol{\varphi}^{(2)}$, $\boldsymbol{\varphi}^{(2,0)}$, and $\delta\boldsymbol{\varphi}^{(2)}$ may be written out explicitly whenever needed for clarity.

Hence, for each $i_1 = 1,\dots,N_\alpha$, the first-order G-differential $\delta R_{i_1}^{(1)}\left(\boldsymbol{\alpha}^0,\boldsymbol{\varphi}^{(2,0)};\delta\boldsymbol{\alpha},\delta\boldsymbol{\varphi}^{(2)}\right)$ of $R_{i_1}^{(1)}\left(\boldsymbol{\alpha},\mathbf{u};\boldsymbol{\psi}^{(1)}\right)$ can be computed in the neighborhood of the point $(\boldsymbol{\alpha}^0, \boldsymbol{\varphi}^{(2,0)})$, in the usual manner, namely:

$$\begin{aligned}\delta R_{i_1}^{(1)}\left(\boldsymbol{\alpha}^0,\boldsymbol{\varphi}^{(2,0)};\delta\boldsymbol{\alpha},\delta\boldsymbol{\varphi}^{(2)}\right) &\triangleq \left\{\frac{d}{d\varepsilon}R_{i_1}^{(1)}\left(\boldsymbol{\alpha}^0+\varepsilon\delta\boldsymbol{\alpha},\mathbf{u}^0+\varepsilon\delta\mathbf{u};\boldsymbol{\psi}^{(1,0)}+\varepsilon\delta\boldsymbol{\psi}^{(1)}\right)\right\}_{\varepsilon=0} \\ &= \left\{\frac{\partial R_{i_1}^{(1)}}{\partial\boldsymbol{\alpha}}\delta\boldsymbol{\alpha}+\frac{\partial R_{i_1}^{(1)}}{\partial\mathbf{u}}\delta\mathbf{u}+\frac{\partial R_{i_1}^{(1)}}{\partial\boldsymbol{\psi}^{(1)}}\delta\boldsymbol{\psi}^{(1)}\right\}_{\left(\boldsymbol{\alpha}^0,\boldsymbol{\varphi}^{(2,0)}\right)}.\end{aligned} \tag{3.32}$$

The above relation can be expressed as

$$\delta R_{i_1}^{(1)}\left(\boldsymbol{\alpha}^0,\boldsymbol{\varphi}^{(2,0)};\delta\boldsymbol{\alpha},\delta\boldsymbol{\varphi}^{(2)}\right) \triangleq \left(\delta R_{i_1}^{(1)}\right)_{direct} + \left(\delta R_{i_1}^{(1)}\right)_{indirect}, \tag{3.33}$$

where $\left(\delta R_{i_1}^{(1)}\right)_{direct}$ denotes the *direct-effect term*, while $\left(\delta R_{i_1}^{(1)}\right)_{indirect}$ denotes the *indirect-effect term*. The direct-effect term $\left(\delta R_{i_1}^{(1)}\right)_{direct}$ is defined as follows:

$$\left(\delta R_{i_1}^{(1)}\right)_{direct} \triangleq \left(\frac{\partial R_{i_1}^{(1)}}{\partial\boldsymbol{\alpha}}\right)_{\left(\boldsymbol{\alpha}^0,\boldsymbol{\varphi}^{(2,0)}\right)}\delta\boldsymbol{\alpha} = \sum_{i_2=1}^{N_\alpha}\frac{\partial R_{i_1}^{(1)}\left(\boldsymbol{\alpha}^0,\boldsymbol{\varphi}^{(2,0)}\right)}{\partial\alpha_{i_2}}\delta\alpha_{i_2}, \quad i_1 = 1,\dots,N_\alpha, \tag{3.34}$$

where

$$\begin{aligned}\frac{\partial R_{i_1}^{(1)}\left(\boldsymbol{\alpha}^0,\boldsymbol{\varphi}^{(2,0)}\right)}{\partial\alpha_{i_2}} &\triangleq \frac{\partial}{\partial\alpha_{i_2}}\left\{\frac{\partial R(\boldsymbol{\alpha},\mathbf{u})}{\partial\alpha_{i_1}}+\left\langle\boldsymbol{\psi}^{(1)},\ \frac{\partial\left[\mathbf{Q}(\boldsymbol{\alpha})-\mathbf{L}(\boldsymbol{\alpha})\mathbf{u}\right]}{\partial\alpha_{i_1}}\right\rangle_{\psi^{(1)}}\right. \\ &\left.-\frac{\partial\hat{P}^{(1)}\left(\boldsymbol{\alpha},\mathbf{u};\boldsymbol{\psi}^{(1)}\right)}{\partial\alpha_{i_1}}\right\}_{\left(\boldsymbol{\alpha}^0,\boldsymbol{\varphi}^{(2,0)}\right)}, \quad i_1, i_2 = 1,\dots,N_\alpha.\end{aligned} \tag{3.35}$$

Since the direct-effect term $\left(\delta R_{i_1}^{(1)}\right)_{direct}$ is independent of variations $\delta\boldsymbol{\varphi}^{(2)} \triangleq (\delta\mathbf{u}; \delta\boldsymbol{\psi}^{(1)})$, it *can be computed immediately* at this stage, without needing any additional large-scale computations.

On the contrary, the indirect-effect term $\left(\delta R_{i_1}^{(1)}\right)_{indirect}$ is defined from Equations 3.32 and 3.33 as follows:

$$\begin{aligned}\left(\delta R_{i_1}^{(1)}\right)_{indirect} &\triangleq \left\{\frac{\partial R_{i_1}^{(1)}}{\partial\boldsymbol{\psi}^{(1)}}\delta\boldsymbol{\psi}^{(1)}+\frac{\partial R_{i_1}^{(1)}}{\partial\mathbf{u}}\delta\mathbf{u}\right\}_{\left(\boldsymbol{\alpha}^0,\boldsymbol{\varphi}^{(2,0)}\right)} \\ &= \left\{\left(\delta\boldsymbol{\psi}^{(1)},\delta\mathbf{u}\right)\begin{pmatrix}\boldsymbol{\Theta}_{1,i_1}^{(2)}\left(\boldsymbol{\alpha},\boldsymbol{\varphi}^{(2)}\right) \\ \boldsymbol{\Theta}_{2,i_1}^{(2)}\left(\boldsymbol{\alpha},\boldsymbol{\varphi}^{(2)}\right)\end{pmatrix}\right\}_{\left(\boldsymbol{\alpha}^0,\boldsymbol{\varphi}^{(2,0)}\right)},\end{aligned} \tag{3.36}$$

where

$$\Theta^{(2)}_{1,i_1}\left(\alpha,\varphi^{(2)}\right) \triangleq \frac{\partial R^{(1)}_{i_1}\left(\alpha,\mathbf{u};\psi^{(1)}\right)}{\partial \psi^{(1)}} = \frac{\partial}{\partial \alpha_{i_1}}\left[\mathbf{Q}(\alpha)-\mathbf{L}(\alpha)\mathbf{u}-\frac{\partial \hat{P}^{(1)}\left(\alpha,\mathbf{u};\psi^{(1)}\right)}{\partial \psi^{(1)}}\right], \tag{3.37}$$

and

$$\begin{aligned}\Theta^{(2)}_{2,i_1}\left(\alpha,\varphi^{(2)}\right) &\triangleq \frac{\partial R^{(1)}_{i_1}\left(\alpha,\mathbf{u};\psi^{(1)}\right)}{\partial \mathbf{u}} \\ &= \frac{\partial}{\partial \alpha_{i_1}}\left[\Theta^{(1)}(\alpha,\mathbf{u})-\Lambda\left(\alpha;\psi^{(1)}\right)-\frac{\partial \hat{P}^{(1)}\left(\alpha,\mathbf{u};\psi^{(1)}\right)}{\partial \mathbf{u}}\right],\end{aligned} \tag{3.38}$$

with

$$\Lambda\left(\alpha,\mathbf{u};\psi^{(1)}\right) \triangleq \left(\Lambda_1,\ldots,\Lambda_{N_u}\right);\quad \Lambda_k\left(\alpha,\mathbf{u};\psi^{(1)}\right) \triangleq \sum_{j=1}^{N_u}\psi^{(1)}_j(\mathbf{x})L_{jk}(\alpha);\ k=1,\ldots,N_u. \tag{3.39}$$

The indirect-effect term, $\left(\delta R^{(1)}_{i_1}\right)_{indirect}$*, cannot be computed at this stage yet, since the vectors of variations* $\delta\mathbf{u}$ and $\delta\psi^{(1)}$ *are not completely known*. The relation between the *vectors of variations* $\delta\mathbf{u}$ and $\delta\psi^{(1)}$ needs to be determined next by noting that the vector of variations $\delta\psi^{(1)}$ will be the solution of the system of operator equations that will result after G-differentiating the 1st-LASS given by Equations 3.28 and 3.29 in the neighborhood ($\alpha^0 + \varepsilon\delta\alpha$, $\mathbf{u}^0 + \varepsilon\delta\mathbf{u}$; $\psi^{(1,0)}+\varepsilon\delta\psi^{(1)}$) of the point ($\alpha^0$, $\varphi^{(2,0)}$). Performing this G-differentiation yields

$$\left[\frac{d}{d\varepsilon}\left\{\mathbf{L}^*\left[\left(\alpha^0+\varepsilon\delta\alpha\right)\right]\left(\psi^{(1,0)}+\varepsilon\delta\psi^{(1)}\right)-\Theta^{(1)}\left(\alpha^0+\varepsilon\delta\alpha,\mathbf{u}^0+\varepsilon\delta\mathbf{u}\right)\right\}\right]_{\varepsilon=0}=\mathbf{0},\ \mathbf{x}\in\Omega_x, \tag{3.40}$$

$$\begin{aligned}&\left\{\frac{d}{d\varepsilon}\left[\mathbf{B}^{(1)}_A\left(\alpha^0+\varepsilon\delta\alpha;\psi^{(1,0)}+\varepsilon\delta\psi^{(1)}\right)\right]\right\}_{\varepsilon=0}= \\ &\left\{\frac{d}{d\varepsilon}\left[\mathbf{B}^*\left(\alpha^0+\varepsilon\delta\alpha\right)\left(\psi^{(1,0)}+\varepsilon\delta\psi^{(1)}\right)-\mathbf{C}^*\left(\alpha^0+\varepsilon\delta\alpha\right)\right]\right\}_{\varepsilon=0}=\mathbf{0},\ \mathbf{x}\in\partial\Omega_x.\end{aligned} \tag{3.41}$$

Carrying out the differentiations and evaluations indicated in Equations 3.40 and 3.41 yields the following system of operator equations:

$$\mathbf{F}^{(2)}_{11}\left(\alpha^0\right)\delta\psi^{(1)}+\mathbf{F}^{(2)}_{12}\left(\alpha^0;\mathbf{u}^0\right)\delta\mathbf{u}=\mathbf{Q}^{(2)}\left(\alpha^0,\varphi^{(2,0)};\delta\alpha\right), \tag{3.42}$$

$$\delta\mathbf{B}^{(1)}_A\left(\alpha^0,\varphi^{(2,0)};\delta\alpha,\delta\psi^{(1)}\right)=\mathbf{0},\ \mathbf{x}\in\partial\Omega_x, \tag{3.43}$$

where the following definitions were used:

$$\mathrm{F}_{11}^{(2)}(\boldsymbol{\alpha}) \triangleq \mathbf{L}^{*}(\boldsymbol{\alpha}); \quad \mathrm{F}_{12}^{(2)}(\boldsymbol{\alpha};\mathbf{u}) \triangleq -\frac{\partial \Theta^{(1)}(\boldsymbol{\alpha},\mathbf{u})}{\partial \mathbf{u}};$$

$$\mathbf{Q}^{(2)}\left(\boldsymbol{\alpha},\boldsymbol{\varphi}^{(2)};\delta\boldsymbol{\alpha}\right) \triangleq \frac{\partial\left[\Theta^{(1)}(\boldsymbol{\alpha},\mathbf{u}) - \mathbf{L}^{*}(\boldsymbol{\alpha})\boldsymbol{\psi}^{(1)}\right]}{\partial \boldsymbol{\alpha}}\delta\boldsymbol{\alpha};$$

$$\frac{\partial \Theta^{(1)}(\boldsymbol{\alpha},\mathbf{u})}{\partial \boldsymbol{\alpha}} \triangleq \begin{pmatrix} \dfrac{\partial^2 R(\mathbf{e})}{\partial u_1\,\partial \alpha_1} & \cdots & \dfrac{\partial^2 R(\mathbf{e})}{\partial u_1\,\partial \alpha_{N_\alpha}} \\ \vdots & \ddots & \vdots \\ \dfrac{\partial^2 R(\mathbf{e})}{\partial u_{N_u}\,\partial \alpha_1} & \cdots & \dfrac{\partial^2 R(\mathbf{e})}{\partial u_{N_u}\,\partial \alpha_{N_\alpha}} \end{pmatrix}; \tag{3.44}$$

$$\frac{\partial \Theta^{(1)}(\boldsymbol{\alpha},\mathbf{u})}{\partial \mathbf{u}} \triangleq \begin{pmatrix} \dfrac{\partial^2 R(\mathbf{e})}{\partial u_1^2} & \cdots & \dfrac{\partial^2 R(\mathbf{e})}{\partial u_1\,\partial u_{N_u}} \\ \vdots & \ddots & \vdots \\ \dfrac{\partial^2 R(\mathbf{e})}{\partial u_{N_u}\,\partial u_1} & \cdots & \dfrac{\partial^2 R(\mathbf{e})}{\partial u_{N_u}^2} \end{pmatrix};$$

$$\delta\mathbf{B}_A^{(1)}\left(\boldsymbol{\alpha},\boldsymbol{\varphi}^{(2)};\delta\boldsymbol{\alpha},\delta\boldsymbol{\psi}^{(1)}\right) \triangleq \mathbf{B}^{*}(\boldsymbol{\alpha})\delta\boldsymbol{\psi}^{(1)} - \frac{\partial\left[\mathbf{B}^{*}(\boldsymbol{\alpha})\boldsymbol{\psi}^{(1)} - \mathbf{C}^{*}(\boldsymbol{\alpha})\right]}{\partial \boldsymbol{\alpha}}\delta\boldsymbol{\alpha}.$$

As Equations 3.42 and 3.43 indicate, the vector of variations $\delta\boldsymbol{\psi}^{(1)}$ will also depend on the parameter variations $\delta\boldsymbol{\alpha}$, just as the vector of variations $\delta\mathbf{u}$ did. To construct a well-posed system of equations for obtaining both $\delta\boldsymbol{\psi}^{(1)}$ and $\delta\mathbf{u}$ in terms of $\delta\boldsymbol{\alpha}$, the 1st-LFSS given by Equations 3.11 and 3.12 is combined together with Equations 3.42 and 3.43 to obtain the following block-matrix system:

$$\begin{pmatrix} \mathbf{F}_{11}^{(2)}\left(\boldsymbol{\alpha}^0\right) & \mathbf{F}_{12}^{(2)}\left(\boldsymbol{\alpha}^0,\mathbf{u}^0\right) \\ \mathbf{0} & \mathbf{F}_{11}^{(1)}\left(\boldsymbol{\alpha}^0\right) \end{pmatrix}\begin{pmatrix} \delta\boldsymbol{\psi}^{(1)} \\ \delta\mathbf{u} \end{pmatrix} = \begin{pmatrix} \mathbf{Q}^{(2)}\left(\boldsymbol{\alpha}^0,\boldsymbol{\varphi}^{(2,0)};\delta\boldsymbol{\alpha}\right) \\ \mathbf{Q}^{(1)}\left(\boldsymbol{\alpha}^0,\mathbf{u}^0;\delta\boldsymbol{\alpha}\right) \end{pmatrix}, \quad \mathbf{x}\in\Omega_x, \tag{3.45}$$

together with the definition $\mathbf{F}_{11}^{(1)}\left(\boldsymbol{\alpha}^0\right) \triangleq \mathbf{L}\left(\boldsymbol{\alpha}^0\right)$ and with the following boundary and/or initial conditions

$$\mathbf{B}_F^{(2)}\left(\boldsymbol{\alpha}^0,\boldsymbol{\varphi}^{(2,0)};\delta\boldsymbol{\alpha},\delta\boldsymbol{\varphi}^{(2)}\right) \triangleq \begin{pmatrix} \delta\mathbf{B}_A^{(1)}\left(\boldsymbol{\alpha}^0,\boldsymbol{\varphi}^{(2,0)};\delta\boldsymbol{\alpha},\delta\boldsymbol{\varphi}^{(2)}\right) \\ \mathbf{B}_F^{(1)}\left(\boldsymbol{\alpha}^0,\mathbf{u}^0;\delta\boldsymbol{\alpha},\delta\mathbf{u}\right) \end{pmatrix} = \mathbf{0}, \ \mathbf{x}\in\partial\Omega_x. \tag{3.46}$$

Equations 3.45 and 3.46 constitute the *second-level forward sensitivity system* (2nd-LFSS). They can be solved, in principle, to obtain the vectors $\delta\mathbf{u}$ and $\delta\boldsymbol{\psi}^{(1)}$ as functions of the vector of parameter variations $\delta\boldsymbol{\alpha}$ and could subsequently be replaced in Equation 3.36 to determine the indirect-effect term $\left(\delta R_{i_1}^{(1)}\right)_{indirect}$. This procedure would be impractical to use for large-scale systems, since it would require $O\left(N_\alpha^2\right)$ large-scale computations.

To circumvent the need for solving 2nd-LFSS, the alternative *Second-Level Adjoint Sensitivity System* (*2nd-LASS*) will be developed in the remaining of this section by using the same guiding principles as those underlying the development of the 1st-LASS. The goal of the 2nd-LASS is to compute the indirect-effect term $\left(\delta R_{i_1}^{(1)}\right)_{indirect}$ in a manner that circumvents the need for computing explicitly the vectors of variations $\delta\boldsymbol{\psi}^{(1)}$ and $\delta\mathbf{u}$ by solving the 2nd-LFSS.

To construct the 2nd-LASS, recall that an element in the Hilbert space $\mathcal{H}_\psi^{(2)}(\Omega_x) \triangleq \mathcal{H}_\psi^{(1)}(\Omega_x) \oplus \mathcal{H}_u(\Omega_x)$ is a two-block vector of the form $\boldsymbol{\Psi}_{i_1}^{(2)} \triangleq \left(\boldsymbol{\psi}_{1,i_1}^{(2)}, \boldsymbol{\psi}_{2,i_1}^{(2)}\right) \in \mathcal{H}_\psi^{(2)}(\Omega_x)$, with $\boldsymbol{\psi}_{1,i_1}^{(2)} \in \mathcal{H}_\psi^{(1)}(\Omega_x)$, $\boldsymbol{\psi}_{2,i_1}^{(2)} \in \mathcal{H}_u(\Omega_x)$. Since Hilbert spaces are self-dual, the spaces $\mathcal{H}_\psi^{(1)}(\Omega_x)$ and $\mathrm{H}_u(\Omega_x)$ actually coincide, but the distinct notation will be kept in this section for clarity. The inner product $\langle \bullet, \bullet \rangle_{\psi^{(2)}}$ between any two vectors $\mathbf{U}_{i_1}^{(2)} \triangleq \left(\mathbf{u}_{1,i_1}^{(2)}, \mathbf{u}_{2,i_1}^{(2)}\right) \in \mathcal{H}_\psi^{(2)}(\Omega_x)$ and $\mathbf{V}_{i_1}^{(2)} \triangleq \left(\mathbf{v}_{1,i_1}^{(2)}, \mathbf{v}_{2,i_1}^{(2)}\right) \in \mathcal{H}_\psi^{(2)}(\Omega_x)$ is defined as follows:

$$\left\langle \mathbf{U}_{i_1}^{(2)}, \mathbf{V}_{i_1}^{(2)} \right\rangle_{\psi^{(2)}} \triangleq \int_{\Omega_x} \left(\mathbf{u}_{1,i_1}^{(2)} \bullet \mathbf{v}_{1,i_1}^{(2)} + \mathbf{u}_{2,i_1}^{(2)} \bullet \mathbf{v}_{2,i_1}^{(2)} \right) d\mathbf{x} = \sum_{j=1}^{2} \int_{\Omega_x} \left(\mathbf{u}_{j,i_1}^{(2)} \bullet \mathbf{v}_{j,i_1}^{(2)} \right) d\mathbf{x}. \tag{3.47}$$

Using the above definition, we construct the inner product of Equation 3.45 with a yet-undefined vector $\boldsymbol{\Psi}_{i_1}^{(2)} \triangleq \left(\boldsymbol{\psi}_{1,i_1}^{(2)}, \boldsymbol{\psi}_{2,i_1}^{(2)}\right) \in \mathcal{H}_\psi^{(2)}(\Omega_x)$ to obtain

$$\begin{aligned}
&\left\langle \left(\boldsymbol{\psi}_{1,i_1}^{(2)}, \boldsymbol{\psi}_{2,i_1}^{(2)}\right), \begin{pmatrix} \mathbf{F}_{11}^{(2)}\left(\boldsymbol{\alpha}^0\right) & \mathbf{F}_{12}^{(2)}\left(\boldsymbol{\alpha}^0; \mathbf{u}^0\right) \\ 0 & \mathbf{F}_{11}^{(1)}\left(\boldsymbol{\alpha}^0\right) \end{pmatrix} \begin{pmatrix} \delta\boldsymbol{\psi}^{(1)} \\ \delta\mathbf{u} \end{pmatrix} \right\rangle_{\psi^{(2)}} \\
&= \left\langle \left(\boldsymbol{\psi}_{1,i_1}^{(2)}, \boldsymbol{\psi}_{2,i_1}^{(2)}\right), \begin{pmatrix} \mathbf{Q}^{(2)}\left(\boldsymbol{\alpha}^0, \boldsymbol{\varphi}^{(2,0)}; \delta\boldsymbol{\alpha}\right) \\ \mathbf{Q}^{(1)}\left(\boldsymbol{\alpha}^0, \mathbf{u}^0; \delta\boldsymbol{\alpha}\right) \end{pmatrix} \right\rangle_{\psi^{(2)}} \\
&= \int_{\Omega_x} \left[\boldsymbol{\psi}_{1,i_1}^{(2)} \bullet \mathbf{Q}^{(2)}\left(\boldsymbol{\alpha}^0, \boldsymbol{\varphi}^{(2,0)}; \delta\boldsymbol{\alpha}\right) + \boldsymbol{\psi}_{2,i_1}^{(2)} \bullet \mathbf{Q}^{(1)}\left(\boldsymbol{\alpha}^0, \mathbf{u}^0; \delta\boldsymbol{\alpha}\right) \right] d\mathbf{x}, \quad \mathbf{x} \in \Omega_x.
\end{aligned} \tag{3.48}$$

The inner product on the left side of Equation 3.48 is now further transformed by using the definition of the adjoint operator to obtain

$$\left\langle \left(\psi_{1,i_1}^{(2)},\psi_{2,i_1}^{(2)}\right), \begin{pmatrix} \mathbf{F}_{11}^{(2)}\left(\boldsymbol{\alpha}^0\right) & \mathbf{F}_{12}^{(2)}\left(\boldsymbol{\alpha}^0,\mathbf{u}^0\right) \\ \mathbf{0} & \mathbf{F}_{11}^{(1)}\left(\boldsymbol{\alpha}^0\right) \end{pmatrix} \begin{pmatrix} \delta\psi^{(1)} \\ \delta\mathbf{u} \end{pmatrix} \right\rangle_{\psi^{(2)}}$$

$$= \left\langle \left(\delta\psi^{(1)},\delta\mathbf{u}\right), \begin{pmatrix} \left[\mathbf{F}_{11}^{(2)}\left(\boldsymbol{\alpha}^0\right)\right]^* & \mathbf{0} \\ \left[\mathbf{F}_{12}^{(2)}\left(\boldsymbol{\alpha}^0,\mathbf{u}^0\right)\right]^* & \left[\mathbf{F}_{11}^{(1)}\left(\boldsymbol{\alpha}^0\right)\right]^* \end{pmatrix} \begin{pmatrix} \psi_{1,i_1}^{(2)} \\ \psi_{2,i_1}^{(2)} \end{pmatrix} \right\rangle_{\psi^{(2)}} \tag{3.49}$$

$$+\left\{P^{(2)}\left[\boldsymbol{\alpha}^0,\boldsymbol{\varphi}^{(2,0)};\psi_{1,i_1}^{(2,0)},\psi_{2,i_1}^{(2,0)};\delta\mathbf{u};\delta\psi^{(1)}\right]\right\}_{\partial\Omega_x}.$$

In the above equation, the quantity $[X]^*$ indicates the "formal adjoint"of the quantity X, and the quantity $\left\{P^{(2)}\left[\boldsymbol{\alpha}^0,\boldsymbol{\varphi}^{(2,0)};\psi_{1,i_1}^{(2,0)},\psi_{2,i_1}^{(2,0)};\delta\mathbf{u};\delta\psi^{(1)}\right]\right\}_{\partial\Omega_x}$ denotes the corresponding bilinear concomitant on $\mathbf{x}\in\partial\Omega_x$. The domains of the adjoint operators appearing in the first term on the right side of Equation 3.49 can now be specified by requiring that the function $\boldsymbol{\Psi}_{i_1}^{(2)} \triangleq \left(\psi_{1,i_1}^{(2)}(\mathbf{x}),\psi_{2,i_1}^{(2)}(\mathbf{x})\right)\in\mathcal{H}_{\psi}^{(2)}(\Omega_x)$ satisfies adjoint boundary conditions denoted as

$$\mathbf{B}_A^{(2)}\left(\boldsymbol{\alpha}^0,\boldsymbol{\varphi}^{(2,0)};\psi_{i_1}^{(2,0)}\right)=\mathbf{0}, \quad \mathbf{x}\in\partial\Omega_x. \tag{3.50}$$

The above adjoint boundary and/or initial conditions for $\boldsymbol{\Psi}_{i_1}^{(2)} \triangleq \left(\psi_{1,i_1}^{(2)},\psi_{2,i_1}^{(2)}\right)\in\mathcal{H}_{\psi}^{(2)}(\Omega_x)$ are determined following the same principles as described in Section 3.2, namely that

1. They must be independent of *unknown* values of $\delta\boldsymbol{\alpha};\delta\mathbf{u};\delta\boldsymbol{\psi}^{(1)}$.
2. The substitution of the boundary and/or initial conditions represented by Equations 3.50 and 3.46 into the expression of $\left\{P^{(2)}\left[\boldsymbol{\alpha}^0,\boldsymbol{\varphi}^{(2,0)};\boldsymbol{\Psi}_{i_1}^{(2,0)};\delta\mathbf{u};\delta\psi^{(1)}\right]\right\}_{\partial\Omega_x}$ must cause all terms containing unknown values of $\delta\mathbf{u}$ and $\delta\boldsymbol{\psi}^{(1)}$ to vanish, which will reduce the bilinear concomitant to an expression that will contain only known values of $\delta\boldsymbol{\alpha}$; this expression will be denoted as $\hat{P}^{(2)}\left(\boldsymbol{\alpha}^0,\boldsymbol{\varphi}^{(2,0)};\boldsymbol{\Psi}_{i_1}^{(2,0)};\delta\boldsymbol{\alpha}\right)$. In general, $\hat{P}^{(2)}\left(\boldsymbol{\alpha}^0,\boldsymbol{\varphi}^{(2,0)};\boldsymbol{\Psi}_{i_1}^{(2,0)};\delta\boldsymbol{\alpha}\right)$ does not automatically vanish as a result of these operations.

Implementing the second-level forward and adjoint boundary and initial conditions, namely Equations 3.50 and 3.46, into Equation 3.49 will transform the latter into the form

$$\left\langle \left(\delta\psi^{(1)},\delta\mathbf{u}\right), \begin{pmatrix} \left[\mathbf{F}_{11}^{(2)}\right]^* & 0 \\ \left[\mathbf{F}_{12}^{(2)}\right]^* & \left[\mathbf{F}_{11}^{(1)}\right]^* \end{pmatrix} \begin{pmatrix} \psi_{1,i_1}^{(2)} \\ \psi_{2,i_1}^{(2)} \end{pmatrix} \right\rangle_{\psi^{(2)}}$$

$$= \left\langle \left(\psi_{1,i_1}^{(2)},\psi_{2,i_1}^{(2)}\right), \begin{pmatrix} \mathbf{F}_{11}^{(2)} & \mathbf{F}_{12}^{(2)} \\ \mathbf{0} & \mathbf{F}_{11}^{(1)} \end{pmatrix} \begin{pmatrix} \delta\psi^{(1)} \\ \delta\mathbf{u} \end{pmatrix} \right\rangle_{\psi^{(2)}} - \hat{P}^{(2)}\left(\boldsymbol{\alpha}^0,\boldsymbol{\varphi}^{(2,0)};\boldsymbol{\Psi}_{i_1}^{(2,0)};\delta\boldsymbol{\alpha}\right). \tag{3.51}$$

The expression containing the variations $\delta\mathbf{u}$ and $\delta\boldsymbol{\psi}^{(1)}$ in the first term on the right side of Equation 3.51 is now replaced by using the right side of Equation 3.45, which transforms Equation 3.51 into the form

$$\left\langle \left(\delta\psi^{(1)},\delta\mathbf{u}\right), \begin{pmatrix} \left[\mathbf{F}_{11}^{(2)}\right]^* & 0 \\ \left[\mathbf{F}_{12}^{(2)}\right]^* & \left[\mathbf{F}_{11}^{(1)}\right]^* \end{pmatrix} \begin{pmatrix} \psi_{1,i_1}^{(2)} \\ \psi_{2,i_1}^{(2)} \end{pmatrix} \right\rangle_{\psi^{(2)}}$$

$$= \left\langle \left(\psi_{1,i_1}^{(2)},\psi_{2,i_1}^{(2)}\right), \begin{pmatrix} \mathbf{Q}^{(2)}\left(\boldsymbol{\alpha}^0,\boldsymbol{\varphi}^{(2,0)};\delta\boldsymbol{\alpha}\right) \\ \mathbf{Q}^{(1)}\left(\boldsymbol{\alpha}^0,\mathbf{u}^0;\delta\boldsymbol{\alpha}\right) \end{pmatrix} \right\rangle_{\psi^{(2)}} \tag{3.52}$$

$$-\hat{P}^{(2)}\left(\boldsymbol{\alpha}^0,\boldsymbol{\varphi}^{(2,0)};\boldsymbol{\Psi}_{i_1}^{(2,0)};\delta\boldsymbol{\alpha}\right).$$

The definition of the second-level adjoint function $\boldsymbol{\Psi}_{i_1}^{(2)}$ will be completed next by requiring that the left side of Equation 3.52 and the right side of Equation 3.36 represent the same functional, namely the *indirect-effect term*, $\left(\delta R_{i_1}^{(1)}\right)_{indirect}$, which yields the following relationship:

$$\left\langle \left(\delta\psi^{(1)},\delta\mathbf{u}\right), \begin{pmatrix} \left[\mathbf{F}_{11}^{(2)}\left(\boldsymbol{\alpha}^0\right)\right]^* & 0 \\ \left[\mathbf{F}_{12}^{(2)}\left(\boldsymbol{\alpha}^0,\mathbf{u}^0\right)\right]^* & \left[\mathbf{F}_{11}^{(1)}\left(\boldsymbol{\alpha}^0\right)\right]^* \end{pmatrix} \begin{pmatrix} \psi_{1,i_1}^{(2)} \\ \psi_{2,i_1}^{(2)} \end{pmatrix} \right\rangle_{\psi^{(2)}} \tag{3.53}$$

$$= \left\langle \left(\delta\psi^{(1)},\delta\mathbf{u}\right), \begin{pmatrix} \boldsymbol{\Theta}_{1,i_1}^{(2)}\left(\boldsymbol{\alpha}^0,\boldsymbol{\varphi}^{(2,0)}\right) \\ \boldsymbol{\Theta}_{2,i_1}^{(2)}\left(\boldsymbol{\alpha}^0,\boldsymbol{\varphi}^{(2,0)}\right) \end{pmatrix} \right\rangle_{\psi^{(2)}}.$$

Equation 3.53 implies that the second-level adjoint function $\boldsymbol{\Psi}_{i_1}^{(2)}(\mathbf{x}) \triangleq \left(\boldsymbol{\psi}_{1,i_1}^{(2)}(\mathbf{x}), \boldsymbol{\Psi}_{2,i_1}^{(2)}(\mathbf{x})\right) \in \mathcal{H}_{\psi}^{(2)}(\Omega_x)$ is the *weak solution* (in the sense of distributions) of the following operator equation:

$$\begin{pmatrix} \mathbf{A}_{11}^{(2)}\left(\boldsymbol{\alpha}^0\right) & \mathbf{0} \\ \mathbf{A}_{21}^{(2)}\left(\boldsymbol{\alpha}^0, \mathbf{u}^0\right) & \mathbf{A}_{22}^{(2)}\left(\boldsymbol{\alpha}^0\right) \end{pmatrix} \begin{pmatrix} \boldsymbol{\psi}_{1,i_1}^{(2)}(\mathbf{x}) \\ \boldsymbol{\psi}_{2,i_1}^{(2)}(\mathbf{x}) \end{pmatrix}$$
$$= \begin{pmatrix} \boldsymbol{\Theta}_{1,i_1}^{(2)}\left(\boldsymbol{\alpha}^0, \boldsymbol{\varphi}^{(2,0)}\right) \\ \boldsymbol{\Theta}_{2,i_1}^{(2)}\left(\boldsymbol{\alpha}^0, \boldsymbol{\varphi}^{(2,0)}\right) \end{pmatrix}, \ \mathbf{x} \in \Omega_x, \ i_1 = 1, \ldots, N_\alpha, \tag{3.54}$$

where the second-level adjoint operators $\mathbf{A}_{ij}^{(2)}$, $i,j = 1,2$, are defined as follows:

$$\mathbf{A}_{11}^{(2)}(\boldsymbol{\alpha}) \triangleq \left[\mathbf{F}_{11}^{(2)}(\boldsymbol{\alpha})\right]^* = \mathbf{L}(\boldsymbol{\alpha}); \ \mathbf{A}_{22}^{(2)}(\boldsymbol{\alpha}) \triangleq \left[\mathbf{F}_{11}^{(1)}(\boldsymbol{\alpha})\right]^* = \mathbf{A}_{11}^{(1)}(\boldsymbol{\alpha}) = \mathbf{L}^*(\boldsymbol{\alpha});$$
$$\mathbf{A}_{21}^{(2)}(\boldsymbol{\alpha}, \mathbf{u}) \triangleq \left[\mathbf{F}_{12}^{(2)}(\boldsymbol{\alpha}, \mathbf{u})\right]^* = -\left[\frac{\partial \boldsymbol{\Theta}^{(1)}(\boldsymbol{\alpha}, \mathbf{u})}{\partial \mathbf{u}}\right]^*. \tag{3.55}$$

The Riesz representation theorem ensures that the relationship in Equation 3.55 holds uniquely. Of course, the second-level adjoint function $\boldsymbol{\Psi}_{i_1}^{(2)}(\mathbf{x}) \triangleq \left(\boldsymbol{\psi}_{1,i_1}^{(2)}(\mathbf{x}), \boldsymbol{\psi}_{2,i_1}^{(2)}(\mathbf{x})\right) \in \mathcal{H}_{\psi}^{(2)}(\Omega_x)$ satisfies the second-level adjoint boundary conditions represented by Equation 3.50. The system of equations comprising Equations 3.54 and 3.50 will be referred to as the *2nd-LASS*, and its solutions, namely the functions $\boldsymbol{\Psi}_{i_1}^{(2)}(\mathbf{x}) \triangleq \left(\boldsymbol{\psi}_{1,i_1}^{(2)}(\mathbf{x}), \boldsymbol{\psi}_{2,i_1}^{(2)}(\mathbf{x})\right) \in \mathcal{H}_{\psi}^{(2)}(\Omega_x)$, will be called the *second-level adjoint functions.*

The results obtained in Equations 3.36, 3.52, 3.53, and 3.54 are now replaced in Equation 3.33 to obtain

$$\delta R_{i_1}^{(1)}\left(\boldsymbol{\alpha}^0, \boldsymbol{\varphi}^{(2,0)}; \delta\boldsymbol{\alpha}, \delta\mathbf{u}, \delta\boldsymbol{\psi}^{(1)}\right) = \sum_{i_2=1}^{N_\alpha} \frac{\partial R_{i_1}^{(1)}\left(\boldsymbol{\alpha}^0, \boldsymbol{\varphi}^{(2,0)}\right)}{\partial \alpha_{i_2}} \delta\alpha_{i_2}$$
$$+ \left\langle \left(\boldsymbol{\psi}_{1,i_1}^{(2)}, \boldsymbol{\psi}_{2,i_1}^{(2)}\right), \begin{pmatrix} \mathbf{Q}^{(2)}\left(\boldsymbol{\alpha}^0, \boldsymbol{\varphi}^{(2,0)}; \delta\boldsymbol{\alpha}\right) \\ \mathbf{Q}^{(1)}\left(\boldsymbol{\alpha}^0, \mathbf{u}^0; \delta\boldsymbol{\alpha}\right) \end{pmatrix} \right\rangle_{\psi^{(2)}} \tag{3.56}$$
$$- \hat{P}^{(2)}\left(\boldsymbol{\alpha}^0, \boldsymbol{\varphi}^{(2,0)}; \boldsymbol{\Psi}_{i_1}^{(2,0)}; \delta\boldsymbol{\alpha}\right), \quad i_1 = 1, \ldots, N_\alpha.$$

Since the quantities $\mathbf{Q}^{(2)}(\boldsymbol{\alpha}^0,\boldsymbol{\varphi}^{(2,0)};\delta\boldsymbol{\alpha})$ and $\hat{P}_2\left(\boldsymbol{\alpha}^0,\boldsymbol{\varphi}^{(2,0)};\boldsymbol{\Psi}_{i_1}^{(2,0)};\delta\boldsymbol{\alpha}\right)$ are both linear in $\delta\boldsymbol{\alpha}$, it follows that they can be expressed in the following forms:

$$\mathbf{Q}^{(2)}\left(\boldsymbol{\alpha}^0,\boldsymbol{\varphi}^{(2,0)};\delta\boldsymbol{\alpha}\right)=\sum_{i_2=1}^{N_\alpha}\left\{\frac{\partial\left[\boldsymbol{\Theta}^{(1)}(\boldsymbol{\alpha},\mathbf{u})-\mathbf{L}^*(\boldsymbol{\alpha})\boldsymbol{\psi}^{(1)}\right]}{\partial\alpha_{i_2}}\delta\alpha_{i_2}\right\}_{\left(\boldsymbol{\alpha}^0,\boldsymbol{\varphi}^{(2,0)}\right)},\tag{3.57}$$

and

$$\hat{P}^{(2)}\left(\boldsymbol{\alpha}^0,\boldsymbol{\varphi}^{(2,0)};\boldsymbol{\psi}^{(2,0)};\delta\boldsymbol{\alpha}\right)=\sum_{i_2=1}^{N_\alpha}\frac{\partial\hat{P}^{(2)}\left(\boldsymbol{\alpha}^0,\boldsymbol{\varphi}^{(2,0)};\boldsymbol{\psi}_{1,i_1}^{(2,0)},\boldsymbol{\psi}_{2,i_1}^{(2,0)}\right)}{\partial\alpha_{i_2}}\delta\alpha_{i_2}.\tag{3.58}$$

Using the relations expressed by Equations 3.57, 3.58, and 3.24 in Equation 3.56, it follows that Equation 3.56 can be expressed in the following form:

$$\delta R_{i_1}^{(1)}\left(\boldsymbol{\alpha}^0,\boldsymbol{\varphi}^{(2,0)};\delta\boldsymbol{\alpha},\delta\mathbf{u},\delta\boldsymbol{\psi}^{(1)}\right)=\sum_{i_2=1}^{N_\alpha}R_{i_1i_2}^{(2)}\left(\boldsymbol{\alpha}^0,\boldsymbol{\varphi}^{(2,0)};\boldsymbol{\psi}_{1,i_1}^{(2,0)},\boldsymbol{\psi}_{2,i_1}^{(2,0)}\right)\delta\alpha_{i_2},\quad i_1=1,\ldots,N_\alpha,\tag{3.59}$$

where

$$R_{i_1i_2}^{(2)}\left(\boldsymbol{\alpha},\boldsymbol{\varphi}^{(2)};\boldsymbol{\psi}_{1,i_1}^{(2)},\boldsymbol{\psi}_{2,i_1}^{(2)}\right)\triangleq\frac{\partial}{\partial\alpha_{i_2}}\left[R_{i_1}^{(1)}\left(\boldsymbol{\alpha},\boldsymbol{\varphi}^{(2)}\right)-\hat{P}^{(2)}\left(\boldsymbol{\alpha},\boldsymbol{\varphi}^{(2)};\boldsymbol{\psi}_{1,i_1}^{(2)},\boldsymbol{\psi}_{2,i_1}^{(2)}\right)\right]$$
$$+\int_{\Omega_x}\left\{\boldsymbol{\psi}_{1,i_1}^{(2)}(\mathbf{x})\bullet\left[\frac{\partial\left[\boldsymbol{\Theta}^{(1)}(\boldsymbol{\alpha},\mathbf{u})-\mathbf{L}^*(\boldsymbol{\alpha})\boldsymbol{\psi}^{(1)}\right]}{\partial\alpha_{i_2}}\right]+\boldsymbol{\psi}_{2,i_1}^{(2)}(\mathbf{x})\bullet\left[\frac{\partial\left[\mathbf{Q}(\boldsymbol{\alpha})-\mathbf{L}(\boldsymbol{\alpha})\mathbf{u}\right]}{\partial\alpha_{i_2}}\right]\right\}d\mathbf{x},$$
$$\text{for }\ i_1,i_2=1,\ldots,N_\alpha.\tag{3.60}$$

As Equation 3.60 indicates, *each of the G-differentials* $\delta R_{i_1}^{(1)}\left(\boldsymbol{\alpha}^0,\boldsymbol{\varphi}^{(2,0)};\delta\boldsymbol{\alpha},\delta\mathbf{u},\delta\boldsymbol{\psi}^{(1)}\right)$ *contains one complete row (running on the index* $i_2=1,\ldots,N_{\boldsymbol{\alpha}}$*) of the mixed second-order partial sensitivities,* $R_{i_1i_2}^{(2)}\left(\boldsymbol{\alpha},\boldsymbol{\varphi}^{(2)};\boldsymbol{\psi}_{1,i_1}^{(2)},\boldsymbol{\psi}_{2,i_1}^{(2)}\right)$, which can be computed exactly and efficiently after solving *only once* the 2nd-LASS, namely Equations 3.54 and 3.50, to obtain the *second-level adjoint functions* $\boldsymbol{\Psi}_{i_1}^{(2)}(\mathbf{x})\triangleq\left(\boldsymbol{\psi}_{1,i_1}^{(2)}(\mathbf{x}),\boldsymbol{\psi}_{2,i_1}^{(2)}(\mathbf{x})\right)\in\mathcal{H}_\psi^{(2)}(\Omega_x)$, for each index $i_1=1,\ldots,N_{\boldsymbol{\alpha}}$. The second-order partial sensitivities are functionals of the form $R_{i_1i_2}^{(2)}\left(\boldsymbol{\alpha},\boldsymbol{\varphi}^{(2)};\boldsymbol{\Psi}_{i_1}^{(2)}\right)\in\mathcal{H}_\alpha\oplus\mathcal{H}_\psi^{(2)}(\Omega_x)\oplus\mathcal{H}_\psi^{(2)}(\Omega_x)=\mathcal{H}_\alpha\oplus2\otimes\mathcal{H}_\psi^{(2)}(\Omega_x)\to\mathcal{F}$.

It is also important to note that the construction and solution of the 2nd-LASS requires very little effort beyond that already invested in solving the original forward system (namely Equations 3.1 and 3.2) to compute the state variable $\mathbf{u}(\mathbf{x})$ and for solving the 1st-LASS (cf. Equations 3.28 and 3.29) to compute the first-level adjoint function $\boldsymbol{\psi}^{(1)}(\mathbf{x})$. This fact becomes readily apparent by recasting the 2nd-LASS, cf. Equation 3.54, into the following form:

$$\mathbf{L}\left(\boldsymbol{\alpha}^0\right)\boldsymbol{\psi}_{1,i_1}^{(2)}(\mathbf{x}) = \boldsymbol{\Theta}_{1,i_1}^{(2)}\left(\boldsymbol{\alpha},\boldsymbol{\varphi}^{(2)}\right), \ \mathbf{x}\in\Omega_x, \ i_1 = 1,\ldots,N_\alpha, \tag{3.61}$$

$$\mathbf{L}^*(\boldsymbol{\alpha})\boldsymbol{\psi}_{2,i_1}^{(2)}(\mathbf{x}) = \boldsymbol{\Theta}_{2,i_1}^{(2)}\left(\boldsymbol{\alpha},\boldsymbol{\varphi}^{(2)}\right) + \left[\frac{\partial\boldsymbol{\Theta}^{(1)}(\boldsymbol{\alpha},\mathbf{u})}{\partial\mathbf{u}}\right]^* \boldsymbol{\psi}_{1,i_1}^{(2)}, \ \mathbf{x}\in\Omega_x, \ i_1 = 1,\ldots,N_\alpha. \tag{3.62}$$

It is now apparent that Equation 3.61 requires the same computational solver and effort as solving the original Equation 3.1 for determining $\mathbf{u}(\mathbf{x})$; only the source on the right side of Equation 3.61 differs from the source on the right side of Equation 3.1. Furthermore, solving Equation 3.62 to obtain the second-level adjoint functions $\boldsymbol{\psi}_{2,i_1}^{(2)}(\mathbf{x})$ is equivalent to solving Equation 3.28, with the same left side as for the 1st-LASS, but with a different source term on the right side of Equation 3.62 by comparison to the right side of Equation 3.28. Thus, *the same computer program and solvers can be used for solving the 2nd-LASS as were already used for solving the original system of Equations 3.1 and 3.2 and for solving the 1st-LASS, namely Equations 3.28 and 3.29. Only the sources for the respective operator equations and boundary conditions for the 2nd-LASS will differ from those used to solve the 1st-LASS and the original system of equations. For this reason, the 2nd-LASS* (which comprises Equations 3.54 and 3.50) *was designated as a* second-level *rather than* second-order *adjoint sensitivity system, since the 2nd-LASS does not involve any explicit second-order G-derivatives of the operators underlying the original system but involves the inversion of the same operators that needed to be inverted for solving the 1st-LASS and the original system of equations.*

Note that the 2nd-LASS is independent of parameter variations $\delta\boldsymbol{\alpha}$, as well as of the variations $\delta\mathbf{u}$ and $\delta\boldsymbol{\psi}^{(1)}$. Thus, *the exact computation of all of the partial second-order sensitivities,* $\left\{\partial^2 R(\boldsymbol{\alpha},\mathbf{u})/\partial\alpha_j\,\partial\alpha_i\right\}_{\left(\boldsymbol{\alpha}^0,\mathbf{u}^0\right)}$, $i, j = 1, \ldots, N_\alpha$, *requires at most* N_α *large-scale (adjoint) computations using the 2nd-LASS*, rather than $O\left(N_\alpha^2\right)$ large-scale computations as would be required by forward methods. Furthermore, in the important particular case *when the response* $R(\mathbf{u},\boldsymbol{\alpha})$ *is linear in* $\mathbf{u}$, that is, when $R(\mathbf{u},\boldsymbol{\alpha}) = r(\boldsymbol{\alpha})\mathbf{u}$, then both the *1st-LASS* and the *2nd-LASS* are independent of the original state function $\mathbf{u}$ and can be solved independently of the original system.

It is also important to note that *by solving the 2nd-LASS* N_α *times, the "off-diagonal" second-order mixed sensitivities* $R_{i_1 i_2}^{(2)}\left(\boldsymbol{\alpha}^0,\boldsymbol{\varphi}^{(2,0)};\boldsymbol{\psi}_{1,i_1}^{(2,0)},\boldsymbol{\psi}_{2,i_1}^{(2,0)}\right)$ *will be computed twice, in two different ways, thereby providing an intrinsic (numerical) verification that the second-level adjoint functions* $\boldsymbol{\psi}_{1,i_1}^{(2)}(\mathbf{x})$ *and* $\boldsymbol{\psi}_{2,i_1}^{(2)}(\mathbf{x})$ *are computed accurately.*

Another important characteristic of using the 2nd-LASS is the flexibility it provides for prioritizing the computation of the second-order sensitivities. The computation of the second-order sensitivities should be based on the priority ranking of the relative magnitudes of the first-order sensitivities: the second-order mixed G-derivatives (sensitivities) corresponding to the largest relative first-order response sensitivity should be computed first; the second largest relative first-order response sensitivity should be considered next, and so on. Computing second-order partial sensitivities that correspond to vanishing first-order sensitivities may also be of interest, since vanishing first-order sensitivities may indicate critical points of the response in the phase space of model parameters. Thus, only the second-order partial sensitivities of the response $R(\mathbf{e})$ that are deemed important will need to be computed; the unimportant ones can be deliberately neglected while knowing the error incurred by neglecting them.

3.4 Concluding Remarks

This chapter has presented the 2nd-ASAM for computing exactly and efficiently all of the first-order and the second-order functional derivatives of system responses (i.e., "system performance parameters") to the system's model parameters. For *one functional-type response* of interest produced by a physical system comprising N_α parameters and N_r responses, the *2nd-ASAM requires one large-scale computation using the 1st-LASS for obtaining all of the first-order sensitivities* $\left\{\partial^2 R(\mathbf{u},\boldsymbol{\alpha})/\partial\alpha_j\,\partial\alpha_i\right\}_{(\mathbf{u}^0,\boldsymbol{\alpha}^0)}$, *$i, j = 1, \ldots, N_\alpha$, followed by at most N_α large-scale computations using the 2nd-LASS for obtaining exactly all of the second-order sensitivities* $\left\{\partial^2 R(\mathbf{u},\boldsymbol{\alpha})/\partial\alpha_j\,\partial\alpha_i\right\}_{(\mathbf{u}^0,\boldsymbol{\alpha}^0)}$, $i, j = 1, \ldots, N_\alpha$. In contradistinction, forward methods would require $\left(N_\alpha^2/2+3N_\alpha/2\right)$ large-scale computations for obtaining all of the first- and second-order sensitivities, for all N_r system responses.

It has also been shown that constructing and solving the 2nd-LASS requires very little additional effort beyond the construction of the 1st-LASS needed for computing the first-order sensitivities. Furthermore, due to the symmetry properties of the second-order sensitivities $\left\{\partial^2 R(\mathbf{u},\boldsymbol{\alpha})/\partial\alpha_j\,\partial\alpha_i\right\}_{(\mathbf{u}^0,\boldsymbol{\alpha}^0)}$, $j = 1, \ldots, N_\alpha$ the 2nd-ASAM encompasses an automatic, inherent, "solution verification" of the correctness and accuracy of the second-level adjoint functions $\boldsymbol{\Psi}_{i1}^{(2)} \triangleq \left(\boldsymbol{\psi}_{i1}^{(2)}, \boldsymbol{\psi}_{i2}^{(2)}\right)$ needed for the efficient and exact computation of the second-order sensitivities.

Since real Hilbert spaces provide the natural mathematical setting for computational purposes, the 2nd-ASAM presented in this chapter is set in *real* (as opposed to complex) Hilbert spaces. This setting does *not* restrict, in any way, the generality of the concepts presented here, since the 2nd-ASAM can be readily set in complex Hilbert spaces by simply changing some terminology, without affecting its substance.

4

Application of the 2nd-ASAM to a Linear Heat Conduction and Convection Benchmark Problem

The adequate transfer of the heat generated in a reactor's fuel rods to the reactor's coolant and the subsequent transport of this heat by the coolant to the primary heat exchanger is of paramount importance for the design and licensing of nuclear reactors. Such heat transport processes are modeled using large-scale three-dimensional thermal-hydraulics numerical computation tools, which comprise many imperfectly known parameters that give rise to uncertainties in the computed results. It is therefore important to develop benchmarks that admit exact solutions, in order to verify, within known uncertainty bands, the results produced by the respective computation tools. The "solution verification" process of such computational tools includes quantification of the accuracy of the sensitivities of the results computed by these tools (e.g., temperature distributions) with respect to the parameters (e.g., conductivities, specific heat capacities, and heat transfer coefficients).

This chapter presents an application of the 2nd-ASAM developed in Chapter 3 to a benchmark problem that models the steady-state heat conduction within a heated cylindrical rod cooled by forced convection to a surrounding liquid flowing along the rod's length. This benchmark problem admits exact analytical solutions for the spatially dependent temperature distributions within the rod and the surrounding coolant, thus providing another readily tractable illustration of the steps involved in the application of the 2nd-ASAM. In particular, this benchmark problem also illustrates the occasional need to consider the *extended adjoint operator*, as has been discussed in Chapter 3.

The development of the benchmark problem presented in this chapter was motivated by the need to verify the numerical results produced by the commercially available three-dimensional thermal-hydraulics code FLUENT Adjoint Solver (ANSYS 2015). This computational tool has been used (Cacuci 2016b, Cacuci et al. 2016a, 2016b) to compute thermal-hydraulics processes within the conceptual design of the G4M reactor (Gen4 Energy, Inc. 2012), as well as for the conceptual design of a test section aimed at analyzing the thermal-hydraulics behavior of a 1/48-segment of the G4M reactor's core.

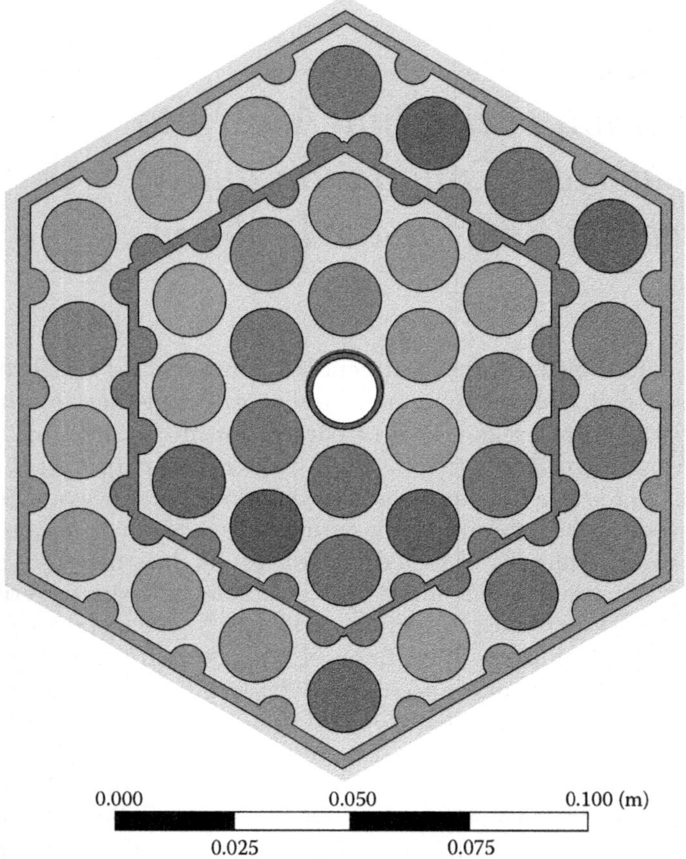

FIGURE 4.1
FLUENT model of the test section comprising 36 electrically heated rods.

The conceptual design of this test section is envisaged to comprise 36 heated rods disposed equally within two rings, as depicted in Figure 4.1, cooled by pumped counter-gravity flow of molten lead–bismuth eutectic (LBE). The thermal-hydraulics conditions envisaged for the test section would be the same as for the G4M reactor design, namely an inlet velocity of 1.65 m/s and an inlet temperature of 673.15 K. Based on the G4M reactor design, the heating conditions within the test section were chosen such that each of the 18 electrically heated inner rods would experience a heat flux $q''_{innner} \cos\left(\frac{\pi z}{\ell}\right)$, $q''_{inner} = 6.46367 \times 10^5 \ \mathrm{W/m^2}$, while each of the 18 electrically heated outer rods would experience a heat flux $q''_{outer} \cos\left(\frac{\pi z}{\ell}\right)$, with $q''_{outer} = 8.53575 \times 10^5 \ \mathrm{W/m^2}$.

Figure 4.2 depicts the temperature distribution computed with FLUENT within the test section. The maximum surface rod temperature, $T_s(z)$, at the location of its maximum value and the coolant temperature at the exit

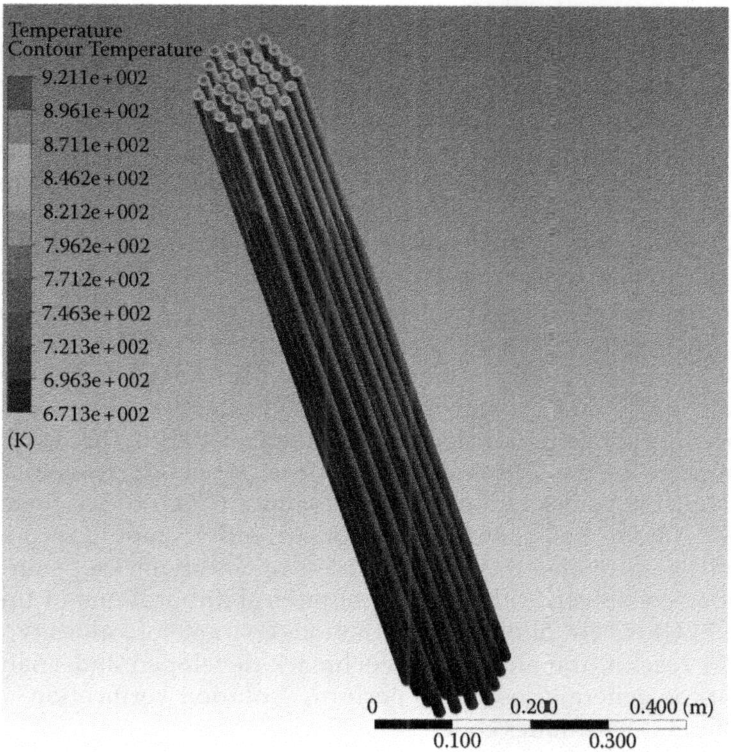

FIGURE 4.2
FLUENT computation of the temperature distribution within the test section's 36 electrically heated rods.

(where it attains its respective maximum) were considered to be the "model responses" subject to sensitivity analysis using the FLUENT Adjoint Solver.

The material presented in this chapter follows the material originally developed by Cacuci (2016d) and is structured as follows: Section 4.1 describes the heat transport benchmark problem, which comprises six thermal-hydraulics parameters and admits an easily tractable closed-form analytical solution. Therefore, for each functional type response, such as the temperature distribution within the heated rod at some specific location within the rod and/or on its surface, there are 6 first-order sensitivities and 21 (i.e., $6 \times 7/2 = 21$) distinct second-order sensitivities. Similarly, there are 6 first-order sensitivities and 21 distinct second-order sensitivities for the coolant temperature, $T_{fl}(z_2)$, at any location within the fluid z_2. The explicit expressions of all first-order sensitivities (to the model parameters) of the temperature distribution within the rod, on the rod's surface, and within the coolant are presented in Section 4.2. Locations of particular importance are those where the rod temperature attains its maximum along the rod's axis or on the rod's surface, as well as the coolant's exit, where the coolant temperature attains its maximum. By applying the 2nd-ASAM, the sensitivities of a temperature response to all (six) model parameters can be obtained after solving the 1st-LASS once. Such an adjoint computation would correspond to large-scale computations for a full-scale system.

Section 4.3 presents the application of the 2nd-ASAM to compute the 21 distinct second-order sensitivities of the temperature distributions to the thermal-hydraulics parameters; all of these second-order sensitivities are obtained using only *six independent* adjoint computations. Section 4.3 also shows that the construction and solution of the 2nd-LASS require very little additional effort beyond the construction of the 1st-LASS needed for computing the first-order sensitivities. Only the terms on the right side (i.e., "sources") of the solver for the 1st-LASS needed to be modified, while the left side of the solver (which solves the differential equations) remained unchanged, in order to solve the 2nd-LASS. This very important property of the 2nd-ASAM opens the possibility to use commercially available computational fluid dynamics (CFD) codes, such as the FLUENT Adjoint Solver (which cannot currently compute second-order sensitivities), provided the user has access to modifying the "sources" for the respective solver. Although the numerical applications in this work use the FLUENT Adjoint Solver with model parameter values typical for the G4M reactor, the analytical benchmark developed and analyzed in this work is generally useful to perform "solution verification" for any other thermal-hydraulics code.

4.1 Heat Transport Benchmark Problem: Mathematical Modeling

The benchmark problem models the steady-state heat conduction within a cylindrical rod of radius a and length (height) ℓ, coupled to the convective heat transfer to a surrounding coolant. The rod is heated by a heat source of the form $q\cos(\pi z/\ell)$, where $q(\mathrm{W \cdot m^{-3}})$ denotes a constant volumetric source and z denotes the coordinate along the rod's axial (customarily, the vertical) direction. Such a heat source is typical of nuclear reactors. The rod's conductivity, $k(\mathrm{W \cdot m^{-1}K^{-1}})$, is considered to be constant throughout the rod. The rod's surface is cooled by forced convection to a surrounding liquid flowing along the rod's length, from the rod's lower end, taken to be located at $z = -\ell/2$, towards the rod's upper end, located at $z = \ell/2$. The heat transfer coefficient, $h(\mathrm{W \cdot m^{-2}K^{-1}})$, from the rod's surface to the coolant is considered to be constant. For this benchmark, the rod's length is typically two orders of magnitude larger than its diameter, so the heat conduction process in the rod's axial direction can be neglected by comparison to the heat conduction in the rod's radial direction. Under these conditions, the steady-state temperature distributions, $T(r, z)$ and $T_{fl}(z)$, within the heated rod and coolant (fluid), respectively, are determined by the following energy conservation balances:

$$\frac{k}{r}\frac{\partial}{\partial r}\left[r\frac{\partial T(r,z)}{\partial r}\right]=-q\cos\frac{\pi z}{\ell},\quad 0\leq r<a,\ -\frac{\ell}{2}\leq z\leq\frac{\ell}{2}, \tag{4.1}$$

$$\frac{\partial T(r,z)}{\partial r}=0,\ \text{at}\ r=0, \tag{4.2}$$

$$-k\frac{\partial T(r,z)}{\partial r}=h\left[T(r,z)-T_{fl}(z)\right],\ \ \text{at}\ r=a, \tag{4.3}$$

$$\frac{dT_{fl}(z)}{dz}=\frac{\pi a^2 q}{Wc_p}\cos\frac{\pi z}{\ell},\quad -\frac{\ell}{2}\leq z\leq\frac{\ell}{2}, \tag{4.4}$$

$$T_{fl}(z)=T_{inlet},\ \ at\ \ z=-\ell/2, \tag{4.5}$$

where W(kg · s^{-1}), denotes the mass flow rate, T_{inlet}(K) denotes the inlet temperature, and c_p(J · kg^{-1}K^{-1}) denotes the coolant's heat capacity. The "system responses" for the heat transport benchmark modeled by Equations 4.1 through 4.5 are typically functionals of the rod and/or coolant temperatures and can be generally represented in the form

$$R\left[T(r,z),T_{fl}(z)\right]\triangleq\int_0^a r\,dr\int_{-\ell/2}^{\ell/2}dz\,F\left[T(r,z),T_{fl}(z)\right], \tag{4.6}$$

where F denotes a user-defined function of the indicated arguments. For example, if the response of interest is the value, $T(r_0, z_0)$, of the heated rod temperature, $T(r, z)$, at some point (r_0, z_0), then this particular type of response is represented in the form

$$T(r_0,z_0)=\int_0^a r\,dr\int_{-\ell/2}^{\ell/2}dz\,T(r,z)\frac{\delta(r-r_0)}{r}\delta(z-z_0). \tag{4.7}$$

On the contrary, if the response of interest is the coolant temperature, $T_{fl}(z_2)$, at some axial point $z = z_2$, then this response can be represented in the form

$$T_{fl}(z_2)=\frac{2}{a^2}\int_0^a r\,dr\int_{-\ell/2}^{\ell/2}dz\,T_{fl}(z)\delta(z-z_2). \tag{4.8}$$

The following correspondences between the general notation introduced in Chapter 3 and the formulation of the above paradigm heat transport problem, described by Equations 4.1 through 4.5, are readily apparent:

$$\mathbf{u}(\mathbf{x}) \triangleq \left[T(r,z), T_{fl}(z)\right], \quad \mathbf{x} \triangleq (r,z), \quad \boldsymbol{\alpha} \triangleq \left(q,k,h,W,c_p,T_{inlet}\right),$$

$$\mathbf{L}(\boldsymbol{\alpha}) \triangleq \begin{pmatrix} \frac{k}{r}\frac{\partial}{\partial r}\left(r\frac{\partial[\ ]}{\partial r}\right) & 0 \\ 0 & \frac{d[\ ]}{dz} \end{pmatrix}, \quad \mathbf{Q}(\boldsymbol{\alpha}) \triangleq \begin{pmatrix} -q\cos\frac{\pi z}{\ell} \\ \frac{\pi a^2 q}{W c_p}\cos\frac{\pi z}{\ell} \end{pmatrix}, \tag{4.9}$$

$$\Omega_x \triangleq \left[r \in (0,a) \times z \in (-\ell/2, \ell/2)\right].$$

For the heat transport benchmark problem under consideration, the base-case solution is obtained by solving Equations 4.1 through 4.5 using the nominal parameter values, $\boldsymbol{\alpha}^0 \equiv \left(q^0,k^0,h^0,W^0,c_p^0,T_{inlet}^0\right)$, to obtain the vector $\mathbf{u}^0(\mathbf{x}) \triangleq \left[T^0(r,z), T_{fl}^0(z)\right]$ of the nominal values $T_{fl}^0(z)$ and $T_l^0(r,z)$ of the respective temperature distributions. The resulting nominal solutions of Equations 4.1 through 4.5 are

$$T^0(r,z) = q^0\left(\frac{a^2-r^2}{4k^0} + \frac{a}{2h^0}\right)\cos\frac{\pi z}{\ell} + T_{fl}^0(z), \quad 0 \le r < a, \ -\frac{\ell}{2} \le z \le \frac{\ell}{2}, \tag{4.10}$$

$$T_{fl}^0(z) = \frac{a^2 \ell q^0}{W^0 c_p^0}\left(\sin\frac{\pi z}{\ell} + 1\right) + T_{inlet}^0, \quad -\frac{\ell}{2} \le z \le \frac{\ell}{2}. \tag{4.11}$$

The specific physical geometry and parameter values considered for benchmarking the computations performed with the Adjoint Solver of the CFD code FLUENT are representative of a fuel rod surrounded by coolant in the preliminary design of the G4M reactor (Gen4 Energy, Inc. 2012), which is envisaged as a small fast reactor cooled by LBE. The dimensions of the G4M reactor's fuel rod are as follows: radius a = 7.705 mm and length (height) ℓ = 1.7 m. The nominal values and relative standard deviations for the G4M reactor's thermal-hydraulics parameters are provided in Table 4.1.

The nominal temperature profiles described by Equations 4.10 and 4.11 are the reference (exact) results. On the other hand, the physical system comprising the heated rod surrounded by the flowing LBE coolant was also modeled with the CFD code FLUENT to obtain the CFD-computed temperature profiles in the rod and coolant. The distributions of the temperatures within the rod obtained by solving Equations 4.10 and 4.11 by using FLUENT are depicted in Figure 4.3. The colors used in this figure, as well as in all subsequent figures

TABLE 4.1

Nominal Values and Relative Standard Deviations for Thermal-hydraulics Parameters Characteristic of the G4M Reactor

Model Parameter	Nominal Value	Relative Standard Deviation
q ($W \cdot m^{-3}$)	2.03×10^8	$\sigma(q) = 15\%$
k ($W \cdot m^{-1}K^{-1}$)	9.382 (at 800 K)	$\sigma(k) = 15\%$
h ($W \cdot m^{-2}K^{-1}$)	34.570	$\sigma(h) = 10\%$
W ($kg \cdot {}^{-1}$)	2.28	$\sigma(W) = 15\%$
c_p ($J \cdot kg^{-1}K^{-1}$)	141.80 (at 800 K)	$\sigma(c_p) = 15\%$
T_{inlet} (K)	673.15	$\sigma(T_{inlet}) = 10\%$

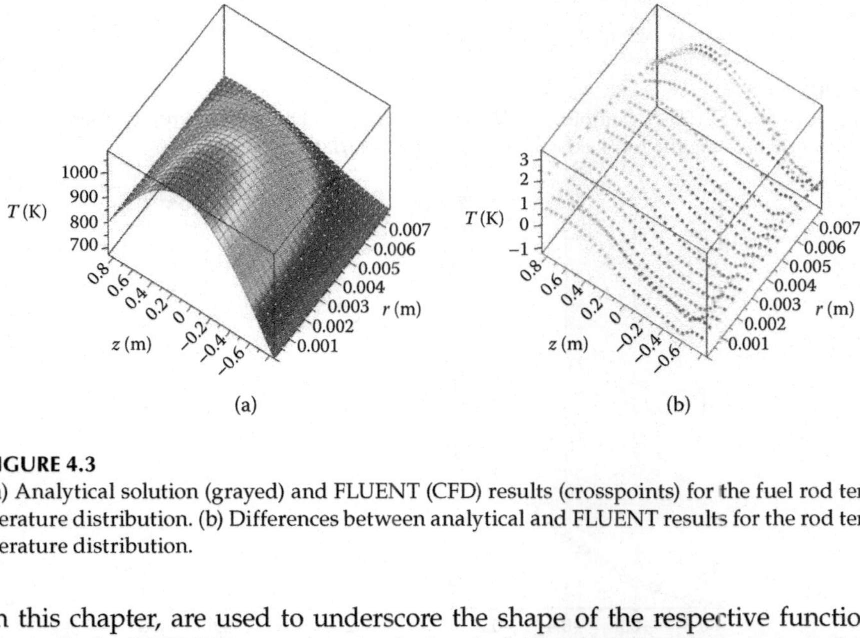

FIGURE 4.3
(a) Analytical solution (grayed) and FLUENT (CFD) results (crosspoints) for the fuel rod temperature distribution. (b) Differences between analytical and FLUENT results for the rod temperature distribution.

in this chapter, are used to underscore the shape of the respective function. Overall, the FLUENT-computed solution is within 0.3% of the exact analytical solution, thus providing the solution verification of the adequacy of the numerical discretization procedure for this benchmark problem.

As has already been discussed in the foregoing, the maximum temperature in the rod is attained along its axis, as depicted in Figure 4.4 (a). Similarly important is the behavior of the temperature distribution on the rod's surface, which is depicted in Figure 4.4 (b).

Figure 4.5 displays the exact and FLUENT-computed coolant temperature distributions along the coolant's flow direction. As before, the FLUENT-computed distribution is in excellent agreement with the exact one, thus providing additional solution verification and confidence that the spatial discretization chosen in FLUENT for this benchmark problem is adequate to yield very accurate results for the temperature distributions.

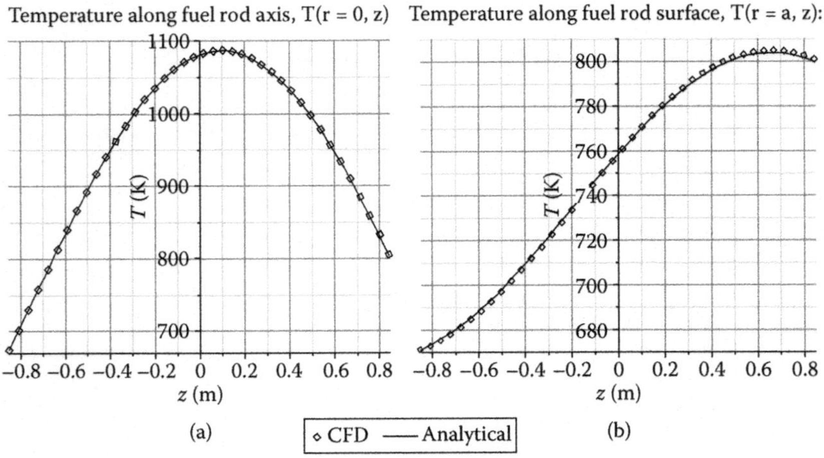

FIGURE 4.4
Analytical and FLUENT (computational fluid dynamics, CFD) computed rod temperature distributions. (a) along the rod's axis; (b) along the rod's surface.

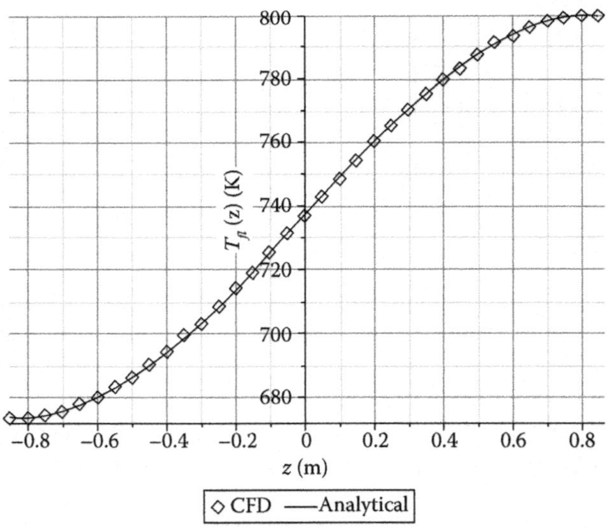

FIGURE 4.5
Analytical and FLUENT (CFD) computed coolant temperature distributions.

4.2 Computation of First-Order Sensitivities

For the paradigm benchmark problem defined by Equations 4.1 through 4.6, the vector of parameter variations is defined as $\delta\boldsymbol{\alpha} \triangleq (\delta q, \delta k, \delta h, \delta W, \delta c_p, \delta T_{inlet})$, where $\delta q, \delta k, \delta h, \delta W, \delta c_p, \delta T_{inlet}$ denote arbitrary variations in the model parameters around the respective nominal values. In turn, these parameter variations induce variations $\delta\mathbf{u} \triangleq [\delta T(r,z), \delta T_{fl}(z)]$ in the state functions, where $\delta T(r,z)$ denotes the corresponding variation in the temperature distribution within the rod, and $\delta T_{fl}(z)$ denotes the corresponding variation in the temperature distribution within the coolant.

Applying the definition of the G-differential to the temperature response considered in Equation 4.6 yields the following expression for the first-order sensitivity $\delta R(\mathbf{e}^0; \mathbf{h})$ of the response $R(\mathbf{e})$ at $\mathbf{e}^0 \triangleq (\mathbf{u}^0, \boldsymbol{\alpha}^0)$ to the variations $\mathbf{h} \triangleq (\delta\mathbf{u}, \delta\boldsymbol{\alpha})$ in the model's state functions and parameters:

$$\delta R(\mathbf{e}^0;\mathbf{h}) \triangleq \frac{d}{d\varepsilon}\left\{R(\mathbf{e}^0+\varepsilon\mathbf{h})\right\}_{\varepsilon=0} = \int_0^a r\,dr \int_{-\ell/2}^{\ell/2} dz \left[\left(\frac{\partial F}{\partial T}\right)_{\mathbf{e}^0} \delta T(r,z) + \left(\frac{\partial F}{\partial T_{fl}}\right)_{\mathbf{e}^0} \delta T_{fl}(z)\right]. \tag{4.12}$$

Note that the expression in Equation 4.12 comprises solely the "indirect-effect term," since there is no "direct-effect term" stemming from the response defined in Equation 4.6. Computing next the G-differentials of Equations 4.1 through 4.5 yields the following set of coupled differential equations:

$$\begin{aligned} &\frac{k^0}{r}\frac{\partial}{\partial r}\left\{r\frac{\partial}{\partial r}[\delta T(r,z)]\right\} = -\frac{(\delta k)}{r}\frac{\partial}{\partial r}\left[r\frac{\partial T^0(r,z)}{\partial r}\right] - (\delta q)\cos\frac{\pi z}{\ell} \\ &= \left[(\delta k)\frac{q^0}{k^0} - (\delta q)\right]\cos\frac{\pi z}{\ell} \triangleq Q(r,z), \quad 0 \le r < a, \; -\frac{\ell}{2} \le z \le \frac{\ell}{2}, \end{aligned} \tag{4.13}$$

$$\frac{\partial}{\partial r}[\delta T(r,z)] = 0, \quad \text{at } r = 0, \tag{4.14}$$

$$\begin{aligned} &-k^0\frac{\partial}{\partial r}[\delta T(r,z)] - h^0[\delta T(r,z) - \delta T_{fl}(z)] = (\delta k)\frac{\partial T^0(r,z)}{\partial r} + (\delta h)[T^0(r,z) - T_{fl}^0(z)] \\ &= \left[-(\delta k)\frac{h^0}{k^0} + (\delta h)\right][T^0(r,z) - T_{fl}^0(z)], \quad \text{at } r = a, \end{aligned} \tag{4.15}$$

$$\frac{d}{dz}[\delta T_{fl}(z)] = \pi a^2\left[\frac{(\delta q)}{W^0 c_p^0} - \frac{q^0(\delta W)}{(W^0)^2 c_p^0} - \frac{q^0(\delta c_p)}{W^0(c_p^0)^2}\right]\cos\frac{\pi z}{\ell} \triangleq Q_{fl}(z), \quad -\frac{\ell}{2} \le z \le \frac{\ell}{2}, \tag{4.16}$$

$$\delta T_{fl}(z) = \delta T_{inlet}, \quad at \;\; z = -\ell/2. \tag{4.17}$$

Equations 4.13 through 4.17 constitute the 1st-LFSS. The Hilbert space appropriate for constructing the 1st-LASS corresponding to this 1st-LFSS comprises the space of all square-integrable two-component vector functions of the form $\mathbf{u}(\mathbf{x}) \equiv [u_1, (r, z), u_2(z)]^\dagger$, endowed with an inner product $\langle \mathbf{u}(\mathbf{x}), \boldsymbol{\psi}(\mathbf{x}) \rangle$ between two functions $\mathbf{u}(\mathbf{x})$ and $\boldsymbol{\psi}(\mathbf{x}) \triangleq [\psi_1, (r, z), \psi_2(z)]^\dagger$, belonging to this Hilbert space, defined as

$$\langle \mathbf{u}(\mathbf{x}), \boldsymbol{\psi}(\mathbf{x}) \rangle \triangleq \int_0^a r\,dr \int_{-\ell/2}^{\ell/2} dz \left[u_1(r,z)\psi_1(r,z) + u_2(z)\psi_2(z) \right]. \tag{4.18}$$

In particular, the adjoint function corresponding to the solution $\delta\mathbf{u}(\mathbf{x}) \triangleq [\delta T(r, z), \delta T_{fl}(z)]$ of the 1st-LFSS will be a two-component vector-valued function denoted as $\boldsymbol{\psi}(\mathbf{x}) = [\Psi(r, z), \Psi_{fl}(z)]$, where $\Psi(r, z)$ and $\Psi_{fl}(z)$ denote the adjoint sensitivity functions that correspond to the forward functions $\delta T(r, z)$ and $\delta T_{fl}(z)$, respectively. Taking the inner product of $\boldsymbol{\psi}(\mathbf{x}) = [\Psi(r, z), \Psi_{fl}(z)]$ with Equations 4.13 and 4.16 and integrating by parts (twice over the variable r and once over the variable z) yields the following relation:

$$\begin{aligned}
&\int_0^a r\,dr \int_{-\ell/2}^{\ell/2} dz \left\{ \Psi(r,z) \frac{k^0}{r} \frac{\partial}{\partial r}\left(r \frac{\partial[\delta T(r,z)]}{\partial r} \right) + \Psi_{fl}(z) \frac{d[\delta T_{fl}(z)]}{dz} \right\} \\
&= \int_0^a r\,dr \left[\Psi_{fl}\left(\frac{\ell}{2}\right) \delta T_{fl}\left(\frac{\ell}{2}\right) - \Psi_{fl}\left(-\frac{\ell}{2}\right) \delta T_{fl}\left(-\frac{\ell}{2}\right) \right] \\
&+ \int_0^a r\,dr \int_{-\ell/2}^{\ell/2} dz \left[\delta T_{fl}(z) \right] \left[-\frac{d\Psi_{fl}(z)}{dz} \right] \\
&+ \int_{-\ell/2}^{\ell/2} dz \left[\Psi(r,z) r k^0 \frac{\partial}{\partial r} \delta T(r,z) - \delta T(r,z) r k^0 \frac{\partial \Psi(r,z)}{\partial r} \right]_{r=a} \\
&- \int_{-\ell/2}^{\ell/2} dz \left[\Psi(r,z) r k^0 \frac{\partial}{\partial r} \delta T(r,z) - \delta T(r,z) r k^0 \frac{\partial \Psi(r,z)}{\partial r} \right]_{r=0} \\
&+ \int_0^a r\,dr \int_{-\ell/2}^{\ell/2} dz \left\{ [\delta T(r,z)] \frac{k^0}{r} \frac{\partial}{\partial r}\left(r \frac{\partial \Psi(r,z)}{\partial r} \right) \right\} \\
&= \int_0^a r\,dr \int_{-\ell/2}^{\ell/2} dz \left[\Psi(r,z) Q(r,z) + \Psi_{fl}(z) Q_{fl}(z) \right].
\end{aligned} \tag{4.19}$$

The following operations are now performed in Equation 4.19: (i) use the boundary condition given in Equation 4.17 to replace the term $\delta T_{fl}(-\ell/2)$, and (ii) use the boundary conditions given in Equations 4.14 and 4.15 to replace the terms $\{\partial[\delta T(r,z)]/\partial r\}_{r=0}$ and $\{\partial[\delta T(r,z)]/\partial r\}_{r=a}$. After performing these operations and regrouping terms, Equation 4.19 can be written in the following form:

$$\begin{aligned}
&\int_0^a r\,dr \int_{-\ell/2}^{\ell/2} dz\left[\Psi(r,z)Q(r,z)+\Psi_{fl}(z)Q_{fl}(z)\right] \\
&= \int_0^a r\,dr \int_{-\ell/2}^{\ell/2} dz\left\{\left[\delta T(r,z)\right]\frac{k^0}{r}\frac{\partial}{\partial r}\left(r\frac{\partial \Psi(r,z)}{\partial r}\right)+\left[\delta T_{fl}(z)\right]\left[-\frac{d\Psi_{fl}(z)}{dz}\right]\right\} \\
&+ \int_{-\ell/2}^{\ell/2} dz\left[\delta T_{fl}(z)h^0\Psi(r,z)r\right]_{r=a} + \int_0^a r\,dr\left[\Psi_{fl}\left(\frac{\ell}{2}\right)\delta T_{fl}\left(\frac{\ell}{2}\right)-\Psi_{fl}\left(-\frac{\ell}{2}\right)\delta T_{inlet}\right] \\
&+ \int_{-\ell/2}^{\ell/2} dz\left\{\Psi(r,z)r\left[(\delta k)\frac{h^0}{k^0}-(\delta h)\right]\left[T^0(r,z)-T^0_{fl}(z)\right]\right\}_{r=a} \\
&+ \int_{-\ell/2}^{\ell/2} dz\left\{\left[\delta T(r,z)\right]\left[-rh^0\Psi(r,z)-rk^0\frac{\partial \Psi(r,z)}{\partial r}\right]\right\}_{r=a} \\
&- \int_{-\ell/2}^{\ell/2} dz\left[-\delta T(r,z)rk^0\frac{\partial \Psi(r,z)}{\partial r}\right]_{r=0}.
\end{aligned} \tag{4.20}$$

The quantity $\int_{-\ell/2}^{\ell/2} dz\left[\delta T_{fl}(z)h^0\Psi(r,z)rk^0\right]_{r=a}$ on the right side of the last equality above contains the function $\delta T_{fl}(z)$ and must therefore be included with the term containing the same function, $\delta T_{fl}(z)$, under the inner product (i.e., double integral over r and z). This situation provides an illustration of the occasional need to consider the *extended adjoint operator*, as has been discussed in Chapter 3.

Applying the criteria discussed in Chapter 3 for selecting the adjoint boundary conditions for the 1st-LASS indicates that the unknown boundary values of the functions $\delta T(r, z)$ and $\delta T_{fl}(z)$ will be eliminated from Equation 4.20 by imposing the following boundary conditions on the adjoint sensitivity functions $\Psi(r, z)$ and $\Psi(z)$:

$$\Psi_{fl}(z) = 0, \quad \text{at} \quad z = \ell/2, \tag{4.21}$$

$$\frac{\partial \Psi(r,z)}{\partial r} = 0, \ \text{at} \ r = 0, \tag{4.22}$$

$$-k^0\frac{\partial\Psi(r,z)}{\partial r}=h^0\Psi(r,z),\quad \text{at } r=a. \tag{4.23}$$

The imposition of the adjoint boundary conditions indicated in Equations 4.21, 4.22, and 4.23 reduces Equation 4.20 to the following form:

$$\begin{aligned}&\int_0^a r\,dr\int_{-\ell/2}^{\ell/2}dz\left[\Psi(r,z)Q(r,z)+\Psi_{fl}(z)Q_{fl}(z)\right]\\&=\int_0^a r\,dr\int_{-\ell/2}^{\ell/2}dz\left\{\left[\delta T(r,z)\right]\frac{k^0}{r}\frac{\partial}{\partial r}\left(r\frac{\partial\Psi(r,z)}{\partial r}\right)+\left[\delta T_{fl}(z)\right]\left[-\frac{d\Psi_{fl}(z)}{dz}+\frac{2}{a}h^0\Psi(r,z)\right]\right\}\\&-\int_0^a r\,dr\Psi_{fl}\left(-\frac{\ell}{2}\right)\left(\delta T_{inlet}\right)+\int_{-\ell/2}^{\ell/2}dz\left\{\Psi(r,z)r\left[(\delta k)\frac{h^0}{k^0}-(\delta h)\right]\left[T^0(r,z)-T_{fl}^0(z)\right]\right\}_{r=a}.\end{aligned}\tag{4.24}$$

The first integral on the right side of Equation 4.24 can be identified with the total response sensitivity represented by Equation 4.12. This identification yields the following relations:

$$\frac{k^0}{r}\frac{\partial}{\partial r}\left[r\frac{\partial\Psi(r,z)}{\partial r}\right]=\left(\frac{\partial F}{\partial T}\right)_{\mathbf{e}^0},\quad 0\le r<a,\ -\frac{\ell}{2}\le z\le\frac{\ell}{2}, \tag{4.25}$$

$$-\frac{\partial\Psi_{fl}(z)}{\partial z}+\frac{2}{a}h^0\Psi(a,z)=\left(\frac{\partial F}{\partial T_{fl}}\right)_{\mathbf{e}^0},\quad -\frac{\ell}{2}\le z\le\frac{\ell}{2}. \tag{4.26}$$

The relations expressed in Equations 4.25 and 4.26 together with the first-level adjoint boundary conditions expressed in Equations 4.21 through 4.23 comprise the 1st-LASS. Replacing in Equation 4.24 the source terms $Q(r, z)$ and $Q_{fl}(z)$ with their definitions provided in Equations 4.13 and 4.16, respectively, it follows from Equation 4.24 that the first-order total response sensitivity $\delta R(\mathbf{e}^0;\mathbf{h})$ can now be expressed in terms of the adjoint sensitivity functions $\Psi_l(r, z)$ and $\Psi_{fl}(z)$ as follows:

$$\begin{aligned}\delta R\left(\mathbf{e}^0;\mathbf{h}\right)&=\left[(\delta k)\frac{q^0}{k^0}-(\delta q)\right]\int_0^a r\,dr\int_{-\ell/2}^{\ell/2}\Psi(r,z)\cos\frac{\pi z}{\ell}dz\\&+\pi a^2\left[\frac{(\delta q)}{W^0c_p^0}-\frac{q^0(\delta W)}{\left(W^0\right)^2c_p^0}-\frac{q^0\left(\delta c_p\right)}{W^0\left(c_p^0\right)^2}\right]\frac{a^2}{2}\int_{-\ell/2}^{\ell/2}\Psi_{fl}(z)\cos\frac{\pi z}{\ell}\\&+\frac{a^2}{2}\Psi_{fl}\left(-\frac{\ell}{2}\right)\left(\delta T_{inlet}\right)-a\left[(\delta k)\frac{h^0}{k^0}-(\delta h)\right]\int_{-\ell/2}^{\ell/2}\Psi(a,z)\left[T^0(a,z)-T_{fl}^0(z)\right]dz.\end{aligned}\tag{4.27}$$

Finally, noting from Equation 4.10 that $\left[T^0(a,z)-T_{fl}^0(z)\right]=\frac{aq^0}{2h^0}\cos\frac{\pi z}{\ell}$ and replacing this result in the last term on the right side of Equation 4.27 yields the following final expression for the first-order sensitivity $\delta R(\mathbf{e}^0;\mathbf{h})$:

$$\delta R\left(\mathbf{e}^0;\mathbf{h}\right)=\left[(\delta k)\frac{q^0}{k^0}-(\delta q)\right]\int_0^a r\,dr\int_{-\ell/2}^{\ell/2}\Psi(r,z)\cos\frac{\pi z}{\ell}dz$$
$$+\pi a^2\left[\frac{(\delta q)}{W^0c_p^0}-\frac{q^0(\delta W)}{\left(W^0\right)^2c_p^0}-\frac{q^0\left(\delta c_p\right)}{W^0\left(c_p^0\right)^2}\right]\frac{a^2}{2}\int_{-\ell/2}^{\ell/2}\Psi_{fl}(z)\cos\frac{\pi z}{\ell} \tag{4.28}$$
$$+\frac{a^2}{2}\Psi_{fl}\left(-\frac{\ell}{2}\right)(\delta T_{inlet})-\left[(\delta k)\frac{h^0}{k^0}-(\delta h)\right]\frac{q^0}{h^0}\frac{a^2}{2}\int_{-\ell/2}^{\ell/2}\Psi(a,z)\cos\frac{\pi z}{\ell}dz.$$

The 1st-LASS, comprising Equations 4.21, 4.22, 4.23, 4.25, and 4.26, is independent of parameter variations and needs to be solved only once for each response of interest. Thus, once the first-level adjoint functions corresponding to the specific response under consideration are available, they are used in Equation 4.28 to obtain very efficiently all of the exact first-order response sensitivities, to all model parameters. For the generic response, R(**e**), defined in Equation 4.6, the partial first-order sensitivities are obtained from Equation 4.28 as follows:

$$S_1\left(\alpha^0,\psi^0\right)\triangleq\left(\frac{\partial R}{\partial q}\right)_{\left(\mathbf{u}^0,\alpha^0,\psi^0\right)}$$
$$=-\int_0^a r\,dr\int_{-\ell/2}^{\ell/2}\Psi^0(r,z)\cos\frac{\pi z}{\ell}dz+\frac{\pi a^4}{2W^0c_p^0}\int_{-\ell/2}^{\ell/2}\Psi_{fl}^0(z)\cos\frac{\pi z}{\ell}, \tag{4.29}$$

$$S_2\left(\alpha^0,\psi^0\right)\triangleq\left(\frac{\partial R}{\partial k}\right)_{\left(\mathbf{u}^0,\alpha^0,\psi^0\right)}$$
$$=\frac{q^0}{k^0}\int_0^a r\,dr\int_{-\ell/2}^{\ell/2}\Psi^0(r,z)\cos\frac{\pi z}{\ell}dz-\frac{q^0}{k^0}\frac{a^2}{2}\int_{-\ell/2}^{\ell/2}\Psi^0(a,z)\cos\frac{\pi z}{\ell}dz, \tag{4.30}$$

$$S_3\left(\alpha^0,\psi^0\right)\triangleq\left(\frac{\partial R}{\partial h}\right)_{\left(\mathbf{u}^0,\alpha^0,\psi^0\right)}=\frac{q^0}{h^0}\frac{a^2}{2}\int_{-\ell/2}^{\ell/2}\Psi^0(a,z)\cos\frac{\pi z}{\ell}dz, \tag{4.31}$$

$$S_4\left(\alpha^0,\psi^0\right)\triangleq\left(\frac{\partial R}{\partial W}\right)_{\left(\mathbf{u}^0,\alpha^0,\psi^0\right)}=-\frac{q^0\pi a^4}{2\left(W^0\right)^2c_p^0}\int_{-\ell/2}^{\ell/2}\Psi_{fl}^0(z)\cos\frac{\pi z}{\ell}, \tag{4.32}$$

$$S_5(\alpha^0,\psi^0) \triangleq \left(\frac{\partial R}{\partial c_p}\right)_{(\mathbf{u}^0,\alpha^0,\psi^0)} = -\frac{q^0\pi a^4}{2W^0\left(c_p^0\right)^2}\int_{-\ell/2}^{\ell/2}\Psi_{fl}^0(z)\cos\frac{\pi z}{\ell}, \tag{4.33}$$

$$S_6(\alpha^0,\psi^0) \triangleq \left(\frac{\partial R}{\partial T_{inlet}}\right)_{(\mathbf{u}^0,\alpha^0,\psi^0)} = \frac{a^2}{2}\Psi_{fl}^0\left(-\frac{\ell}{2}\right). \tag{4.34}$$

4.2.1 Computation of First-Order Sensitivities of the Heated Rod Temperature, $T(r, z)$, at an Arbitrary Location (r_0, z_0)

When the response of interest is the value of the heated rod temperature, $T(r, z)$, at some location (r_0, z_0), the generic form of the sensitivity $\delta R(\mathbf{e}^0; \mathbf{h})$ given in Equation 4.12 takes on the following particular form:

$$\delta T(r_0,z_0) = \int_0^a r\,dr\int_{-\ell/2}^{\ell/2}\delta T(r,z)\frac{\delta(r-r_0)}{r}\delta(z-z_0)\,dz, \tag{4.35}$$

which indicates that $(\partial F/\partial T)_{\mathbf{e}^0} = \frac{\delta(r-r_0)}{r}\delta(z-z_0)$ and $(\partial F/\partial T_{fl})_{\mathbf{e}^0} = 0$ for this response. Replacing these particular expressions for the generic source terms in Equations 4.25 and 4.26, respectively, and solving the resulting 1st-LASS yields the following expressions for the adjoint functions $\Psi(r,z)$ and $\Psi_{fl}(z)$, respectively:

$$\Psi(r,z) = \delta(z-z_0)\left[-\frac{1}{ah^0} + H(r-r_0)\frac{1}{k^0}\ln\frac{r}{r_0} - \frac{1}{k^0}\ln\frac{a}{r_0}\right],\ 0\le r,r_0\le a,\ -\frac{\ell}{2}\le z,z_0\le\frac{\ell}{2},$$

$$\Psi_{fl}(z) = \frac{2}{a^2}\left[1-H(z-z_0)\right] = \frac{2}{a^2}H(z_0-z),\ -\frac{\ell}{2}\le z,z_0\le\frac{\ell}{2}. \tag{4.36}$$

Replacing the results obtained in Equation 4.36 into Equation 4.28 yields

$$\begin{aligned}
&\delta T(r_0,z_0) = \\
&\left[(\delta k)\frac{q^0}{k^0} - (\delta q)\right]\int_0^a r\,dr\int_{-\ell/2}^{\ell/2}\delta(z-z_0)\left[-\frac{1}{ah^0} + H(r-r_0)\frac{1}{k^0}\ln\frac{r}{r_0} - \frac{1}{k^0}\ln\frac{a}{r_0}\right]\cos\frac{\pi z}{\ell}dz \\
&+\pi a^2\left[\frac{\delta q}{W^0c_p^0} - \frac{q^0(\delta W)}{(W^0)^2c_p^0} - \frac{q^0(\delta c_p)}{W^0\left(c_p^0\right)^2}\right]\frac{a^2}{2}\int_{-\ell/2}^{\ell/2}\frac{2}{a^2}H(z_0-z)\cos\frac{\pi z}{\ell}dz + \\
&+\left[(\delta k)\frac{h^0}{k^0} - (\delta h)\right]\frac{q^0}{h^0}\frac{a^2}{2}\int_{-\ell/2}^{\ell/2}\frac{\delta(z-z_0)}{ah^0}\cos\frac{\pi z}{\ell}dz + (\delta T_{inlet}).
\end{aligned} \tag{4.37}$$

Recalling that $\int r \ln r\, dr = r^2 \ln r/2 - r^2/4$ and carrying out the integrations in Equation 4.37 yields

$$\delta T(r_0,z_0)=\left[(\delta k)\frac{q^0}{k^0}-(\delta q)\right]\left(-\frac{a}{2h^0}+\frac{r_0^2-a^2}{4k^0}\right)\cos\frac{\pi z_0}{\ell}$$
$$+a^2\ell\left[\frac{\delta q}{W^0 c_p^0}-\frac{q^0(\delta W)}{(W^0)^2 c_p^0}-\frac{q^0(\delta c_p)}{W^0(c_p^0)^2}\right]\left(\sin\frac{\pi z_0}{\ell}+1\right) \quad (4.38)$$
$$+\left[(\delta k)\frac{h^0}{k^0}-(\delta h)\right]\frac{q^0}{(h^0)^2}\frac{a}{2}\cos\frac{\pi z_0}{\ell}+(\delta T_{inlet}).$$

Identifying the expressions that multiply the various parameter variations in Equation 4.38 yields the following first-order sensitivities of the heated rod temperature, $T(r_0, z_0)$, at an arbitrary location (r_0, z_0), to the model parameters:

$$\frac{\partial T(r_0,z_0)}{\partial q}=\left(\frac{a^2-r_0^2}{4k^0}+\frac{a}{2h^0}\right)\cos\frac{\pi z_0}{\ell}+\frac{a^2\ell}{W^0 c_p^0}\left(\sin\frac{\pi z_0}{\ell}+1\right), \quad (4.39)$$

$$\frac{\partial T(r_0,z_0)}{\partial k}=\frac{q^0\left(r_0^2-a^2\right)}{4\left(k^0\right)^2}\cos\frac{\pi z_0}{\ell}, \quad (4.40)$$

$$\frac{\partial T(r_0,z_0)}{\partial h}=-\frac{q^0 a}{2\left(h^0\right)^2}\cos\frac{\pi z_0}{\ell}, \quad (4.41)$$

$$\frac{\partial T(r_0,z_0)}{\partial W}=-\frac{q^0 a^2\ell}{\left(W^0\right)^2 c_p^0}\left(\sin\frac{\pi z_0}{\ell}+1\right), \quad (4.42)$$

$$\frac{\partial T(r_0,z_0)}{\partial c_p}=-\frac{q^0 a^2\ell}{W^0\left(c_p^0\right)^2}\left(\sin\frac{\pi z_0}{\ell}+1\right), \quad (4.43)$$

$$\frac{\partial T(r_0,z_0)}{\partial T_{inlet}}=1. \quad (4.44)$$

The sensitivities of the rod temperature distribution are computed using Equations 4.39 through 4.44, evaluated at the nominal parameter values provided in Table 4.1 as functions of (r_0, z_0), $0 \le r_0 \le a$, $-\frac{l}{2} \le z_0 \le \frac{l}{2}$. The space-dependent sensitivities thus obtained are depicted in Figures 4.6 through 4.11 as functions of (r_0, z_0). Note that the superscript 0, which indicates nominal values, is omitted in these figures (as in all other figures presenting sensitivities, to follow), for notational simplicity.

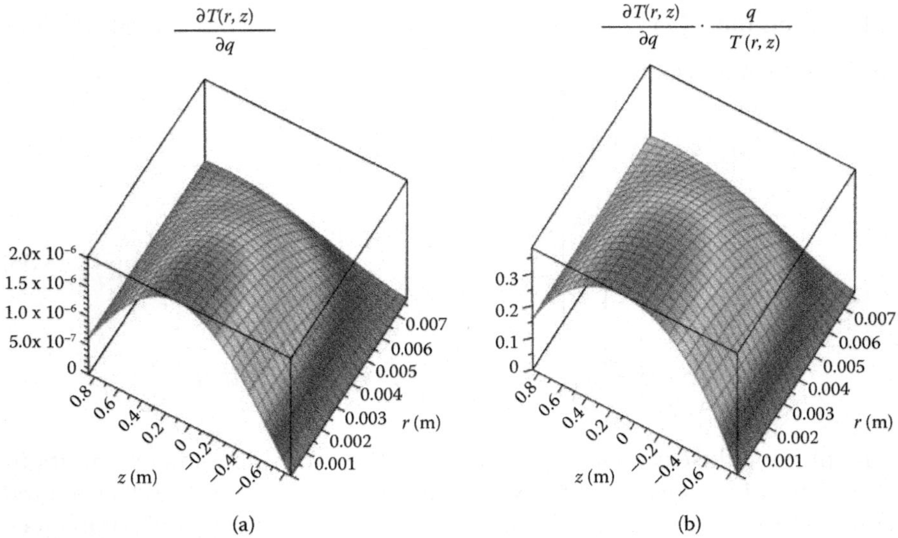

FIGURE 4.6
Spatially dependent absolute (a) and relative (b) sensitivities of the rod temperature to the heat source strength q.

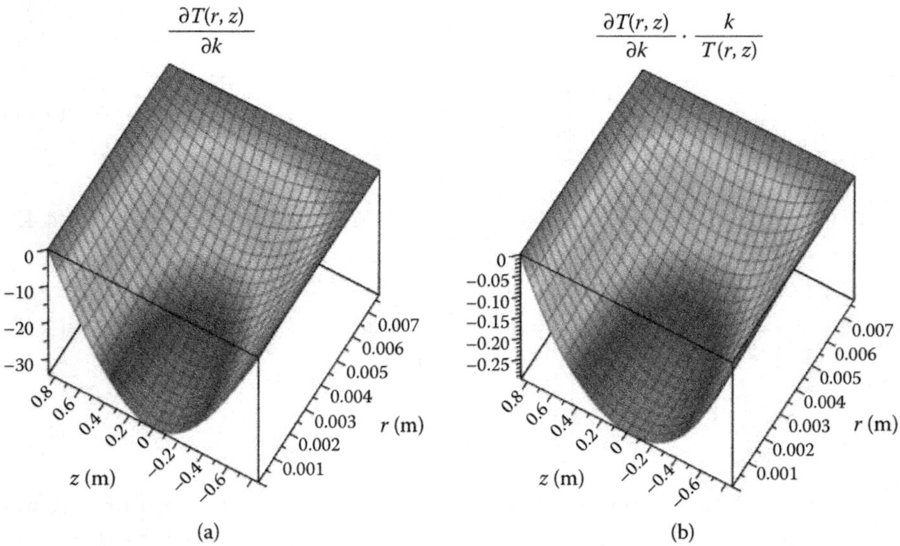

FIGURE 4.7
Spatially dependent absolute (a) and relative (b) sensitivities of the rod temperature to the fuel conductivity k.

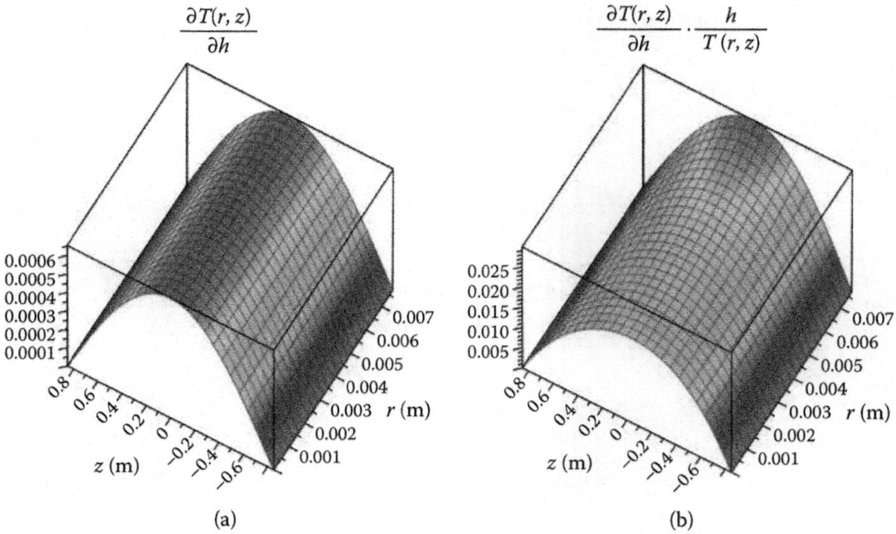

FIGURE 4.8
Spatially dependent absolute (a) and relative (b) sensitivities of the rod temperature to the heat transfer coefficient h.

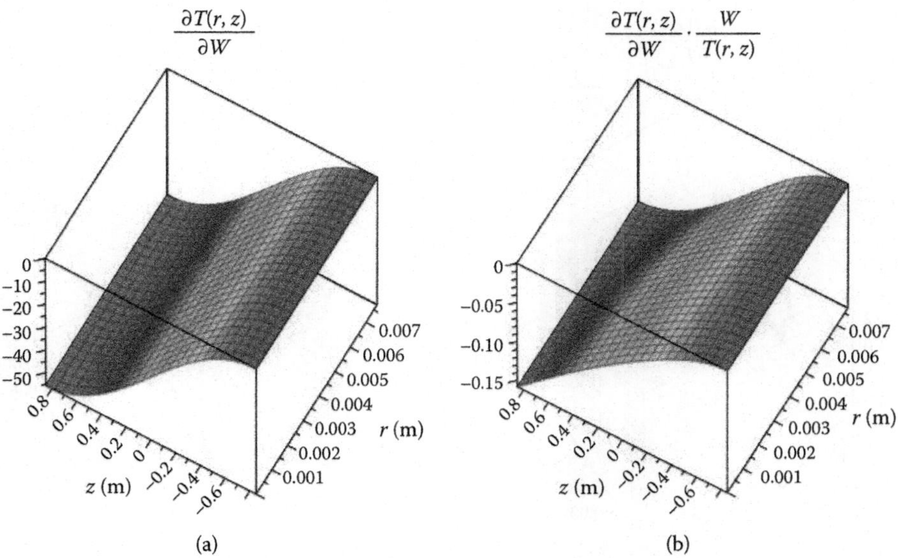

FIGURE 4.9
Spatially dependent absolute (a) and relative (b) sensitivities of the rod temperature to the mass flow rate W.

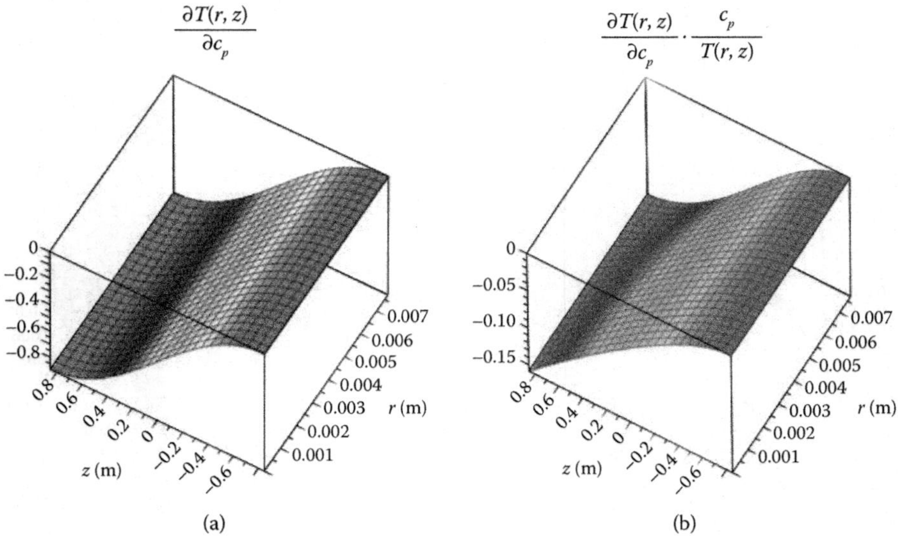

FIGURE 4.10
Spatially dependent absolute (a) and relative (b) sensitivities of the rod temperature to the coolant's heat capacity c_p.

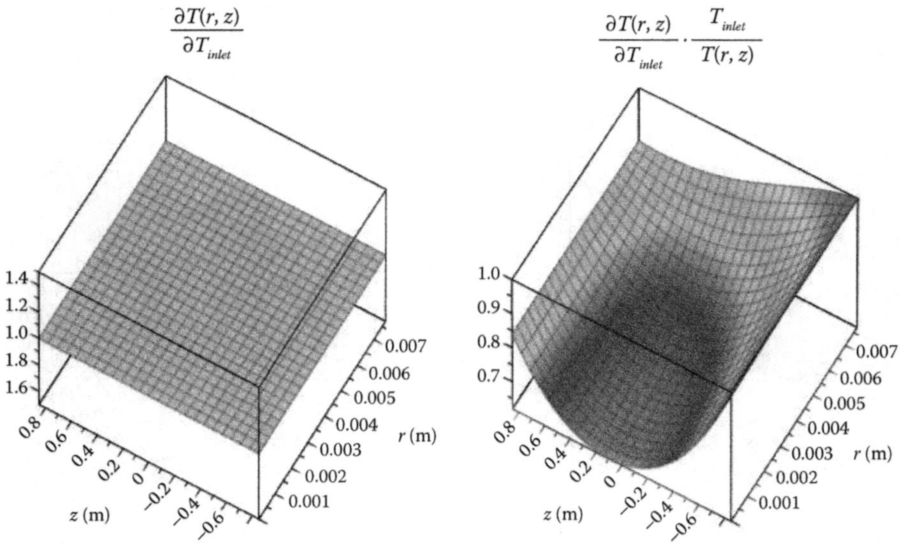

FIGURE 4.11
Spatially dependent absolute (a) and relative (b) sensitivities of the rod temperature to the inlet temperature T_{inlet}.

4.2.2 Computation of First-Order Sensitivities of the Heated Rod Temperature, $T^{max}(z_{max})$, at the Location z_{max}

Of particular interest is the location where $T(r, z)$ attains its maximum value within the fuel rod; this location is implicitly defined by the requirement that the partial derivatives of $T(r, z)$ with respect to r and z, respectively, vanish simultaneously. In view of the boundary condition specified by Equation 4.2, it follows that the maximum fuel temperature is represented by the functional

$$T^{max}(\alpha) \triangleq T(0, z_{max}) = \int_0^a r\,dr \int_{-\ell/2}^{\ell/2} dz\, T(r,z)\frac{\delta(r)}{r}\delta(z - z_{max}), \tag{4.45}$$

where

$$z_{max} = \frac{\ell}{\pi}\arctan\left[\frac{2a\ell}{Wc_p}\left(\frac{a}{2k}+\frac{1}{h}\right)^{-1}\right]. \tag{4.46}$$

To compute the sensitivities of the maximum value of the heated rod temperature, $T^{max}(z_{max})$, Equation 4.45 indicates that the sources for the 1st-LASS will be $\left(\partial F/\partial T\right)_{e^0} = \frac{\delta(r)}{r}\delta(z - z_{max})$ and $\left(\partial F/\partial T_{fl}\right)_{e^0} = 0$. Replacing these expressions for the source terms in Equations 4.25 and 4.26, respectively, and solving the resulting 1st-LASS yields the following expressions for the adjoint functions $\Psi(r, z)$ and $\Psi_{fl}(z)$, respectively:

$$\begin{aligned} \Psi^{max}(r,z) &= -\delta(z - z_{max})\left(\frac{1}{ah^0}+\frac{1}{k^0}\ln\frac{a}{r_0}\right),\ 0 \le r \le a,\ -\frac{\ell}{2} \le z, z_{max} \le \frac{\ell}{2}, \\ \Psi_{fl}^{max}(z) &= \frac{2}{a^2}\left[1 - H(z - z_{max})\right] = \frac{2}{a^2}H(z_{max} - z),\ -\frac{\ell}{2} \le z, z_{max} \le \frac{\ell}{2}. \end{aligned} \tag{4.47}$$

Replacing the above results for $\Psi^{max}(r, z)$ and $\Psi_{fl}^{max}(z)$ into Equation 4.28 yields the following expression of the sensitivity $\delta T^{max}(z_{max})$:

$$\begin{aligned} \delta T^{max}(z_{max}) = &-\left[(\delta k)\frac{q^0}{k^0} - (\delta q)\right]\left(\frac{1}{ah^0}+\frac{1}{k^0}\ln\frac{a}{r_0}\right)\int_0^a r\,dr \int_{-\ell/2}^{\ell/2} \delta(z - z_{max})\cos\frac{\pi z}{\ell}dz \\ &+\pi a^2\left[\frac{\delta q}{W^0 c_p^0} - \frac{q^0(\delta W)}{(W^0)^2 c_p^0} - \frac{q^0(\delta c_p)}{W^0 (c_p^0)^2}\right]\frac{a^2}{2}\int_{-\ell/2}^{\ell/2}\frac{2}{a^2}H(z_{max} - z)\cos\frac{\pi z}{\ell}dz \\ &+\left[(\delta k)\frac{h^0}{k^0} - (\delta h)\right]\frac{q^0}{h^0}\frac{a^2}{2}\int_{-\ell/2}^{\ell/2}\frac{\delta(z - z_{max})}{ah^0}\cos\frac{\pi z}{\ell}dz + (\delta T_{inlet}). \end{aligned} \tag{4.48}$$

Carrying out the integrations in Equation 4.48 and identifying the expressions multiplying the various parameter variations yields the following first-order sensitivities of the maximum rod temperature, $\delta T^{max}(z_{max})$:

$$\frac{\partial T^{max}(z_{max})}{\partial q} = \left(\frac{a^2}{4k^0} + \frac{a}{2h^0}\right)\cos\frac{\pi z_{max}}{\ell} + \frac{a^2 \ell}{W^0 c_p^0}\left(\sin\frac{\pi z_{max}}{\ell} + 1\right), \tag{4.49}$$

$$\frac{\partial T^{max}(z_{max})}{\partial k} = -\frac{a^2 q^0}{4\left(k^0\right)^2}\cos\frac{\pi z_{max}}{\ell}, \tag{4.50}$$

$$\frac{\partial T^{max}(z_{max})}{\partial h} = -\frac{q^0 a}{2\left(h^0\right)^2}\cos\frac{\pi z_{max}}{\ell}, \tag{4.51}$$

$$\frac{\partial T^{max}(z_{max})}{\partial W} = -\frac{q^0 a^2 \ell}{\left(W^0\right)^2 c_p^0}\left(\sin\frac{\pi z_{max}}{\ell} + 1\right), \tag{4.52}$$

$$\frac{\partial T^{max}(z_{max})}{\partial c_p} = -\frac{q^0 a^2 \ell}{W^0 \left(c_p^0\right)^2}\left(\sin\frac{\pi z_{max}}{\ell} + 1\right), \tag{4.53}$$

$$\frac{\partial T^{max}(z_{max})}{\partial T_{inlet}} = 1. \tag{4.54}$$

Recall from Equation 4.45 that $T(r, z)$ attains its maximum value, T^{max}, at the location $r = 0$, $z_{max} = \frac{\ell}{\pi}\arctan\left[\frac{2a\ell}{Wc_p}\left(\frac{a}{2k} + \frac{1}{h}\right)^{-1}\right]$. Using the nominal parameter values shown in Table 4.1, in Equations 4.10 and 4.46, respectively, yields $T^{max} = 1086.025$ K, which is attained at the location ($r = 0$ m, $z_{max} = 0.098646$ m). The relative sensitivities of T^{max}, computed for the nominal parameter values given in Table 4.1, are presented below.

$$\begin{aligned}
&\frac{\partial T^{max}}{\partial q}\frac{q}{T^{max}} = 0.38021; \quad \frac{\partial T^{max}}{\partial k}\frac{k}{T^{max}} = -0.2908; \quad \frac{\partial T^{max}}{\partial h}\frac{h}{T^{max}} = -0.02048; \\
&\frac{\partial T^{max}}{\partial W}\frac{W}{T^{max}} = -0.06893; \quad \frac{\partial T^{max}}{\partial c_p}\frac{c_p}{T^{max}} = -0.06893; \quad \frac{\partial T^{max}}{\partial T_{inlet}}\frac{T_{inlet}}{T^{max}} = -0.61980.
\end{aligned} \tag{4.55}$$

The results presented in Equation 4.55 indicate that the most influential parameter on T^{max} is T_{inlet}, since it displays the largest (in absolute value) relative sensitivity. The relative sensitivity is negative, indicating that an increase in T_{inlet} will decrease the maximum temperature T^{max}, as would also be expected on physical grounds. The relative sensitivities of T^{max} to k, c_p, W, and h are also negative, indicating that increases in any of these model parameters would lead to a decrease in T^{max}; again, this behavior is expected to occur based on physical considerations. Notably, the relative sensitivities of T^{max} to c_p and W are equal, indicating that these parameters have equal influence on T^{max}. The only positive relative sensitivity is that of T^{max} to q, which indicates that an increase in the heat source q would cause the maximum temperature, T^{max}, to increase, which is to be expected on physical grounds.

The influence of the first-order sensitivities on propagating the uncertainties in the respective model parameters to produce uncertainties in the rod temperature response can be evaluated by considering the model parameters (generically denoted as α_i, $i = 1,\ldots,6$) to be uncorrelated and using the following particular form of Equation 1.19 from Chapter 1:

$$\text{var}(r_k) = \sum_{i=1}^{N_\alpha} \left(\frac{\partial r_k}{\partial \alpha_i} \right)^2 \sigma_i^2 . \tag{4.56}$$

Using the sensitivities depicted in Figures 4.6 through 4.11 together with the parameter standard deviations presented in Table 4.1 in the "sandwich formula" given in Equation 4.56 yields the spatially dependent (first-order) standard deviation, denoted as $dev_{FO}(T)$ and depicted in Figure 4.12, of the rod temperature, $T(r, z)$. In particular, the results depicted in Figure 4.12 indicate that the largest standard deviation in the rod temperature is the standard deviation of T^{max}, which has the value $dev_{FO}(T^{max}) = 104.25\text{K}$.

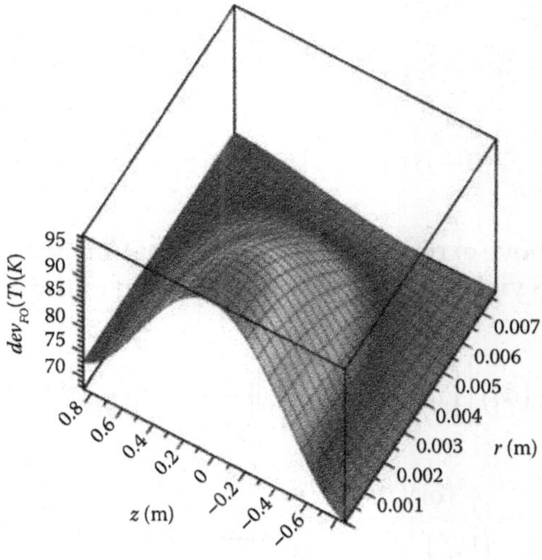

FIGURE 4.12
Spatially dependent distribution of the absolute standard deviation, $dev_{FO}(T)$, computed using first-order sensitivities, of the fuel rod temperature $T(r, z)$.

4.2.3 Computation of First-Order Sensitivities of the Heated Rod Temperature, $T_s(z_1)$, at an Arbitrary Location z_1

Another response of particular interest is the surface fuel temperature, which is defined mathematically as the functional

$$T_s(z_1) \equiv T(a,z_1) = \int_0^a r\,dr \int_{-\ell/2}^{\ell/2} dz\,T(r,z)\frac{\delta(r-a)}{r}\delta(z-z_1), \quad -\ell/2 \le z_1 \le \ell/2. \tag{4.57}$$

In particular, it can be readily shown that $T_s(z)$ attains its maximum value, T_s^{max}, at the location z_s, where

$$T_s^{max} = \int_0^a r\,dr \int_{-\ell/2}^{\ell/2} dz\,T(r,z)\frac{\delta(r-a)}{r}\delta(z-z_s), \tag{4.58}$$

with

$$z_s = \frac{\ell}{\pi}\arctan\left(\frac{2a\ell h}{Wc_p}\right). \tag{4.59}$$

The adjoint functions for computing the sensitivities of the rod surface temperature are obtained by setting $(\partial F/\partial T)_{e^0} = \frac{\delta(r-a)}{r}\delta(z-z_1)$ and $(\partial F/\partial T_{fl})_{e^0} = 0$, in Equations 4.25 and 4.26, respectively, and solving the resulting 1st-LASS. This yields the following expressions for the adjoint functions $\Psi^s(a,z)$ and $\Psi_{fl}^s(z)$, respectively:

$$\begin{aligned}
&\Psi^s(a,z) = \delta(z-z_1)\frac{1}{ah^0}, \quad -\frac{\ell}{2} \le z, z_1 \le \frac{\ell}{2}, \\
&\Psi_{fl}^s(z) = \frac{2}{a^2}\left[1 - H(z-z_1)\right] = \frac{2}{a^2}H(z_1-z), \quad -\frac{\ell}{2} \le z, z_1 \le \frac{\ell}{2}.
\end{aligned} \tag{4.60}$$

Replacing the above expressions for the adjoint functions $\Psi^s(a,z)$ and $\Psi_{fl}^s(z)$ into Equation 4.28 yields the following expression of the sensitivity $\delta T_s(z_1)$:

$$\begin{aligned}
\delta T_s(z_1) = &\left[(\delta k)\frac{q^0}{k^0} - (\delta q)\right]\left[\int_0^a r\,dr \int_{-\ell/2}^{\ell/2} \delta(z-z_1)\left(-\frac{1}{ah^0}\right)\cos\frac{\pi z}{\ell}dz\right. \\
&+ \pi a^2\left[\frac{\delta q}{W^0 c_p^0} - \frac{q^0(\delta W)}{(W^0)^2 c_p^0} - \frac{q^0(\delta c_p)}{W^0 (c_p^0)^2}\right]\frac{a^2}{2}\int_{-\ell/2}^{\ell/2}\frac{2}{a^2}H(z_1-z)\cos\frac{\pi z}{\ell}dz \\
&+ \left[(\delta k)\frac{h^0}{k^0} - (\delta h)\right]\frac{q^0}{h^0}\frac{a^2}{2}\int_{-\ell/2}^{\ell/2}\frac{\delta(z-z_1)}{ah^0}\cos\frac{\pi z}{\ell}dz + (\delta T_{inlet}).
\end{aligned} \tag{4.61}$$

Carrying out the integrations in Equation 4.61 and identifying the expressions multiplying the respective parameter variations yields the following first-order sensitivities of the heated rod temperature, $\delta T_s(z_1)$, at an arbitrary location $z = z_1$:

$$\frac{\partial T_s(z_1)}{\partial q} = \frac{a}{2h^0}\cos\frac{\pi z_1}{\ell} + \frac{a^2\ell}{W^0 c_p^0}\left(\sin\frac{\pi z_1}{\ell}+1\right), \tag{4.62}$$

$$\frac{\partial T_s(z_1)}{\partial k} = 0, \tag{4.63}$$

$$\frac{\partial T_s(z_1)}{\partial h} = -\frac{q^0 a}{2\left(h^0\right)^2}\cos\frac{\pi z_1}{\ell}, \tag{4.64}$$

$$\frac{\partial T_s(z_1)}{\partial W} = -\frac{q^0 a^2 \ell}{\left(W^0\right)^2 c_p^0}\left(\sin\frac{\pi z_1}{\ell}+1\right), \tag{4.65}$$

$$\frac{\partial T_s(z_1)}{\partial c_p} = -\frac{q^0 a^2 \ell}{W^0\left(c_p^0\right)^2}\left(\sin\frac{\pi z_1}{\ell}+1\right), \tag{4.66}$$

$$\frac{\partial T_s(z_1)}{\partial T_{inlet}} = 1. \tag{4.67}$$

As Equations 4.39 through 4.44 indicate, the sensitivities of $T(r_0, z_0)$ are continuous functions in (r_0, z_0). In principle, therefore, substituting $r_0 = 0$ and $z_0 = z_1$ in Equations 4.39 through 4.44 reduces the resulting expressions *formally* to Equations 4.62 through 4.67. Of course, such a simple substitution is not possible when the analytical expressions of the various sensitivities are not available. In practice, the sensitivities of $T(r_0, z_0)$ can be computed only by numerically integrating Equation 4.37, while the sensitivities of $\delta T_s(z_1)$ must be computed by numerically integrating Equation 4.61. In particular, the sensitivities of the maximum fuel surface temperature are obtained by computing the adjoint functions $\Psi^s(a, z)$ and $\Psi^s_{fl}(z)$ at the particular location $z = z_s$, replacing the resulting adjoint functions $\Psi^s(a, z_s)$ and $\Psi^s_{fl}(z_s)$ in Equation 4.28, and carrying out the respective integrations. This sequence of operations would yield the sensitivities of the maximum (rod) surface temperature, $\delta T_s(z_s)$, which will have the same formal expressions as those shown in Equations 4.62 through 4.67, with z_1 replaced by z_s.

The sensitivities of the rod's surface temperature distribution are computed using Equations 4.62 through 4.67 evaluated at the nominal parameter values provided in Table 4.1. The space-dependent sensitivities thus obtained are depicted in Figures 4.13 through 4.17.

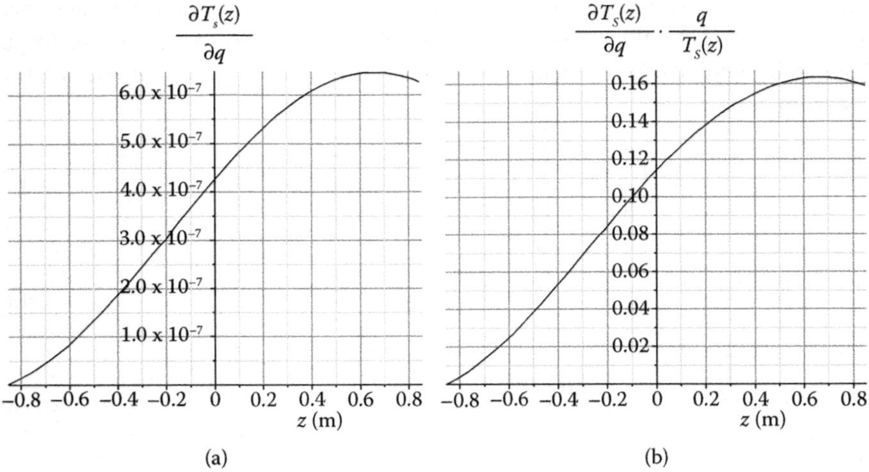

FIGURE 4.13
Spatially dependent absolute (a) and relative (b) sensitivities of the rod's surface temperature to the heat source strength q.

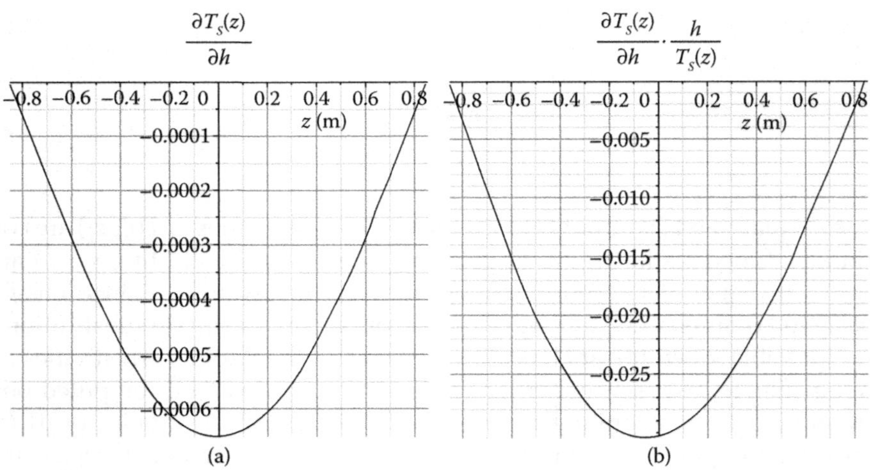

FIGURE 4.14
Spatially dependent absolute (a) and relative (b) sensitivities of the rod's surface temperature to the heat transfer coefficient h.

Recall from Equations 4.58 and 4.59 that the rod's surface temperature attains its maximum value, $T_s^{\max}$, at the spatial location $z_s = \frac{\ell}{\pi}\arctan\left(\frac{2a\ell h}{Wc_p}\right)$. Using the nominal parameter values presented in Table 4.1 yields $z_s = 0.66445$ m, where the corresponding maximum rod surface temperature is obtained as $T_s^{\max} = 803.8\,K$.

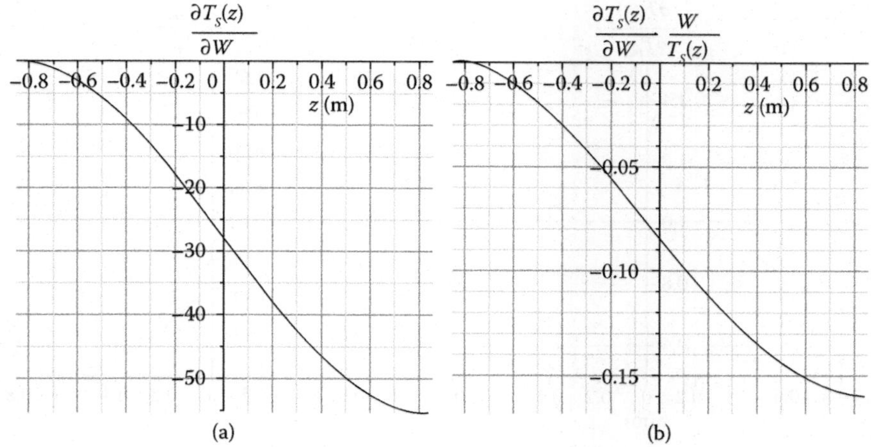

FIGURE 4.15
Spatially dependent absolute (a) and relative (b) sensitivities of the rod's surface temperature to the mass flow rate W.

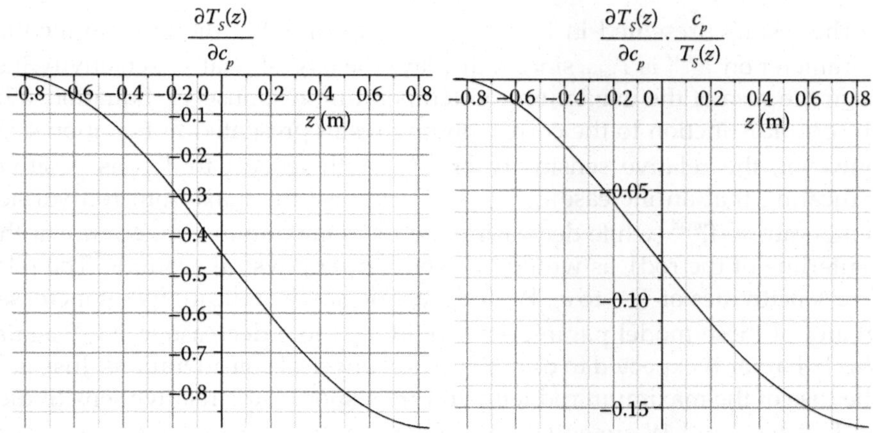

FIGURE 4.16
Spatially dependent absolute (a) and relative (b) sensitivities of the rod's surface temperature to the coolant's heat capacity c_p.

The numerical values of the relative sensitivities of T_s^{max} are as follows:

$$\frac{\partial T_s^{max}}{\partial q}\frac{q}{T_s^{max}} = 0.162546; \quad \frac{\partial T_s^{max}}{\partial k}\frac{\mathrm{k}}{T_s^{max}} = 0; \quad \frac{\partial T_s^{max}}{\partial h}\frac{\mathrm{h}}{T_s^{max}} = -0.009446;$$
$$\frac{\partial T_s^{max}}{\partial W}\frac{\mathrm{W}}{T_s^{max}} = -0.15310; \quad \frac{\partial T_s^{max}}{\partial c_p}\frac{c_p}{T_s^{max}} = -0.15310; \quad \frac{\partial T_s^{max}}{\partial T_{inlet}}\frac{T_{inlet}}{T_s^{max}} = 0.837454. \tag{4.68}$$

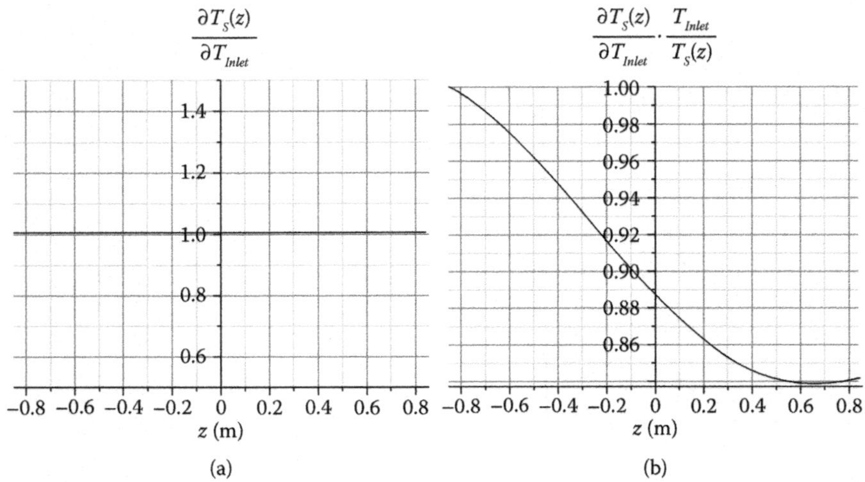

FIGURE 4.17
Spatially dependent absolute (a) and relative (b) sensitivities of the rod's surface temperature to the inlet temperature T_{inlet}.

The results presented in Equation 4.68 indicate that the most influential parameter on $T_s^{\max}$ is T_{inlet}, since it displays the largest relative sensitivity, just as in the case of the maximum rod temperature response, cf. Equation 4.55. In contradistinction to the corresponding result presented in Equation 4.55, however, the relative sensitivity of $T_s^{\max}$ with respect to T_{inlet} is positive, indicating that an increase in T_{inlet} will increase the maximum rod surface temperature, $T_s^{\max}$ (while decreasing the rod's maximum temperature at the centerline of the rod), as would be expected on physical grounds. The relative sensitivities of T^{max} to c_p, W, and h are negative, indicating that increases in any of these model parameters would lead to a decrease in $T_s^{\max}$; again, this behavior is expected to occur based on physical considerations. Just as in the case of the maximum rod temperature response, the relative sensitivities of $T_s^{\max}$ to c_p and W are equal, indicating that these parameters have equal influence on $T_s^{\max}$. The relative sensitivity of $T_s^{\max}$ to q is positive, which indicates that an increase in the heat source q would cause the maximum temperature, $T_s^{\max}$, to increase, which is to be expected on physical grounds. Also as expected on physical grounds, the relative sensitivity of T^{max} to k is zero, thus confirming that the maximum rod surface temperature does not depend on the rod's internal thermal conductivity.

Using the sensitivities depicted in Figures 4.13 through 4.17 together with the parameter standard deviations presented in Table 4.1 in the "sandwich formula" given in Equation 4.56 yields the spatially dependent (first-order) standard deviation, denoted as $dev_{FO}(T_s)$, of the rod's surface temperature, $T_s(z)$, which is depicted in Figure 4.18. In particular, the standard deviation of $T_s^{\max}$ is $dev_{FO}\left(T_s^{max}\right) = 74.8\ K$.

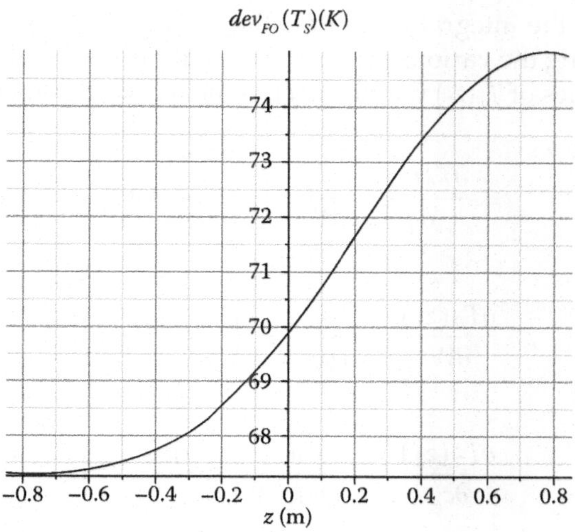

FIGURE 4.18
Spatially dependent distribution of the absolute standard deviation, $dev_{FO}(T_s)$, computed using first-order sensitivities, of the fuel rod's surface temperature.

4.2.4 Computation of First-Order Sensitivities of the Coolant Temperature

When the response of interest is the coolant temperature $T_{fl}(z_2)$ at some axial point $z = z_2$, as defined in Equation 4.8, the respective response sensitivity takes on the form

$$\delta T_{fl}(z_2) = \frac{2}{a^2}\int_0^a r\,dr \int_{-\ell/2}^{\ell/2} \delta T_{fl}(z)\delta(z-z_2)dz, \tag{4.69}$$

which indicates that $(\partial F/\partial T)_{e^0} = 0$ and $(\partial F/\partial T_{fl})_{e^0} = 2a^{-2}\delta(z-z_2)$ for this response. Replacing these expressions for the source terms in Equations 4.25 and 4.26, respectively, and solving the resulting 1st-LASS yields the following expressions for the adjoint functions $\Psi(r, z)$ and $\Psi_{fl}(z)$, respectively:

$$\Psi(r,z) \equiv 0,\quad \Psi_{fl}(z) = 2a^{-2}[1-H(z-z_2)] = 2a^{-2}H(z_2-z),\ -\frac{\ell}{2} \le z, z_2 \le \frac{\ell}{2}. \tag{4.70}$$

Replacing the results from Equation 4.70 into Equation 4.28 yields the following expression of the sensitivity $\delta T_{fl}(z_2)$:

$$\delta T_{fl}(z_2) = \pi a^2\left[\frac{\delta q}{W^0 c_p^0} - \frac{q^0(\delta W)}{(W^0)^2 c_p^0} - \frac{q^0(\delta c_p)}{W^0 (c_p^0)^2}\right]\frac{a^2}{2}\int_{-\ell/2}^{\ell/2} \frac{2}{a^2}H(z_2-z)\cos\frac{\pi z}{\ell}dz + (\delta T_{inlet}). \tag{4.71}$$

Carrying out the integration in Equation 4.71 and identifying the expressions multiplying the various parameter variations yields the following first-order sensitivities of $T_{fl}(z_2)$ with respect to the model parameters:

$$\frac{\partial T_{fl}(z_2)}{\partial q} = \frac{a^2 \ell}{W^0 c_p^0}\left(\sin\frac{\pi z_2}{\ell} + 1\right), \tag{4.72}$$

$$\frac{\partial T_{fl}(z_2)}{\partial W} = -\frac{q^0 a^2 \ell}{\left(W^0\right)^2 c_p^0}\left(\sin\frac{\pi z_2}{\ell} + 1\right), \tag{4.73}$$

$$\frac{\partial T_{fl}(z_2)}{\partial c_p} = -\frac{q^0 a^2 \ell}{W^0 \left(c_p^0\right)^2}\left(\sin\frac{\pi z_2}{\ell} + 1\right), \tag{4.74}$$

$$\frac{\partial T_{fl}(z_2)}{\partial T_{inlet}} = 1. \tag{4.75}$$

Note, in particular, that $T_{fl}(z)$ attains its maximum value, $T_{fl}^{\max}$, at $z_{\max} = \ell/2$. The sensitivities of the coolant temperature distribution are computed using Equations 4.72 through 4.75 evaluated at the nominal parameter values provided in Table 4.1. The space-dependent sensitivities thus obtained are depicted in Figures 4.19 through 4.22.

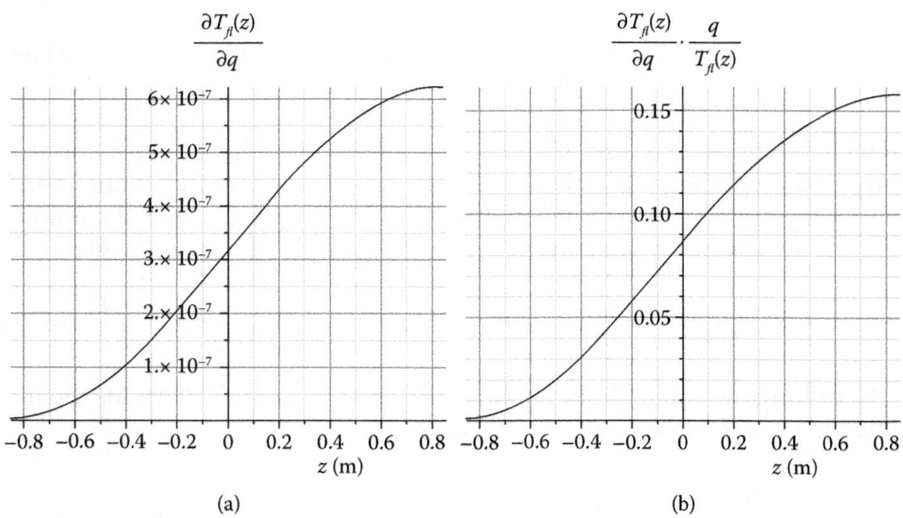

FIGURE 4.19
Spatially dependent absolute (a) and relative (b) sensitivities of the coolant temperature to the heat source strength q.

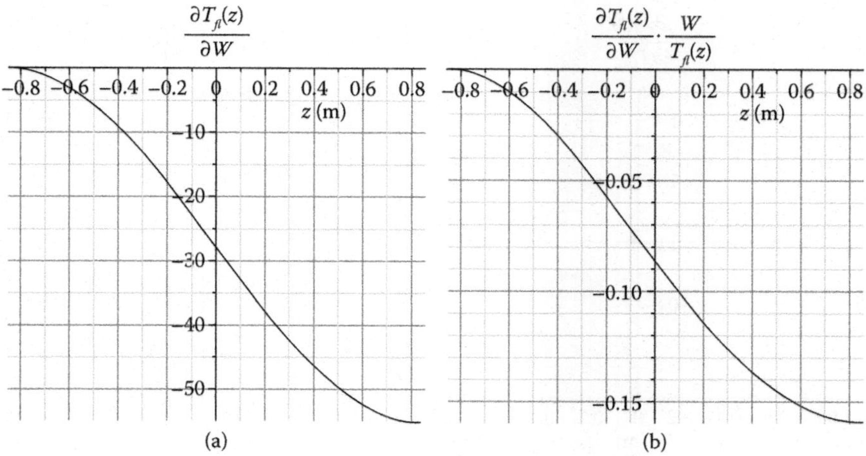

FIGURE 4.20
Spatially dependent absolute (a) and relative (b) sensitivities of the coolant temperature to the mass flow rate W.

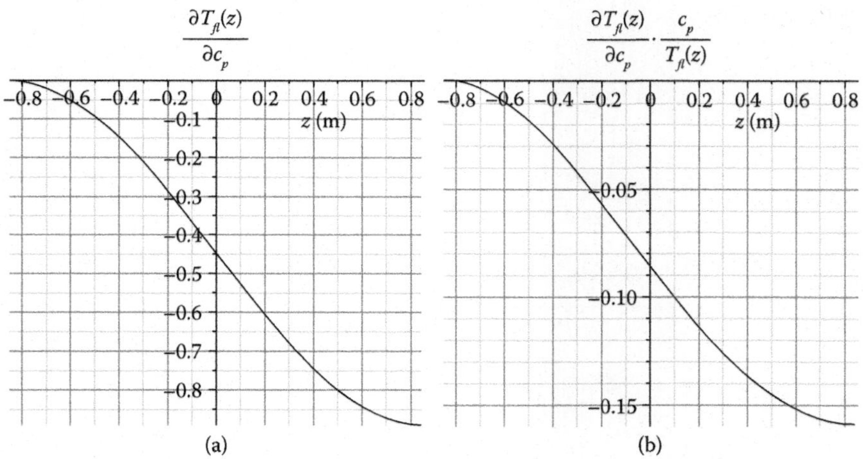

FIGURE 4.21
Spatially dependent absolute (a) and relative (b) sensitivities of the coolant temperature to the coolant's heat capacity c_p.

Recalling that $T_{fl}(z)$ attains its maximum value, T_{fl}^{max}, at $z_{max} = \ell/2$ and using the nominal parameter values presented in Table 4.1 yields $T_{fl}^{max} = 799.9\,K$. The values of the relative sensitivities of T_{fl}^{max} are as follows:

$$\frac{\partial T_{fl}^{max}}{\partial q}\frac{q}{T_{fl}^{max}} = 0.15844; \quad \frac{\partial T_{fl}^{max}}{\partial W}\frac{W}{T_{fl}^{max}} = -0.15844;$$
$$\frac{\partial T_{fl}^{max}}{\partial c_p}\frac{c_p}{T_{fl}^{max}} = -0.15844; \quad \frac{\partial T_{fl}^{max}}{\partial T_{inlet}}\frac{T_{inlet}}{T_{fl}^{max}} = 0.84155. \tag{4.76}$$

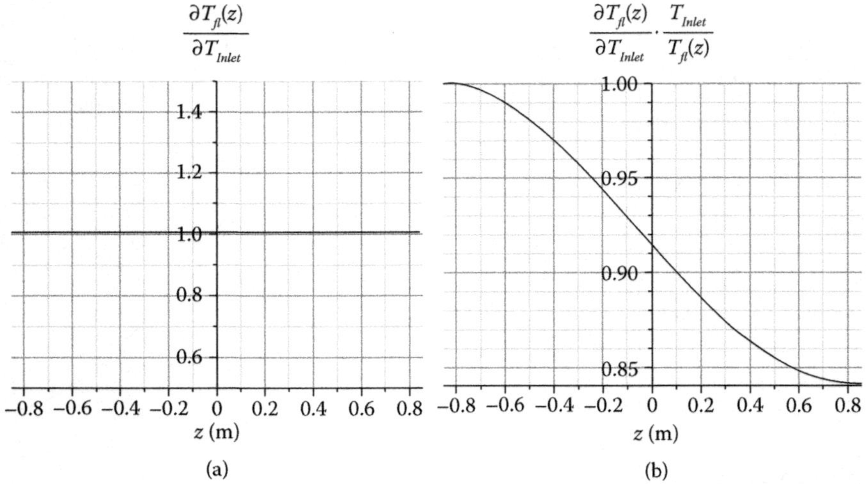

FIGURE 4.22
Spatially dependent absolute (a) and relative (b) sensitivities of the coolant temperature to the inlet temperature T_{inlet}.

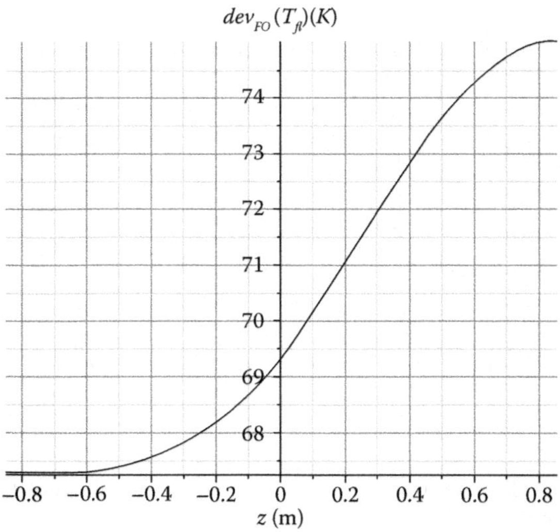

FIGURE 4.23
Spatially dependent distribution of the absolute standard deviation, $dev_{FO}(T_{fl})$, computed using first-order sensitivities of the coolant temperature.

Using the sensitivities depicted in Figures 4.19 through 4.22 together with the (parameter) standard deviations presented in Table 4.1 in the sandwich formula given in Equation 4.56 yields the spatially dependent (first-order) standard deviation, denoted as $dev_{FO}(T_{fl})$, of the coolant temperature, $T_{fl}(z)$. The spatial behavior of $dev_{FO}(T_{fl})$ is depicted in Figure 4.23. In particular, the standard deviation of T_{fl}^{max} is obtained as $dev_{FO}\left(T_{fl}^{max}\right) = 74.9\ K$.

TABLE 4.2
Parameter Importance Ranking for T^{max}, T_s^{max}, T_{fl}^{max}

Response	Parameter Importance Ranking by ARS		
T^{max}	T_{inlet}(ARS = 0.62)	q (ARS = 0.380)	k (ARS = 0.291)
T_s^{max}	T_{inlet}(ARS = 0.837)	q (ARS = 0.1625)	$W = c_p$ (ARS = 0.153)
T_{fl}^{max}	T_{inlet}(ARS = 0.842)	$W = c_p = q$ (ARS = 0.1584)	

ARS: absolute value of relative sensitivity.

Although the relative importance of the model parameters to each of the responses has already been ranked and discussed for each of the temperature responses *separately*, it is of interest to highlight the parameters' importance ranking *across* all three of the most important responses, namely T^{max}, T_s^{max}, and T_{fl}^{max}. The parameter rankings for these responses are presented in Table 4.2, which underscores the fact that the inlet temperature, T_{inlet}, has the most significant overall impact on all responses. The second most important parameter, overall, is the heat flux, q. The importance of the model parameters, k, c_p, and W follow in Positions 3 through 5, respectively, while the least important parameter to all responses is the heat transfer coefficient, h, which (for this reason) has been omitted from Table 4.2.

4.2.5 Verification of the ANSYS/FLUENT Adjoint Solver

The results presented in Sections 4.2.1 through 4.2.4 provide a benchmark problem for verifying the capabilities and accuracy of the FLUENT Adjoint Solver (ANSYS, 2015) provided within ANSYS release 16.0. Ideally, one would wish to verify the adjoint functions computed by the FLUENT Adjoint Solver directly by comparing the FLUENT results to the exact adjoint functions underlying the 1st-LASS provided in Sections 4.2.1 through 4.2.4. Such a direct comparison and, hence, a direct "solution verification" of the FLUENT Adjoint Solver is not possible, however, because the FLUENT Adjoint Solver does not provide access to the adjoint functions it computes. Therefore, the "next best" path for verifying the results produced by the FLUENT Adjoint Solver is to compare temperature sensitivities computed using the FLUENT Adjoint Solver to the corresponding temperature sensitivities presented in Sections 4.2.1 through 4.2.4. The relative sensitivities of the rod surface temperature and coolant temperature, respectively, to the inlet LBE flow temperature and velocity, respectively, were computed using the FLUENT Adjoint Solver for both the full reactor design as well as for the envisaged test section. These results are presented in Table 4.3; they indicate that the sensitivities for the test section design are close to the corresponding sensitivities for the G4M reactor core design, as well as to those computed for the benchmark problem. Such results illustrate the importance of developing heat transport benchmarks for verifying the computational tools used for designing experimental facilities and full-scale power plants.

TABLE 4.3

Relative Sensitivities Computed Using the FLUENT Adjoint Solver

Relative Sensitivities	FLUENT Adjoint Solver G4M Reactor Core	FLUENT Adjoint Solver Test Section	FLUENT Adjoint Solver Benchmark	Exact (analytical) benchmark
$\frac{\partial T_s^{max}}{\partial V_{inlet}} \frac{V_{inlet}}{T_s^{max}}$	–0.15	–0.11	–0.1524	–0.1531
$\frac{\partial T_s^{max}}{\partial T_{inlet}} \frac{T_{inlet}}{T_s^{max}}$	0.84	0.84	0.836	0.8374
$\frac{\partial T_{fl}^{max}}{\partial V_{inlet}} \frac{V_{inlet}}{T_{fl}^{max}}$	–0.12	–0.11	–0.15840	–0.15845
$\frac{\partial T_{fl}^{max}}{\partial T_{inlet}} \frac{T_{inlet}}{T_{fl}^{max}}$	0.87	0.89	0.8415	0.8415

4.3 Applying the 2nd-ASAM to Compute the Second-Order Sensitivities and Uncertainties for the Heat Transport Benchmark Problem

The results obtained in Chapter 3 will be particularized in this section to derive the specific expressions of the second-order sensitivities of the temperature distributions at any location within the heated rod, on the rod's surface, and within the coolant. Spatial points of particularly important interest are the locations where maximum values are attained by the temperatures within the rod, on the rod's surface, and within the coolant. Recall that the 1st-LASS comprises Equations 4.21, 4.22, 4.23, 4.25, and 4.26. Computing the G-differential of the 1st-LASS will yield the following system of equations to be satisfied by the two-component vector of variations $\delta\boldsymbol{\psi} \triangleq [\delta\psi(r, z), \delta\Psi_{fl}(z)]$:

$$\frac{k^0}{r}\frac{\partial}{\partial r}\left[r\frac{\partial}{\partial r}\delta\Psi(r,z)\right] = -\frac{\delta k}{r}\frac{\partial}{\partial r}\left[r\frac{\partial \Psi^0(r,z)}{\partial r}\right], \quad 0 \le r < a, \; -\frac{\ell}{2} \le z \le \frac{\ell}{2}, \tag{4.77}$$

$$-\frac{d}{dz}\delta\Psi_{fl}(z) + \frac{2}{a}h^0\left[\delta\Psi(a,z)\right] = -\frac{2}{a}(\delta h)\Psi^0(a,z), \quad -\frac{\ell}{2} \le z \le \frac{\ell}{2}, \tag{4.78}$$

$$\delta\Psi_{fl}(z) = 0, \quad \text{at } z = \ell/2, \tag{4.79}$$

$$\frac{\partial}{\partial r}\delta\Psi(r,z) = 0, \text{ at } r = 0, \tag{4.80}$$

$$-k^0\frac{\partial}{\partial r}\delta\Psi(a,z)-h^0\delta\Psi(a,z)=(\delta k)\frac{\partial\Psi^0(a,z)}{\partial r}+(\delta h)\Psi^0(a,z). \quad (4.81)$$

The expressions of the first-order sensitivities in Equations 4.29 through 4.34 depend only on the first-level adjoint function $\boldsymbol{\psi}\triangleq[\Psi(r,z),\Psi_{fl}(z)]$ and on model parameters but do *not* depend on $\mathbf{u}(\mathbf{x})\triangleq[T(r,z),T_{fl}(z)]$. Therefore, the second-order sensitivities will depend only on variations $\delta\boldsymbol{\psi}\triangleq[\delta\Psi(r,z),\delta\Psi_{fl}(z)]$ and $\delta\boldsymbol{\alpha}\triangleq(\delta q,\delta k,\delta h,\delta W,\delta c_p,\delta T_{inlet})$ but will *not* depend on variations $\delta\mathbf{u}(\mathbf{x})\triangleq[\delta T(r,z),\delta T_{fl}(z)]$. It therefore follows from the general theory presented in Chapter 3, Section 3.3, that the 2nd-LFSS is provided by Equations 4.77 through 4.81, and consequently the general second-level adjoint function $\Psi^{(2)}_{i_1}(\mathbf{x})\triangleq\left(\boldsymbol{\psi}^{(2)}_{1,i_1}(\mathbf{x}),\boldsymbol{\psi}^{(2)}_{2,i_1}(\mathbf{x})\right)\in\mathcal{H}^{(2)}_{\Psi}(\Omega_x)$ will reduce to a two-component vector of the form $\Psi^{(2)}_{i}(r,z)\triangleq\left[\Psi^{(2,i)}(r,z),\Psi^{(2,i)}_{fl}(z)\right]^{\dagger}\in\mathcal{H}^{(2)}_{\Psi}(\Omega_x)$, which will be the solution of the following generic form of the 2nd-LASS for the heat transport benchmark problem considered in this chapter:

$$\frac{k^0}{r}\frac{\partial}{\partial r}\left[r\frac{\partial\Psi^{(2,i)}(r,z)}{\partial r}\right]=\frac{\partial}{\partial\Psi}S_i\left(\mathbf{u}^0,\boldsymbol{\psi}^0,\boldsymbol{\alpha}^0\right),\quad 0\le r<a,\ -\frac{\ell}{2}\le z\le\frac{\ell}{2}, \quad (4.82)$$

$$\frac{\partial\Psi^{(2,i)}(r,z)}{\partial r}=0,\ \text{ at } r=0, \quad (4.83)$$

$$-k^0\frac{\partial\Psi^{(2,i)}(r,z)}{\partial r}=h^0\left[\Psi^{(2,i)}(r,z)-\Psi^{(2,i)}_{fl}(z)\right],\ \text{ at } r=a, \quad (4.84)$$

$$\frac{d\Psi^{(2,i)}_{fl}(z)}{dz}=\frac{\partial}{\partial\Psi_{fl}}S_i\left(\mathbf{u}^0,\boldsymbol{\psi}^0,\boldsymbol{\alpha}^0\right),\quad -\frac{\ell}{2}\le z\le\frac{\ell}{2}, \quad (4.85)$$

$$\Psi^{(2,i)}_{fl}(z)=0,\ \text{ at } z=-\ell/2. \quad (4.86)$$

4.3.1 Computation of the Second-Order Sensitivities and Uncertainties of the Heated Rod Temperature, $T(r, z)$, at an Arbitrary Location (r_0, z_0)

For the response $T(r_0, z_0)$, the first-order sensitivities given generically in Equations 4.29 through 4.34 take on the following particular expressions:

$$\left(\frac{\partial T(r_0,z_0)}{\partial q}\right)_{(\mathbf{u}^0,\boldsymbol{\alpha}^0,\boldsymbol{\psi}^0)}=-\int_0^a r\,dr\int_{-\ell/2}^{\ell/2}\Psi^0(r,z)\cos\frac{\pi z}{\ell}dz+\frac{\pi a^2}{W^0c_p^0}\int_0^a r\,dr\int_{-\ell/2}^{\ell/2}dz\Psi^0_{fl}(z)\cos\frac{\pi z}{\ell}, \quad (4.87)$$

$$\left(\frac{\partial T(r_0,z_0)}{\partial k}\right)_{(\mathbf{u}^0,\alpha^0,\psi^0)} = \frac{q^0}{k^0}\int_0^a r\,dr \int_{-\ell/2}^{\ell/2} \Psi^0(r,z)\left[1-\frac{a^2}{2}\frac{\delta(r-a)}{r}\right]\cos\frac{\pi z}{\ell}dz, \quad (4.88)$$

$$\left(\frac{\partial T(r_0,z_0)}{\partial h}\right)_{(\mathbf{u}^0,\alpha^0,\psi^0)} = \frac{q^0}{h^0}\frac{a^2}{2}\int_0^a r\,dr \int_{-\ell/2}^{\ell/2} \Psi^0(r,z)\frac{\delta(r-a)}{r}\cos\frac{\pi z}{\ell}dz, \quad (4.89)$$

$$\left(\frac{\partial T(r_0,z_0)}{\partial W}\right)_{(\mathbf{u}^0,\alpha^0,\psi^0)} = -\frac{q^0\pi a^2}{\left(W^0\right)^2 c_p^0}\int_0^a r\,dr \int_{-\ell/2}^{\ell/2} \Psi_{fl}^0(z)\cos\frac{\pi z}{\ell}dz, \quad (4.90)$$

$$\left(\frac{\partial T(r_0,z_0)}{\partial c_p}\right)_{(\mathbf{u}^0,\alpha^0,\psi^0)} = -\frac{q^0\pi a^2}{W^0\left(c_p^0\right)^2}\int_0^a r\,dr \int_{-\ell/2}^{\ell/2} \Psi_{fl}^0(z)\cos\frac{\pi z}{\ell}dz, \quad (4.91)$$

$$\left(\frac{\partial T(r_0,z_0)}{\partial T_{inlet}}\right)_{(\mathbf{u}^0,\alpha^0,\psi^0)} = 1, \quad (4.92)$$

where the expressions of the first-level adjoint sensitivity functions $\Psi^0(r, z)$ and $\Psi_{fl}^0(z)$ are provided in Equation 4.36.

4.3.1.1 Computation of the Second-Order Response Sensitivities $\partial^2 T(r_0, z_0)/(\partial\alpha_1\, \partial\alpha_j), \alpha_1 \equiv q$, and $j = 1, \ldots, N_\alpha = 6$

Taking the G-differential of Equation 4.87 yields

$$\delta\left[\frac{\partial T(r_0,z_0)}{\partial q}\right] = \left\{\delta\left[\frac{\partial T(r_0,z_0)}{\partial q}\right]\right\}_{direct} + \left\{\delta\left[\frac{\partial T(r_0,z_0)}{\partial q}\right]\right\}_{indirect}, \quad (4.93)$$

where

$$\left\{\delta\left[\frac{\partial T(r_0,z_0)}{\partial q}\right]\right\}_{direct} \triangleq -\pi a^2\left(\frac{\delta W}{\left(W^0\right)^2 c_p^0} + \frac{\delta c_p}{W^0\left(c_p^0\right)^2}\right)\int_0^a r\,dr \int_{-\ell/2}^{\ell/2} \Psi_{fl}^0(z)\cos\frac{\pi z}{\ell}dz, \quad (4.94)$$

$$\left\{\delta\left[\frac{\partial T(r_0,z_0)}{\partial q}\right]\right\}_{indirect} \triangleq -\int_0^a r\,dr \int_{-\ell/2}^{\ell/2} \delta\Psi(r,z)\cos\frac{\pi z}{\ell}dz + \frac{\pi a^2}{W^0 c_p^0}\int_0^a r\,dr \int_{-\ell/2}^{\ell/2} \delta\Psi_{fl}(z)\cos\frac{\pi z}{\ell}dz. \quad (4.95)$$

The direct-effect term can be evaluated now by inserting the expression of $\Psi_{fl}^{0}(z)$ from Equation 4.36 into Equation 4.94 and performing the integrations to obtain

$$\left\{\delta\left[\frac{\partial T(r_0,z_0)}{\partial q}\right]\right\}_{direct}=-(\delta W)\frac{a^2\ell}{\left(W^0\right)^2c_p^0}\left(\sin\frac{\pi z_0}{\ell}+1\right)-\left(\delta c_p\right)\frac{a^2\ell}{W^0\left(c_p^0\right)^2}\left(\sin\frac{\pi z_0}{\ell}+1\right). \tag{4.96}$$

On the other hand, the indirect-effect term in Equation 4.93 provides the source terms for Equations 4.82 and 4.85, respectively. Consequently, for this particular form of the indirect-effect term, Equations 4.82 through 4.86 take on the following particular forms for the particular *second-level adjoint function* $\Psi_1^{(2)}(r,z)\equiv\left[\Psi^{(2,1)}(r,z),\,\Psi_{fl}^{(2,1)}(z)\right]$:

$$\frac{k^0}{r}\frac{\partial}{\partial r}\left[r\frac{\partial\Psi^{(2,1)}(r,z)}{\partial r}\right]=-\cos\frac{\pi z}{\ell},\quad 0\le r<a,\ -\frac{\ell}{2}\le z\le\frac{\ell}{2}, \tag{4.97}$$

$$\frac{\partial\Psi^{(2,1)}(r,z)}{\partial r}=0,\ \text{ at } r=0, \tag{4.98}$$

$$-k^0\frac{\partial\Psi^{(2,1)}(r,z)}{\partial r}=h^0\left[\Psi^{(2,1)}(r,z)-\Psi_{fl}^{(2,1)}(z)\right],\ \text{ at } r=a, \tag{4.99}$$

$$\frac{d\Psi_{fl}^{(2,1)}(z)}{dz}=\frac{\pi a^2}{W^0c_p^0}\cos\frac{\pi z}{\ell},\quad -\frac{\ell}{2}\le z\le\frac{\ell}{2}, \tag{4.100}$$

$$\Psi_{fl}^{(2,1)}(z)=0,\quad \text{at } z=-\ell/2. \tag{4.101}$$

Comparing next Equations 4.97 through 4.101 to Equations 4.1 through 4.5 reveals that $\Psi^{(2,1)}(r,z)=\left\{T^0(r,z)\right\}_{q^0=1}$ and $\Psi_{fl}^{(2,1)}(z)=\left\{T_{fl}^0(z)\right\}_{T_{inlet}^0=0}$. Consequently, the second-level adjoint functions $\Psi^{(2,1)}(r,z)$ and $\Psi_{fl}^{(2,1)}(z)$ can be obtained directly by setting $q^0=1$ and $T_{inlet}^0=0$ in Equations 4.10 and 4.11, without actually needing to solve Equations 4.97 through 4.101. Thus, the resulting expressions for the second-level adjoint functions $\Psi^{(2,1)}(r,z)$ and $\Psi_{fl}^{(2,1)}(z)$ are

$$\begin{aligned}\Psi^{(2,1)}(r,z)&=\left(\frac{a^2-r^2}{4k^0}-\frac{a}{2h^0}\right)\cos\frac{\pi z}{\ell}+\Psi_{fl}^{(2,1)}(z),\quad 0\le r<a,\ -\frac{\ell}{2}\le z\le\frac{\ell}{2},\\ \Psi_{fl}^{(2,1)}(z)&=\frac{a^2\ell}{W^0c_p^0}\left(\sin\frac{\pi z}{\ell}+1\right),\quad -\frac{\ell}{2}\le z\le\frac{\ell}{2}.\end{aligned} \tag{4.102}$$

In terms of the second-level adjoint functions $\Psi^{(2,1)}(r, z)$ and $\Psi_{fl}^{(2,1)}(z)$, the indirect-effect term defined in Equation 4.95 takes on the following expression:

$$\left\{\delta\left[\frac{\partial T(r_0,z_0)}{\partial q}\right]\right\}_{indirect} = \frac{\delta k}{k^0}\left[\Psi^{(2,1)}(a,z)-\Psi^{(2,1)}(r_0,z_0)\right] + \frac{\delta h}{h^0}\left[\Psi_{fl}^{(2,1)}(z_0)-\Psi^{(2,1)}(a,z_0)\right]. \tag{4.103}$$

The expressions obtained in Equation 4.102 can now be inserted into Equation 4.103 to obtain

$$\left\{\delta\left[\frac{\partial T(r_0,z_0)}{\partial q}\right]\right\}_{indirect} = (\delta h)\frac{a}{2\left(h^0\right)^2}\cos\frac{\pi z_0}{\ell} + (\delta k)\frac{\left(r_0^2-a^2\right)}{4\left(k^0\right)^2}\cos\frac{\pi z_0}{\ell}. \tag{4.104}$$

Identifying the coefficients multiplying the respective arbitrary parameter variations in Equations 4.96 and 4.104, it follows that the expressions of the respective second-order sensitivities of $T(r_0, z_0)$ are

$$\frac{\partial^2 T(r_0,z_0)}{\partial q^2} = 0, \tag{4.105}$$

$$\frac{\partial^2 T(r_0,z_0)}{\partial q\,\partial h} = -\frac{a}{2\left(h^0\right)^2}\cos\frac{\pi z_0}{\ell}, \tag{4.106}$$

$$\frac{\partial^2 T(r_0,z_0)}{\partial q\,\partial k} = \frac{\left(r_0^2-a^2\right)}{4\left(k^0\right)^2}\cos\frac{\pi z_0}{\ell}, \tag{4.107}$$

$$\frac{\partial^2 T(r_0,z_0)}{\partial q\,\partial W} = -\frac{a^2\ell}{\left(W^0\right)^2 c_p^0}\left(\sin\frac{\pi z_0}{\ell}+1\right), \tag{4.108}$$

$$\frac{\partial^2 T(r_0,z_0)}{\partial q\,\partial c_p} = -\frac{a^2\ell}{W^0\left(c_p^0\right)^2}\left(\sin\frac{\pi z_0}{\ell}+1\right), \tag{4.109}$$

$$\frac{\partial^2 T(r_0,z_0)}{\partial q\,\partial T_{inlet}} = 0. \tag{4.110}$$

4.3.1.2 Computation of the Second-Order Response Sensitivities $\partial^2 T(r_0, z_0)/(\partial\alpha_2\, \partial\alpha_j)$, $\alpha_2 \equiv k$, and $j = 1, \ldots, N_\alpha = 6$

Taking the G-differential of Equation 4.88 yields

$$\delta\left[\frac{\partial T(r_0,z_0)}{\partial k}\right]=\left\{\delta\left[\frac{\partial T(r_0,z_0)}{\partial k}\right]\right\}_{direct}+\left\{\delta\left[\frac{\partial T(r_0,z_0)}{\partial k}\right]\right\}_{indirect}, \quad (4.111)$$

where

$$\left\{\delta\left[\frac{\partial T(r_0,z_0)}{\partial k}\right]\right\}_{direct} \triangleq \left(\frac{\delta q}{k^0}-\frac{q^0\delta k}{\left(k^0\right)^2}\right)\int_0^a r\,dr\int_{-\ell/2}^{\ell/2}\Psi^0(r,z)\left(1-\frac{a^2}{2}\frac{\delta(r-a)}{r}\right)\cos\frac{\pi z}{\ell}dz, \quad (4.112)$$

$$\left\{\delta\left[\frac{\partial T(r_0,z_0)}{\partial q}\right]\right\}_{indirect} \triangleq \frac{q^0}{k^0}\int_0^a r\,dr\int_{-\ell/2}^{\ell/2}\left(1-\frac{a^2}{2}\frac{\delta(r-a)}{r}\right)\delta\Psi(r,z)\cos\frac{\pi z}{\ell}dz. \quad (4.113)$$

The direct-effect term in Equation 4.112 can be evaluated now by inserting the expression of $\Psi^0(r, z)$ from Equation 4.36 into Equation 4.112 and performing the respective integrations to obtain

$$\left\{\delta\left[\frac{\partial T(r_0,z_0)}{\partial k}\right]\right\}_{direct} \equiv \left(\frac{\delta q}{k^0}-\frac{q^0\delta k}{\left(k^0\right)^2}\right)\frac{\left(r_0^2-a^2\right)}{4k^0}\cos\frac{\pi z_0}{\ell}. \quad (4.114)$$

On the other hand, the indirect-effect term in Equation 4.113 provides the source terms for Equations 4.82 and 4.85, respectively. Consequently, for this particular form of the indirect-effect term, the 2nd-LASS given by Equations 4.82 through 4.86 take on the following particular forms for the particular *second-level adjoint function* $\Psi_2^{(2)} \triangleq \left[\Psi^{(2,2)}(r,z),\, \Psi_{fl}^{(2,2)}(z)\right]$:

$$\frac{k^0}{r}\frac{\partial}{\partial r}\left[r\frac{\partial\Psi^{(2,2)}(r,z)}{\partial r}\right]=\frac{q^0}{k^0}\left(1-\frac{a^2}{2}\frac{\delta(r-a)}{r}\right)\cos\frac{\pi z}{\ell},\ 0\le r<a,\ -\frac{\ell}{2}\le z\le\frac{\ell}{2}, \quad (4.115)$$

$$\frac{\partial\Psi^{(2,2)}(r,z)}{\partial r}=0,\ \text{at } r=0, \quad (4.116)$$

$$-k^0\frac{\partial\Psi^{(2,2)}(r,z)}{\partial r}=h^0\left[\Psi^{(2,2)}(r,z)-\Psi_{fl}^{(2,2)}(z)\right],\ \text{at } r=a, \quad (4.117)$$

$$\frac{d\Psi_{fl}^{(2,2)}(z)}{dz}=0,\quad -\frac{\ell}{2}\le z\le\frac{\ell}{2}, \quad (4.118)$$

$$\Psi_{fl}^{(2,2)}(z)=0,\ \text{at } z=-\ell/2. \quad (4.119)$$

The solution to Equations 4.115 through 4.119 can be readily obtained as

$$\Psi^{(2,2)}(r,z)=\frac{q^0}{\left(k^0\right)^2}\left[\frac{r^2-a^2}{4}-\frac{a^2}{2}H(r-a)\ln\frac{r}{a}\right]\cos\frac{\pi z}{\ell},\ 0\le r<a,\ -\frac{\ell}{2}\le z\le\frac{\ell}{2},$$

$$\Psi_{fl}^{(2,2)}(z)\equiv 0,\quad -\frac{\ell}{2}\le z\le\frac{\ell}{2}. \tag{4.120}$$

In terms of the second-level adjoint functions $\Psi^{(2,2)}(r, z)$ and $\Psi_{fl}^{(2,2)}(z)$, the indirect-effect term defined in Equation 4.113 takes on the expression

$$\begin{aligned}\left\{\delta\left[\frac{\partial T(r_0,z_0)}{\partial k}\right]\right\}_{indirect}&=\frac{\delta k}{k^0}\left[\Psi^{(2,2)}(a,z)-\Psi^{(2,2)}(r_0,z_0)\right]+\frac{\delta h}{h^0}\left[\Psi_{fl}^{(2,2)}(z_0)-\Psi^{(2,2)}(a,z_0)\right]\\&=(\delta k)\frac{q^0\left(a^2-r_0^2\right)}{4\left(k^0\right)^3}\cos\frac{\pi z_0}{\ell}.\end{aligned} \tag{4.121}$$

Adding Equations 4.114 and 4.121, and identifying the coefficients multiplying the respective parameter variations, yields the following expressions for the respective second-order sensitivities of $T(r_0, z_0)$:

$$\frac{\partial^2 T(r_0,z_0)}{\partial k\partial q}=\frac{\left(r_0^2-a^2\right)}{4\left(k^0\right)^2}\cos\frac{\pi z_0}{\ell}, \tag{4.122}$$

$$\frac{\partial^2 T(r_0,z_0)}{\partial k^2}=\frac{q^0\left(a^2-r_0^2\right)}{2\left(k^0\right)^3}\cos\frac{\pi z_0}{\ell}, \tag{4.123}$$

$$\frac{\partial^2 T(r_0,z_0)}{\partial k\partial h}=\frac{\partial^2 T(r_0,z_0)}{\partial k\partial W}=\frac{\partial^2 T(r_0,z_0)}{\partial k\partial c_p}=\frac{\partial^2 T(r_0,z_0)}{\partial k\partial T_{inlet}}=0. \tag{4.124}$$

4.3.1.3 Computation of the Second-Order Response Sensitivities $\partial^2 T(r_0, z_0)/(\partial\alpha_3\, \partial\alpha_j)$, $\alpha_3 \equiv h$, and $j = 1, \ldots, N_\alpha = 6$

Taking the G-differential of Equation 4.89 yields

$$\delta\left[\frac{\partial T(r_0,z_0)}{\partial h}\right]=\left\{\delta\left[\frac{\partial T(r_0,z_0)}{\partial h}\right]\right\}_{direct}+\left\{\delta\left[\frac{\partial T(r_0,z_0)}{\partial h}\right]\right\}_{indirect}, \tag{4.125}$$

where

$$\left\{\delta\left[\frac{\partial T(r_0,z_0)}{\partial h}\right]\right\}_{direct} \triangleq \left(\frac{\delta q}{h^0} - \frac{q^0 \delta h}{(h^0)^2}\right)\frac{a^2}{2}\int_0^a r\,dr \int_{-\ell/2}^{\ell/2} \Psi^0(r,z)\frac{\delta(r-a)}{r}\cos\frac{\pi z}{\ell}dz, \tag{4.126}$$

$$\left\{\delta\left[\frac{\partial T(r_0,z_0)}{\partial h}\right]\right\}_{indirect} \triangleq \frac{q^0}{h^0}\frac{a^2}{2}\int_0^a r\,dr \int_{-\ell/2}^{\ell/2} \delta\Psi(r,z)\frac{\delta(r-a)}{r}\cos\frac{\pi z}{\ell}dz. \tag{4.127}$$

The direct-effect term in Equation 4.126 can be evaluated now by inserting the expression of $\Psi^0(r,z)$ from Equation 4.36 into Equation 4.126 and performing the respective integrations to obtain

$$\left\{\delta\left[\frac{\partial T(r_0,z_0)}{\partial h}\right]\right\}_{direct} \triangleq \left(-\frac{\delta q}{(h^0)^2} + \frac{q^0 \delta h}{(h^0)^3}\right)\frac{a}{2}\cos\frac{\pi z_0}{\ell}. \tag{4.128}$$

On the other hand, the indirect-effect term in Equation 4.127 provides the source terms for Equations 4.82 and 4.85, respectively. Consequently, for this particular form of the indirect-effect term, Equations 4.82 and 4.86 take on the following particular forms for the particular *second-level adjoint function* $\Psi_3^{(2)} \triangleq \left[\Psi^{(2,3)}(r,z), \Psi_{fl}^{(2,3)}(z)\right]$:

$$\frac{k^0}{r}\frac{\partial}{\partial r}\left[r\frac{\partial \Psi^{(2,3)}(r,z)}{\partial r}\right] = \frac{q^0}{h^0}\frac{a^2}{2}\frac{\delta(r-a)}{r}\cos\frac{\pi z}{\ell}, \quad 0 \le r < a, \ -\frac{\ell}{2} \le z \le \frac{\ell}{2}, \tag{4.129}$$

$$\frac{\partial \Psi^{(2,3)}(r,z)}{\partial r} = 0, \text{ at } r = 0, \tag{4.130}$$

$$-k^0\frac{\partial \Psi^{(2,3)}(r,z)}{\partial r} = h^0\left[\Psi^{(2,3)}(r,z) - \Psi_{fl}^{(2,3)}(z)\right], \text{ at } r = a, \tag{4.131}$$

$$\frac{d\Psi_{fl}^{(2,3)}(z)}{dz} = 0, \quad -\frac{\ell}{2} \le z \le \frac{\ell}{2}, \tag{4.132}$$

$$\Psi_{fl}^{(2,3)}(z) = 0, \text{ at } z = -\ell/2. \tag{4.133}$$

Solving Equations 4.129 through 4.133 yields

$$\Psi^{(2,3)}(r,z)=\frac{q^0}{h^0}\frac{a^2}{2}\left[\frac{1}{ah^0}-\frac{1}{k^0}H(r-a)\ln\frac{r}{a}\right]\cos\frac{\pi z}{\ell},\ 0\le r<a,\ -\frac{\ell}{2}\le z\le\frac{\ell}{2},$$
$$\Psi_{fl}^{(2,3)}(z)\equiv 0,\ -\frac{\ell}{2}\le z\le\frac{\ell}{2}. \tag{4.134}$$

In terms of the second-level adjoint functions $\Psi^{(2,3)}(r, z)$ and $\Psi_{fl}^{(2,3)}(z)$, the indirect-effect term defined in Equation 4.127 takes on the expression

$$\left\{\delta\left[\frac{\partial T(r_0,z_0)}{\partial h}\right]\right\}_{indirect}=\frac{\delta k}{k^0}\left[\Psi^{(2,3)}(a,z)-\Psi^{(2,3)}(r_0,z_0)\right]+\frac{\delta h}{h^0}\left[\Psi_{fl}^{(2,3)}(z_0)-\Psi^{(2,3)}(a,z_0)\right]$$
$$=\frac{q^0\delta h}{\left(h^0\right)^3}\frac{a}{2}\cos\frac{\pi z_0}{\ell}. \tag{4.135}$$

Adding Equations 4.128 and 4.135 and identifying the coefficients multiplying the respective arbitrary parameter variations, it follows that the expressions of the respective second-order sensitivities of $T(r_0, z_0)$ are

$$\frac{\partial^2 T(r_0,z_0)}{\partial h\,\partial q}=-\frac{a}{2\left(h^0\right)^2}\cos\frac{\pi z_0}{\ell}, \tag{4.136}$$

$$\frac{\partial^2 T(r_0,z_0)}{\partial h^2}=\frac{q^0 a}{\left(h^0\right)^3}\cos\frac{\pi z_0}{\ell}, \tag{4.137}$$

$$\frac{\partial^2 T(r_0,z_0)}{\partial h\,\partial k}=\frac{\partial^2 T(r_0,z_0)}{\partial h\,\partial W}=\frac{\partial^2 T(r_0,z_0)}{\partial h\,\partial c_p}=\frac{\partial^2 T(r_0,z_0)}{\partial h\,\partial T_{inlet}}=0. \tag{4.138}$$

4.3.1.4 Computation of the Second-Order Response Sensitivities $\partial^2 T(r_0, z_0)/(\partial\alpha_4\,\partial\alpha_j)$, $\alpha_4 \equiv W$, and $j = 1, \ldots, N_\alpha = 6$

Taking the G-differential of the expression of the first-order sensitivity $\partial T(r_0, z_0)/\partial W$ given in Equation 4.90 indicates that the indirect-effect term of this G-differential will depend only on the variation $\delta\Psi_{fl}(z)$. However, Equation 4.36 indicates that $\Psi_{fl}(z)$ is independent of the model parameters, which implies that $\delta\Psi_{fl}(z) \equiv 0$. Hence, the respective indirect-effect term will be identically zero, and consequently the second-order sensitivities corresponding to $\partial T(r_0, z_0)/\partial W$ can be obtained directly by computing the direct-effect term, inserting the expression of $\Psi_{fl}^0(z)$ from Equation 4.36 into the resulting integral and performing the respective integration over z. This process yields

$$\frac{\partial^2 T(r_0,z_0)}{\partial W \partial q} = -\frac{a^2 \ell}{\left(W^0\right)^2 c_p^0}\left(\sin\frac{\pi z_0}{\ell}+1\right), \tag{4.139}$$

$$\frac{\partial^2 T(r_0,z_0)}{\partial W^2} = \frac{2q^0 a^2 \ell}{\left(W^0\right)^3 c_p^0}\left(\sin\frac{\pi z_0}{\ell}+1\right), \tag{4.140}$$

$$\frac{\partial^2 T(r_0,z_0)}{\partial W \partial c_p} = \frac{q^0 a^2 \ell}{\left(W^0\right)^2 \left(c_p^0\right)^2}\left(\sin\frac{\pi z_0}{\ell}+1\right), \tag{4.141}$$

$$\frac{\partial^2 T(r_0,z_0)}{\partial W \partial k} = \frac{\partial^2 T(r_0,z_0)}{\partial W \partial h} = \frac{\partial^2 T(r_0,z_0)}{\partial W \partial T_{inlet}} = 0. \tag{4.142}$$

4.3.1.5 Computation of Second-Order Response Sensitivities $\partial^2 T(r_0, z_0)/(\partial\alpha_5\, \partial\alpha_j)$, $\alpha_5 \equiv c_p$, and $j = 1, \ldots, N_\alpha = 6$

The G-differential of the sensitivity $\partial T(r_0, z_0)/\partial c_p$ given in Equation 4.91 can be computed similarly to the computation of the G-differential of $\partial T(r_0, z_0)/\partial W$, as discussed above in subsection 4.3.1.4, to obtain

$$\frac{\partial^2 T(r_0,z_0)}{\partial c_p \partial q} = -\frac{a^2 \ell}{W^0 \left(c_p^0\right)^2}\left(\sin\frac{\pi z_0}{\ell}+1\right), \tag{4.143}$$

$$\frac{\partial^2 T(r_0,z_0)}{\partial c_p \partial W} = \frac{q^0 a^2 \ell}{\left(W^0\right)^2 \left(c_p^0\right)^2}\left(\sin\frac{\pi z_0}{\ell}+1\right), \tag{4.144}$$

$$\frac{\partial^2 T(r_0,z_0)}{\partial c_p^2} = \frac{2q^0 a^2 \ell}{W^0 \left(c_p^0\right)^3}\left(\sin\frac{\pi z_0}{\ell}+1\right), \tag{4.145}$$

$$\frac{\partial^2 T(r_0,z_0)}{\partial c_p \partial k} = \frac{\partial^2 T(r_0,z_0)}{\partial c_p \partial h} = \frac{\partial^2 T(r_0,z_0)}{\partial c_p \partial T_{inlet}} = 0. \tag{4.146}$$

4.3.1.6 Computation of Second-Order Response Sensitivities $\partial^2 T(r_0, z_0)/(\partial\alpha_6\, \partial\alpha_j)$, $\alpha_6 \equiv T_{inlet}$, and $j = 1, \ldots, N_\alpha = 6$

As shown in Equation 4.92, the first-order sensitivity $\left(\frac{\partial T(r_0, z_0)}{\partial T_{inlet}}\right)_{(\mathbf{u}^0, \alpha^0, \psi^0)} = 1.$ Taking the G-differential of this expression directly yields the following results:

$$\frac{\partial^2 T(r_0, z_0)}{\partial T_{inlet}\, \partial q} = \frac{\partial^2 T(r_0, z_0)}{\partial T_{inlet}\, \partial k} = \frac{\partial^2 T(r_0, z_0)}{\partial T_{inlet}\, \partial h} = \frac{\partial^2 T(r_0, z_0)}{\partial T_{inlet}\, \partial W} = \frac{\partial^2 T(r_0, z_0)}{\partial T_{inlet}\, \partial c_p} = \frac{\partial^2 T(r_0, z_0)}{\partial (T_{inlet})^2} = 0. \tag{4.147}$$

4.3.1.7 Quantitative Comparison of Second-Order Sensitivities of the Rod Temperature Distribution to G4M Reactor Model Parameters

Since the location (r_0, z_0) is arbitrary within the heated rod, the second-order sensitivities of $T(r_0, z_0)$ are plotted in Figures 4.24 through 4.32 as functions of (r_0, z_0), $0 \le r_0 \le a$, $-\ell/2 \le z_0 \le \ell/2$. For notational simplicity, the units of the absolute sensitivities and the subscript/superscript 0 are omitted in the figure captions. The relative sensitivities have no units, of course.

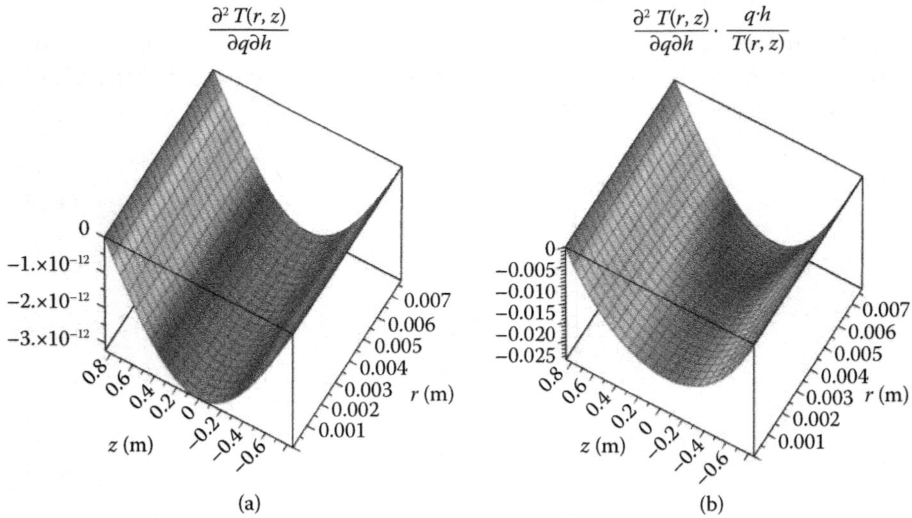

FIGURE 4.24
Absolute (a) and relative (b) second-order sensitivities of the heated rod temperature distribution with respect to the heat source strength and heat transfer coefficient.

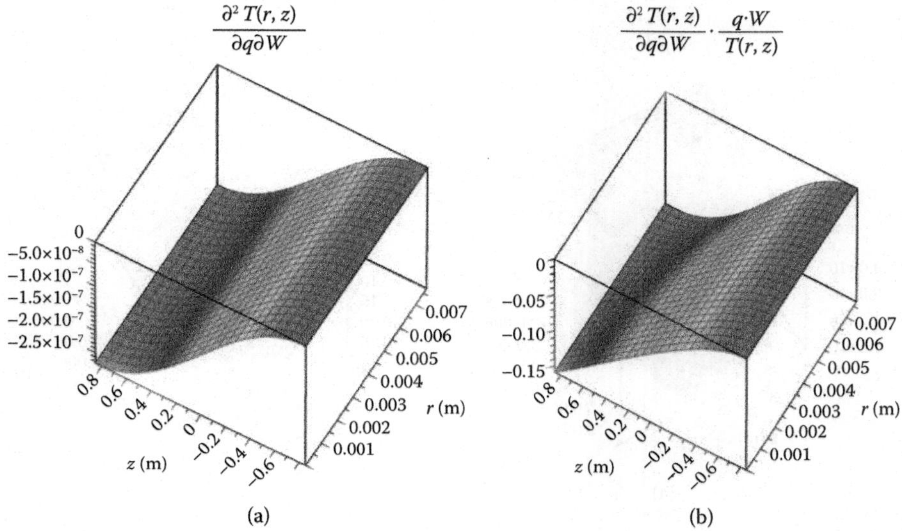

FIGURE 4.25
Absolute (a) and relative (b) second-order sensitivities of the heated rod temperature distribution with respect to the heat source strength and mass flow rate.

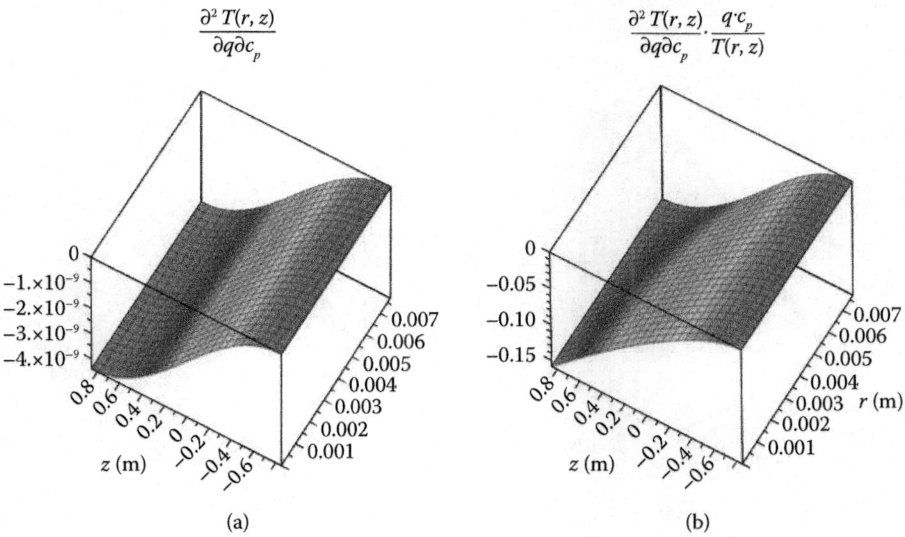

FIGURE 4.26
Absolute (a) and relative (b) second-order sensitivities of the heated rod temperature distribution with respect to the heat source strength and coolant's heat capacity.

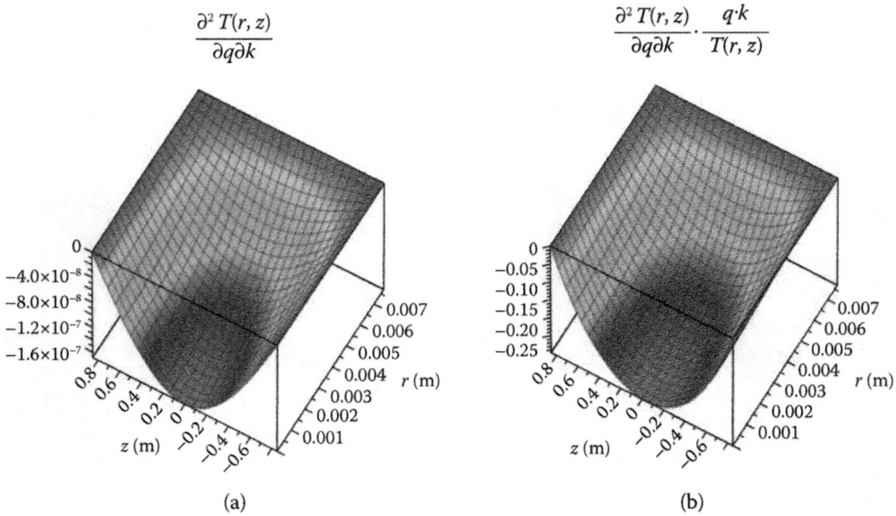

FIGURE 4.27
Absolute (a) and relative (b) second-order sensitivities of the heated rod temperature distribution with respect to the heat source strength and rod's conductivity.

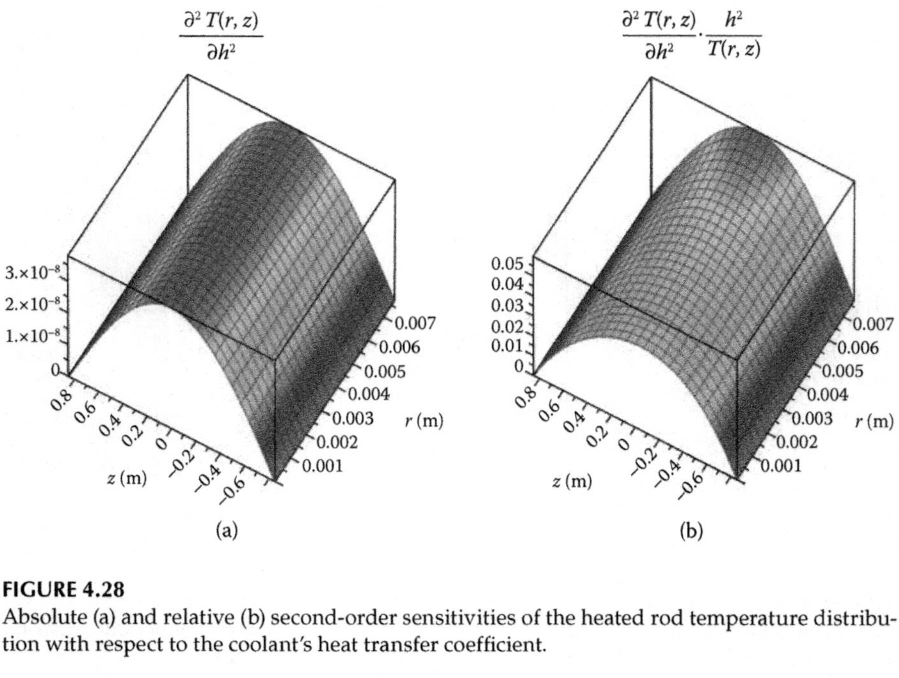

FIGURE 4.28
Absolute (a) and relative (b) second-order sensitivities of the heated rod temperature distribution with respect to the coolant's heat transfer coefficient.

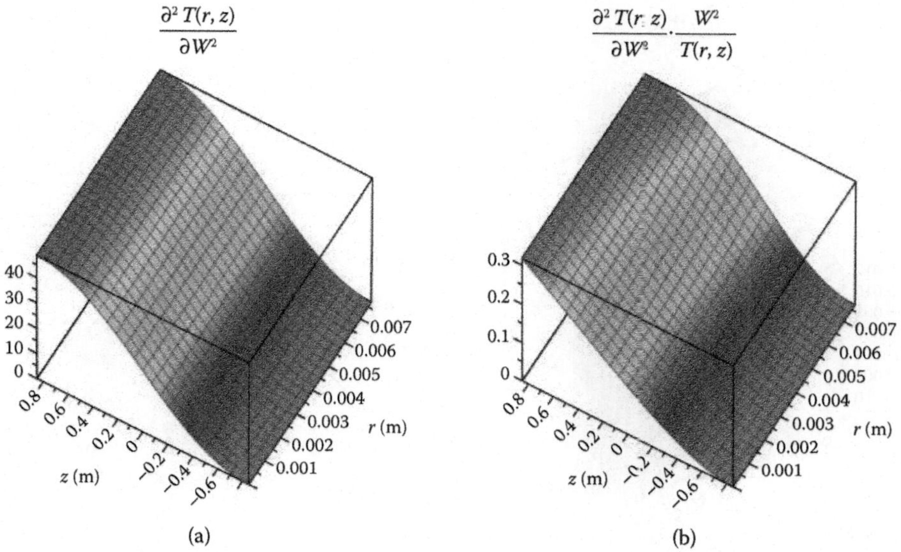

FIGURE 4.29
Absolute (a) and relative (b) second-order sensitivities of the heated rod temperature distribution with respect to the mass flow rate.

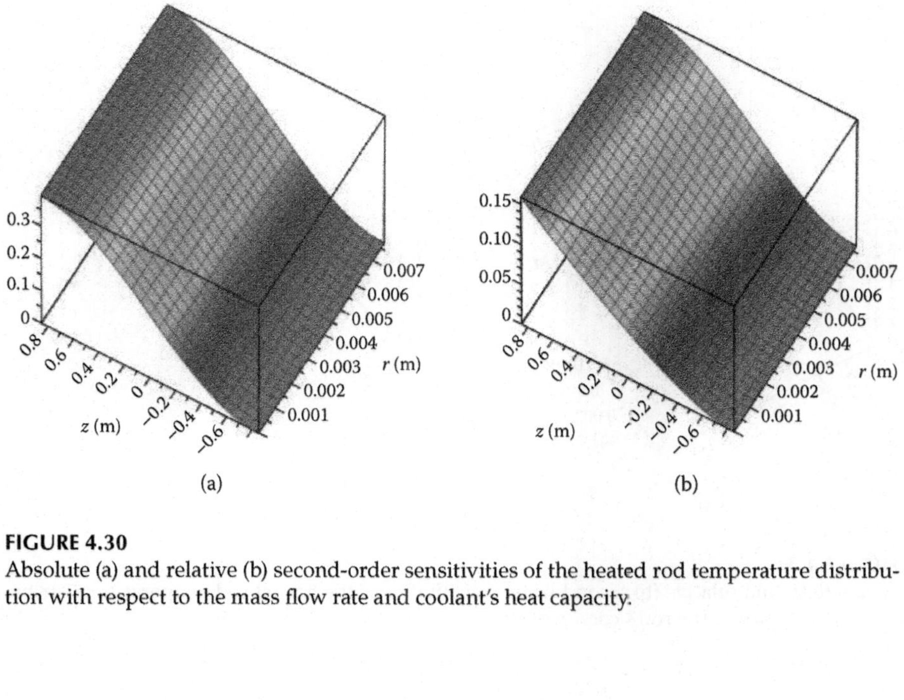

FIGURE 4.30
Absolute (a) and relative (b) second-order sensitivities of the heated rod temperature distribution with respect to the mass flow rate and coolant's heat capacity.

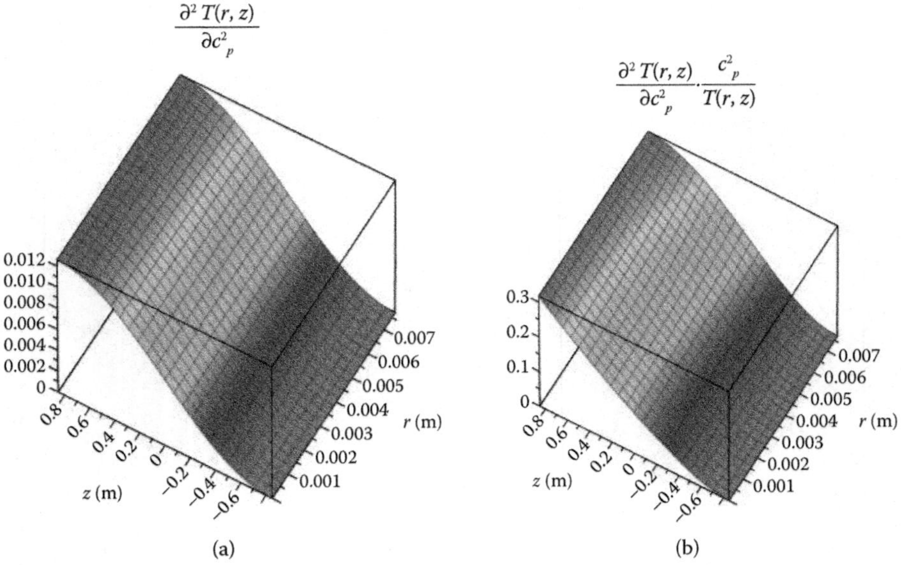

FIGURE 4.31
Absolute (a) and relative (b) second-order sensitivities of the heated rod temperature distribution with respect to the coolant's heat capacity.

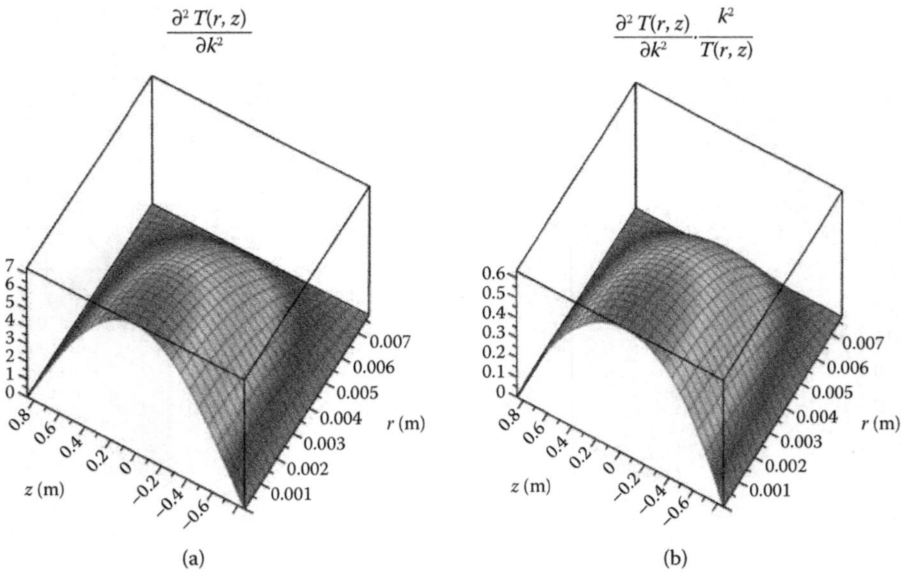

FIGURE 4.32
Absolute (a) and relative (b) second-order sensitivities of the heated rod temperature distribution with respect to the rod's conductivity.

The relative sensitivities plotted on the right sides in Figures 4.24 through 4.32 enable the relative ranking of the respective second-order sensitivities in influencing the response $T(r_0, z_0)$. It is clear from these figures that the influence of these sensitivities varies significantly according to the actual location (r_0, z_0). The largest value, of ca. 0.6, is attained by the second-order sensitivities of the heated rod temperature distribution with respect to the rod's conductivity, as seen in Figure 4.32. Although the largest first-order sensitivity of $T(r_0, z_0)$ was with respect to the inlet temperature, T_{inlet}, as shown in Table 4.2, the second-order sensitivity of $T(r_0, z_0)$ with respect to T_{inlet} is identically zero.

4.3.1.8 Quantitative Contributions of Second-Order Sensitivities to the Uncertainty in the Rod Temperature Distribution for G4M Reactor Model Parameters

To quantify the effects of the first- and second-order response sensitivities for the benchmark problem analyzed in this chapter, it suffices to consider response correlations up to third-order (skewness), for the simple case when (i) the parameters are uncorrelated and normally distributed, with known means $\alpha^0 \equiv \left(q^0, k^0, h^0, W^0, c_p^0, T_{inlet}^0\right)$ and known standard deviations $\boldsymbol{\sigma} \equiv [\sigma(q), \sigma(k), \sigma(h), \sigma(W), \sigma(c_p), \sigma(T_{inlet})]$ and (ii) only the first- and second-order response sensitivities are available.

Replacing the above sensitivities into Equations 1.18 through 1.20 yields the following results:

1. Contributions of the second-order sensitivities to the *second-order effects*, denoted as $E_{SO}[T(r_0, z_0)]$, in the expected value of $T(r_0, z_0)$:

$$E_{SO}\left[T(r_0,z_0)\right]= \frac{1}{2}\left[\frac{\partial^2 T(r_0,z_0)}{\partial h^2}\sigma_h^2+\frac{\partial^2 T(r_0,z_0)}{\partial W^2}\sigma_W^2+\frac{\partial^2 T(r_0,z_0)}{\partial c_p^2}\sigma_{c_p}^2+\frac{\partial^2 T(r_0,z_0)}{\partial k^2}\sigma_k^2\right]. \quad (4.148)$$

2. Contributions of the second-order sensitivities to the *second-order effects*, denoted as $\mathrm{var}_{SO}[T(r_0, z_0)]$, of the variance of the distribution of $T(r_0, z_0)$:

$$\mathrm{var}_{SO}\left[T(r_0,z_0)\right]=\frac{1}{2}\left(\frac{\partial^2 T(r_0,z_0)}{\partial h^2}\right)^2\sigma_h^4+\frac{1}{2}\left(\frac{\partial^2 T(r_0,z_0)}{\partial W^2}\right)^2\sigma_W^4 +\frac{1}{2}\left(\frac{\partial^2 T(r_0,z_0)}{\partial c_p^2}\right)^2\sigma_{c_p}^4+\frac{1}{2}\left(\frac{\partial^2 T(r_0,z_0)}{\partial k^2}\right)^2\sigma_k^4. \quad (4.149)$$

3. Contributions of the second-order sensitivities to the *third-order moment*, denoted as $\mu_3[T(r_0, z_0)]$, of the distribution of $T(r_0, z_0)$:

$$\mu_3\left[\mathrm{T}(r_0,z_0)\right]=3\left(\frac{\partial T(r_0,z_0)}{\partial h}\right)^2\frac{\partial^2 T(r_0,z_0)}{\partial h^2}\sigma_h^4$$

$$+3\left(\frac{\partial T(r_0,z_0)}{\partial W}\right)^2\frac{\partial^2 T(r_0,z_0)}{\partial W^2}\sigma_W^4+3\left(\frac{\partial T(r_0,z_0)}{\partial c_p}\right)^2\frac{\partial^2 T(r_0,z_0)}{\partial c_p^2}\sigma_{c_p}^4 \quad (4.150)$$

$$+3\left(\frac{\partial T(r_0,z_0)}{\partial k}\right)^2\frac{\partial^2 T(r_0,z_0)}{\partial k^2}\sigma_k^4.$$

4. Contributions of the second-order sensitivities to the *skewness* of the distribution of $T(r_0, z_0)$, defined as

$$\gamma_1\left[T(r_0,z_0)\right]=\frac{\mu_3\left[\mathrm{T}(r_0,z_0)\right]}{\left[var_{FO}T(r_0,z_0)+var_{SO}T(r_0,z_0)\right]^{3/2}} \quad .(4.151)$$

The absolute second-order sensitivities, plotted on the left sides of Figures 4.24 through 4.32, were used in Equation 4.148 together with the parameter standard deviations given in Table 4.1 to compute the respective second-order contributions, $E_{SO}[T(r_0, z_0)]$, to the expected value of the rod temperature distribution $T(r_0, z_0)$. As can be seen from Figure 4.33, the second-order contributions, $E_{SO}[T(r_0, z_0)]$, are less than 1% of the corresponding nominally computed values of $T(r_0, z_0)$.

The absolute second-order sensitivities, plotted on the left sides of Figures 4.24 through 4.32, were also used in Equation 4.149 together with the parameter standard deviations given in Table 4.1 to compute the respective second-order contributions, $\mathrm{var}_{SO}[T(r_0, z_0)]$, to the variance of the rod temperature distribution $T(r_0, z_0)$. As can be seen from Figure 4.34, the second-order contributions, $\mathrm{var}_{SO}[T(r_0, z_0)]$, have values that are about 1% of the corresponding values of the first-order contributions, $\mathrm{var}_{FO}[T(r_0, z_0)]$, thereby adding relatively little to the total standard deviation (uncertainty), $Std.Dev.[T(r_0, z_0)] = \{\mathrm{var}_{FO}[T(r_0, z_0)] + \mathrm{var}_{SO}[T(r_0, z_0)]\}^{1/2}$, of the rod temperature distribution $T(r_0, z_0)$.

The absolute second-order sensitivities, plotted on the left sides of Figures 4.24 through 4.32, were also used in Equation 4.150 and 4.151, together with the parameter standard deviations given in Table 4.1, to compute the third-order moment, $\mu_3[T(r_0, z_0)]$, and skewness, $\gamma_1[T(r_0, z_0)]$, respectively, of the rod temperature distribution $T(r_0, z_0)$. Both $\mu_3[T(r_0, z_0)]$ and $\gamma_1[T(r_0, z_0)]$ are plotted as functions of (r_0, z_0) in Figure 4.35. This figure indicates that $\gamma_1[T(r_0, z_0)] > 0$, everywhere; consequently, the distribution of $T(r_0, z_0)$ is *not Gaussian* but is asymmetrically skewed towards the right of its expected value, indicating that temperatures higher than the mean temperature are more likely to occur than

Fuel rod temperature $T^0(r, z)$(K)

$$E_{SO}(T) = \frac{1}{2}\cdot\left[\frac{\partial^2 T(r,z)}{\partial h^2}\cdot\sigma_h^2 + \frac{\partial^2 T(r,z)}{\partial W^2}\cdot\sigma_W^2 + \frac{\partial^2 T(r,z)}{\partial C_p^2}\cdot\sigma^2_{Cp} + \frac{\partial^2 T(r,z)}{\partial k^2}\cdot\sigma_k^2\right]$$

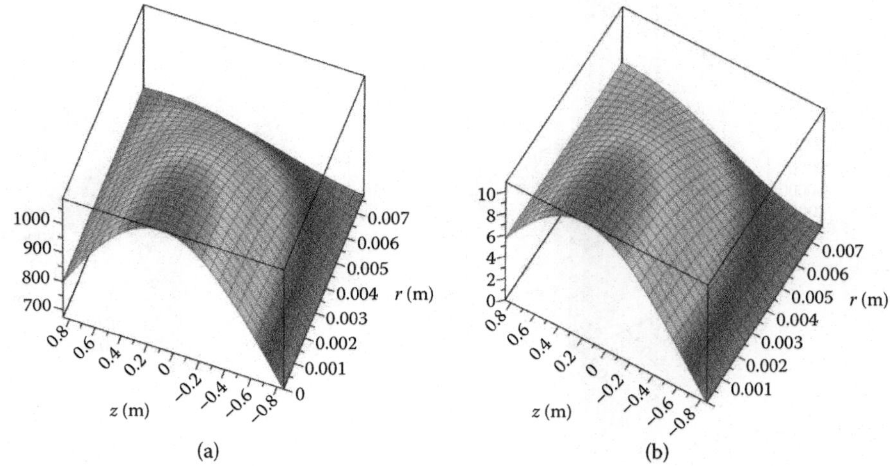

FIGURE 4.33
Heated rod nominal temperature distribution (a) and second-order contributions to its spatially dependent expected value (b).

$$var_{FO}(T) = \left[\frac{\partial T(r,z)}{\partial q}\right]^2\cdot\sigma_q^2 + \left[\frac{\partial T(r,z)}{\partial h}\right]^2\cdot\sigma_h^2 + \left[\frac{\partial T(r,z)}{\partial W}\right]^2\cdot\sigma_W^2 + \left[\frac{\partial T(r,z)}{\partial C_p}\right]\cdot\sigma^2_{Cp} + \left[\frac{\partial T(r,z)}{\partial k}\right]\cdot\sigma_k^2 + \left[\frac{\partial T(r,z)}{\partial k_{Tinlet}}\right]\cdot\sigma^2_{Tinlet}$$

$$var_{SO}(T) = \frac{1}{2}\cdot\left(\left[\frac{\partial^2 T(r,z)}{\partial h^2}\right]^2\cdot\sigma_h^2 + \left[\frac{\partial^2 T(r,z)}{\partial W^2}\right]^2\cdot\sigma_W^2 + \left[\frac{\partial^2 T(r,z)}{\partial C_p^2}\right]^2\cdot\sigma^4_{C_p} + \left[\frac{\partial^2 T(r,z)}{\partial k^2}\right]^2\cdot\sigma_k^2\right]$$

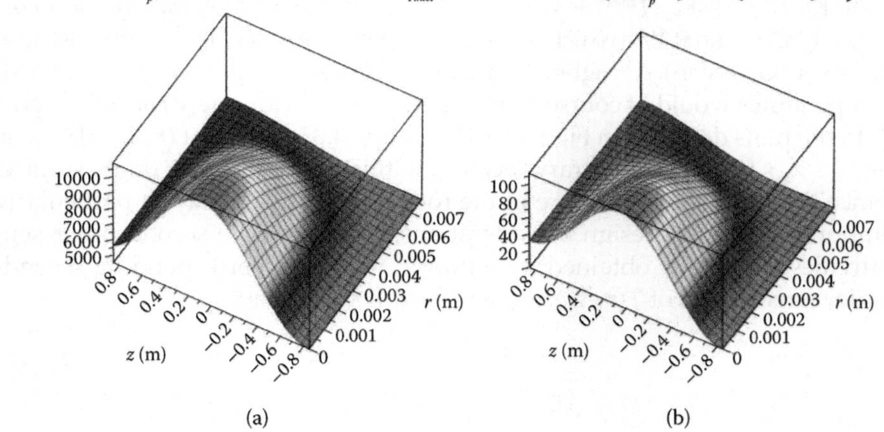

FIGURE 4.34
Contributions of first-order sensitivities (a) and second-order sensitivities (b) to the spatially dependent variance of the heated rod temperature distribution.

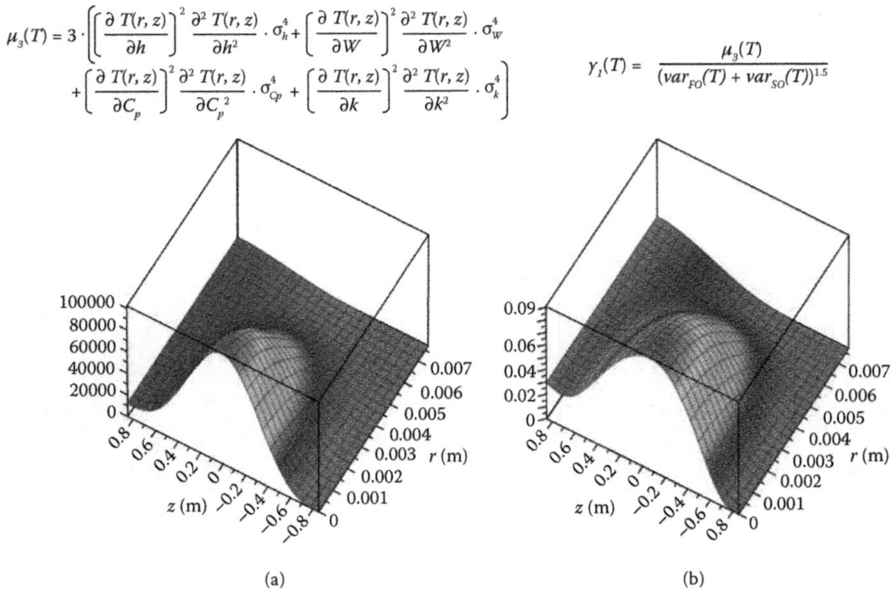

FIGURE 4.35
Third-order moment (a) and skewness (b) of the heated rod temperature distribution.

temperatures lower than the mean temperature. Thus, *the conventional metrics that are valid for Gaussian distributions* (namely, that about 66% of the distribution is contained within one standard deviation, about 90% is contained within two standard deviations, and about 99.9% is contained within three standard deviations) *are not applicable to the temperature distribution* $T(r_0, z_0)$. In particular, because of the distributions' positive skewness, temperatures higher than 1410 K [$= T^{max} + E_{SO}(T^{max}) + 3 \times dev_{FO}(T^{max})$] would have a probability of occurrence larger than 0.1% (which would have been the probability if the distribution were Gaussian). A "higher than Gaussian" probability of occurrence of high temperatures would of course be undesirable for reactor safety consideration.

In the plots depicted in Figures 4.24 through 4.35, the point (r_0, z_0), $0 \le r_0 \le a$, $-\ell/2 \le z_0 \le \ell/2$, is an arbitrary location within the fuel rod or on its surface. Since the surface temperature of the rod, $T_s(z) = T(a, z)$, plays a particularly important role for design considerations, the respective second-order sensitivities, which are obtained by setting $r_0 = a$ in the corresponding second-order sensitivities of $T(r_0, z_0)$, are given explicitly below:

$$\frac{\partial^2 T_s(z)}{\partial q \partial h} = -\frac{a}{2(h^0)^2} \cos\frac{\pi z}{\ell}, \tag{4.152}$$

$$\frac{\partial^2 T_s(z)}{\partial q \partial W} = -\frac{a^2 \ell}{(W^0)^2 c_p^0} \left(\sin\frac{\pi z}{\ell} + 1 \right), \tag{4.153}$$

$$\frac{\partial^2 T_s(z)}{\partial q \, \partial c_p} = -\frac{a^2 \ell}{W^0 \left(c_p^0\right)^2} \left(\sin\frac{\pi z}{\ell} + 1 \right), \tag{4.154}$$

$$\frac{\partial^2 T_s(z)}{\partial h^2} = \frac{q^0 a}{\left(h^0\right)^3} \cos\frac{\pi z}{\ell}, \tag{4.155}$$

$$\frac{\partial^2 T_s(z)}{\partial W^2} = \frac{2 q^0 a^2 \ell}{\left(W^0\right)^3 c_p^0} \left(\sin\frac{\pi z}{\ell} + 1 \right), \tag{4.156}$$

$$\frac{\partial^2 T_s(z)}{\partial W \, \partial c_p} = \frac{q^0 a^2 \ell}{\left(W^0\right)^2 \left(c_p^0\right)^2} \left(\sin\frac{\pi z}{\ell} + 1 \right), \tag{4.157}$$

$$\frac{\partial^2 T_s(z)}{\partial c_p^2} = \frac{2 q^0 a^2 \ell}{W^0 \left(c_p^0\right)^3} \left(\sin\frac{\pi z}{\ell} + 1 \right). \tag{4.158}$$

Although the second-order sensitivities of the surface rod temperature $T_s(z_1)$ are actually included in Figures 4.24 through 4.32, they are plotted separately in Figures 4.36 through 4.42 as functions of z_1, $-\ell/2 \le z_1 \le \ell/2$, for enhanced clarity. For notational simplicity, the units of the absolute second-order sensitivities and the subscript 1 for the z_1-coordinate are omitted in the figure captions.

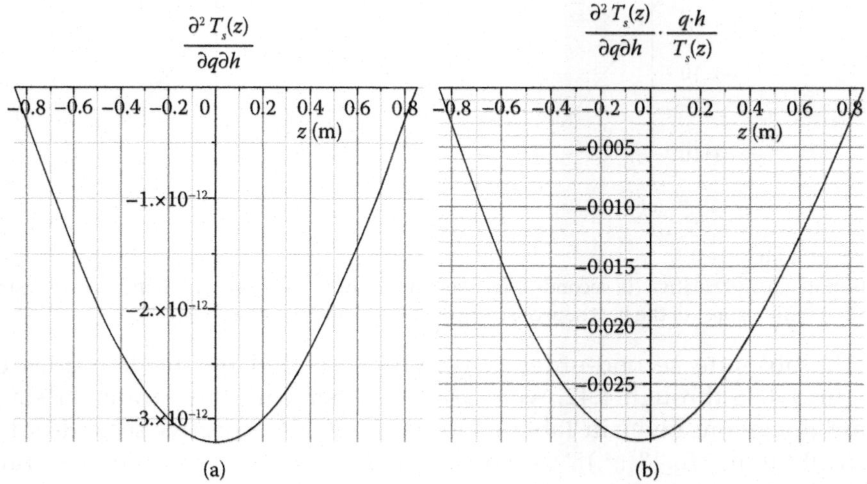

FIGURE 4.36
Absolute (a) and relative (b) second-order sensitivities of the rod's surface temperature distribution with respect to the heat source strength and heat transfer coefficient.

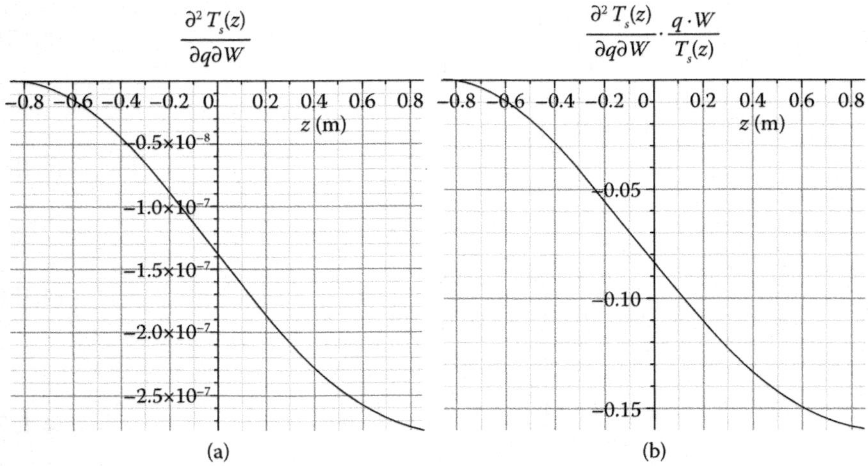

FIGURE 4.37
Absolute (a) and relative (b) second-order sensitivities of the rod's surface temperature distribution with respect to the heat source strength and mass flow rate.

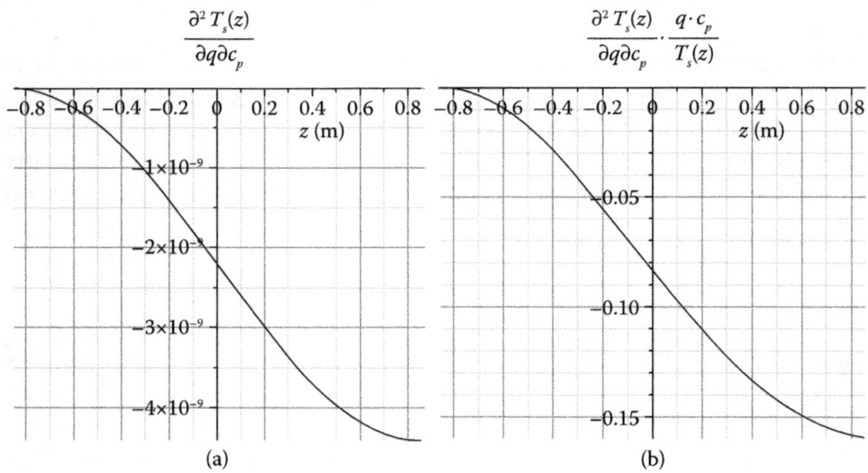

FIGURE 4.38
Absolute (a) and relative (b) second-order sensitivities of the rod's surface temperature distribution with respect to the heat source strength and coolant's heat capacity.

The absolute second-order sensitivities, plotted on the left sides of Figures 4.36 through 4.42, were used together with the parameter standard deviations given in Table 4.1 to compute the respective second-order contributions, $E_{SO}[T_s(z_1)]$, to the expected value of the rod temperature distribution $T_s(z_1)$. The expression of $E_{SO}[T_s(z_1)]$ is

$$\mathrm{E}_{SO}\left[\mathrm{T}_s(z_1)\right]=\frac{1}{2}\left[\frac{\partial^2 T_s(z_1)}{\partial h^2}\sigma_h^2+\frac{\partial^2 T_s(z_1)}{\partial W^2}\sigma_W^2+\frac{\partial^2 T_s(z_1)}{\partial c_p^2}\sigma_{c_p}^2\right], \quad (4.159)$$

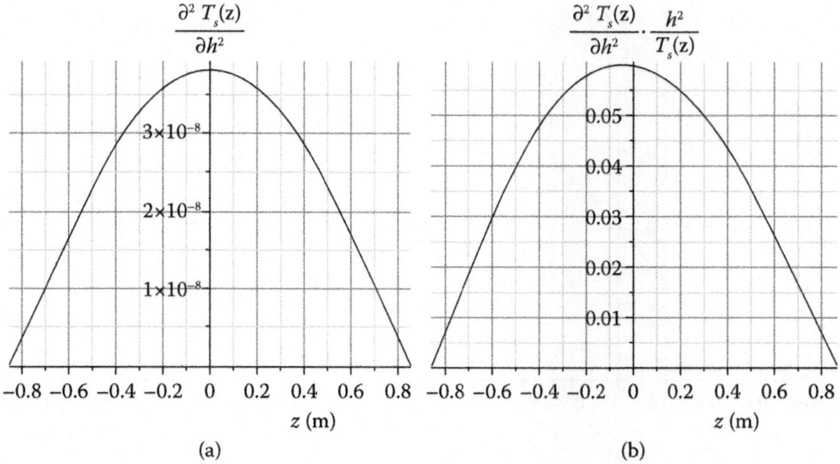

FIGURE 4.39
Absolute (a) and relative (b) second-order sensitivities of the rod's surface temperature distribution with respect to the coolant's heat transfer coefficient.

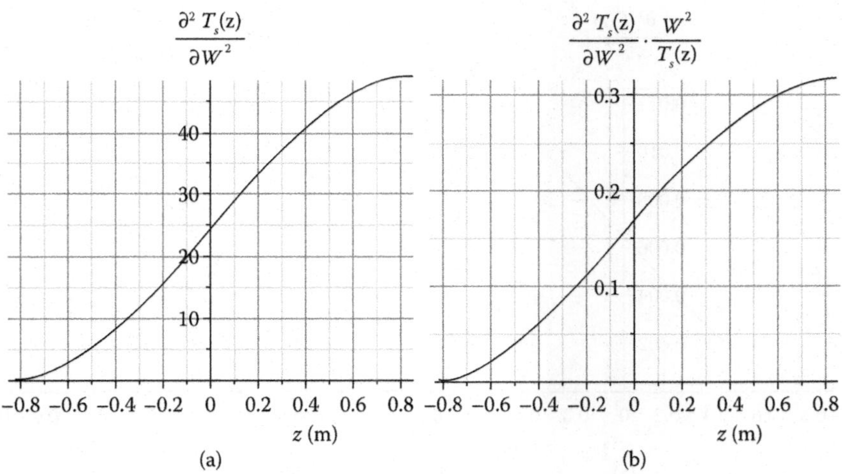

FIGURE 4.40
Absolute (a) and relative (b) second-order sensitivities of the rod's surface temperature distribution with respect to the mass flow rate.

and is plotted on the Figure 4.43 (b), along with the spatial behavior of the surface temperature $T_s(z_1)$. This figure shows that the values of $E_{SO}[T_s(z_1)]$ are less than 1% of the corresponding nominally computed values of $T_s(z_1)$.

The second-order sensitivities plotted on the left sides of Figures 4.36 through 4.42 were also used to compute the respective second-order

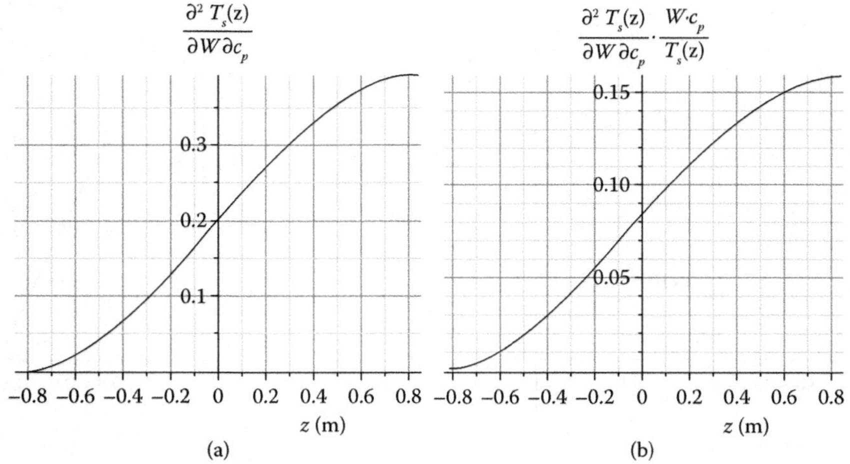

FIGURE 4.41
Absolute (a) and relative (b) second-order sensitivities of the rod's surface temperature distribution with respect to the mass flow rate and coolant's heat capacity.

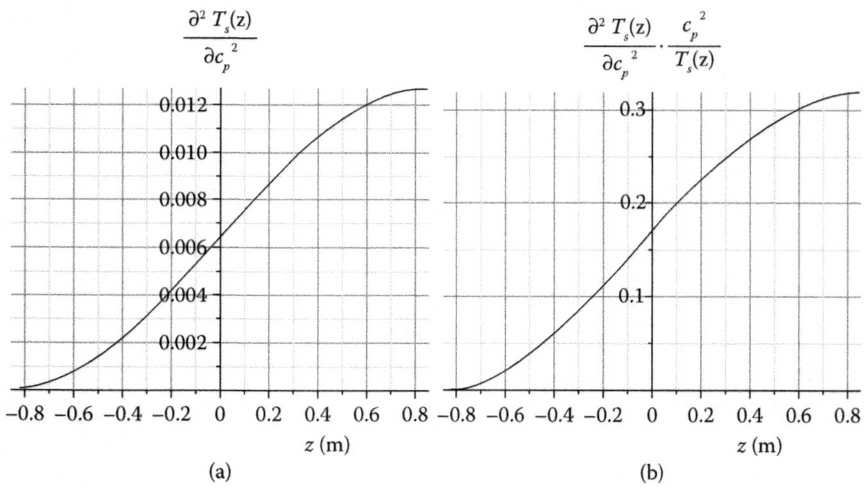

FIGURE 4.42
Absolute (a) and relative (b) second-order sensitivities of the rod's surface temperature distribution with respect to the coolant's heat capacity.

contributions, $\mathrm{var}_{SO}[T_s(z_1)]$, to the variance of the rod's surface temperature distribution $T_s(z_1)$, which takes on the following particular form:

$$\mathrm{var}_{SO}\left[T_s(z_1)\right]=\frac{1}{2}\left[\left(\frac{\partial^2 T_s(z_1)}{\partial h^2}\right)^2\sigma_h^4+\left(\frac{\partial^2 T_s(z_1)}{\partial W^2}\right)^2\sigma_W^4+\left(\frac{\partial^2 T_s(z_1)}{\partial c_p^2}\right)^2\sigma_{c_p}^4\right]. \quad (4.160)$$

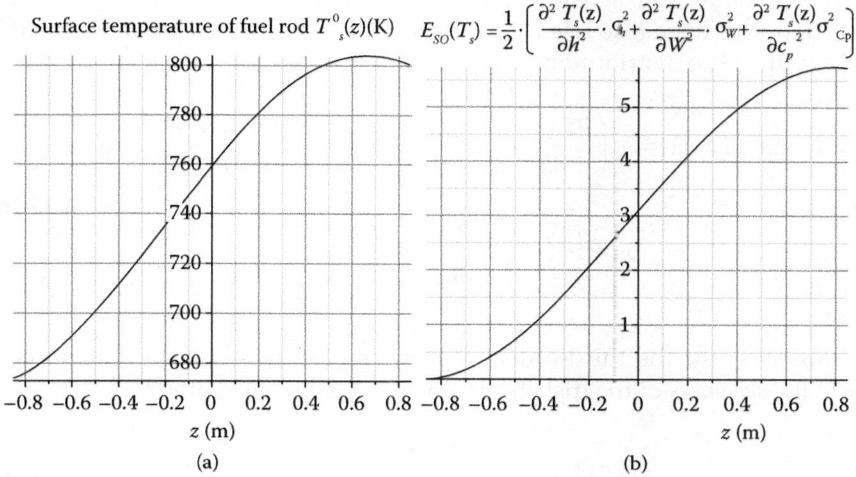

FIGURE 4.43
Rod's surface nominal temperature distribution (a). Second-order contributions to the spatially dependent expected value of the rod's surface temperature distribution (b).

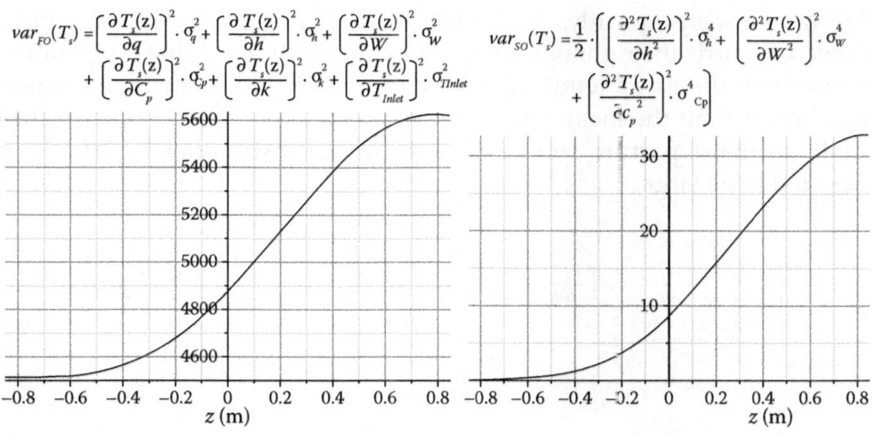

FIGURE 4.44
Contributions of first-order sensitivities (a) and second-order sensitivities (b) to the spatially dependent variance of the rod's surface temperature distribution.

The second-order contributions shown in Equation 4.160 are plotted in Figure 4.44, along with the corresponding values of the first-order contributions, $\mathrm{var}_{FO}[T_s(z_1)]$. Figure 4.44 indicates that the second-order contributions have values that are less than 0.6% of the corresponding values of the first-order contributions, $\mathrm{var}_{FO}[T_s(z_1)]$, thereby adding relatively little to the total standard deviation (uncertainty), $Std.Dev.[T_s(z_1)] = \{\mathrm{var}_{FO}[T_s(z_1)] + \mathrm{var}_{SO}[T_s(z_1)]\}^{1/2}$, of the rod's surface temperature distribution $T_s(z_1)$.

The second-order sensitivities plotted on the left sides of Figures 4.36 through 4.42 were also used to compute the *third-order moment*, denoted as $\mu_3[T_s(z_1)]$, of the distribution of $T_s(z_1)$, namely

$$\mu_3\left[T_s(z_1)\right]=3\left[\left(\frac{\partial T_s(z_1)}{\partial h}\right)^2\frac{\partial^2 T_s(z_1)}{\partial h^2}\sigma_h^4+\left(\frac{\partial T_s(z_1)}{\partial W}\right)^2\frac{\partial^2 T_s(z_1)}{\partial W^2}\sigma_W^4\right.$$
$$\left.+\left(\frac{\partial T_s(z_1)}{\partial c_p}\right)^2\frac{\partial^2 T_s(z_1)}{\partial c_p^2}\sigma_{c_p}^4\right] \quad (4.161)$$

Subsequently, the third-order moment, $\mu_3[T_s(z_1)]$, has been used to compute the *skewness* of the distribution of $T_s(z_1)$, which is defined as

$$\gamma_1\left[T_s(z_1)\right]=\frac{\mu_3\left[T_s(z_1)\right]}{\left[var_{FO}T_s(z_1)+var_{SO}T_s(z_1)\right]^{3/2}}. \quad (4.162)$$

Both the third-order moment, $\mu_3[T_s(z)]$, and the skewness, $\gamma_1[T_s(z_1)]$, are plotted as functions of z in Figure 4.45. This figure indicates that $\gamma_1[T_s(z_1)] > 0$, throughout; consequently, the distribution of $T_s(z)$ is *not Gaussian* but is skewed towards the right of its expected value, indicating that temperatures higher than the mean temperature are more likely to occur than temperatures lower than the mean. However, the skewness $\gamma_1[T_s(z)] > 0$ is relatively small, so the departure from normality of the distribution of $T_s(z)$ is not expected to be large.

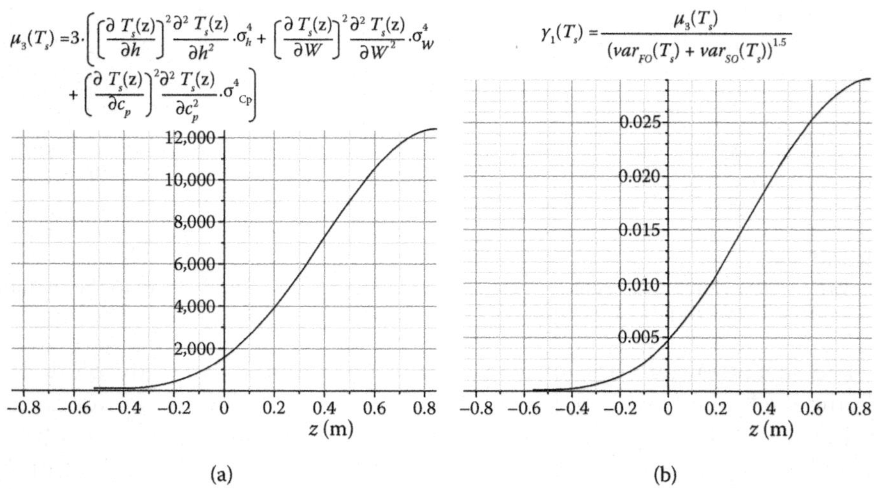

FIGURE 4.45
Third-order moment (a) and skewness (b) of the rod's surface temperature distribution.

4.3.2 Computation of Second-Order Sensitivities of the Coolant Temperature, $T_{fl}(z)$

The solution of the 1st-LASS used for computing the first-order sensitivities of coolant temperature response, $T_{fl}(z_2)$, at some axial point $z = z_2$, was provided in Equation 4.70, which is reproduced, for convenience, below:

$$\Psi^0(r,z) \equiv 0,\ \Psi_{fl}^0(z) = 2a^{-2}\left[1 - H(z - z_2)\right] = 2a^{-2}H(z_2 - z),\ -\frac{\ell}{2} \le z, z_2 \le \frac{\ell}{2}. \quad (4.70)$$

Replacing $\Psi^0(r, z) \equiv 0$ in the general expressions of the first-order sensitivities given in Equations 4.29 through 4.34 yields the following expressions:

$$\left(\frac{\partial T_{fl}(z_2)}{\partial q}\right)_{(\mathbf{u}^0,\alpha^0,\psi^0)} = \frac{\pi a^4}{2W^0 c_p^0} \int_{-\ell/2}^{\ell/2} \Psi_{fl}^0(z) \cos\frac{\pi z}{\ell}, \quad (4.163)$$

$$\left(\frac{\partial T_{fl}(z_2)}{\partial k}\right)_{(\mathbf{u}^0,\alpha^0,\psi^0)} \equiv 0, \quad (4.164)$$

$$\left(\frac{\partial T_{fl}(z_2)}{\partial h}\right)_{(\mathbf{u}^0,\alpha^0,\psi^0)} \equiv 0, \quad (4.165)$$

$$\left(\frac{\partial T_{fl}(z_2)}{\partial W}\right)_{(\mathbf{u}^0,\alpha^0,\psi^0)} = -\frac{q^0 \pi a^4}{2\left(W^0\right)^2 c_p^0} \int_{-\ell/2}^{\ell/2} \Psi_{fl}^0(z) \cos\frac{\pi z}{\ell}, \quad (4.166)$$

$$\left(\frac{\partial T_{fl}(z_2)}{\partial c_p}\right)_{(\mathbf{u}^0,\alpha^0,\psi^0)} = -\frac{q^0 \pi a^4}{2W^0 \left(c_p^0\right)^2} \int_{-\ell/2}^{\ell/2} \Psi_{fl}^0(z) \cos\frac{\pi z}{\ell}, \quad (4.167)$$

$$\left(\frac{\partial T_{fl}(z_2)}{\partial T_{inlet}}\right)_{(\mathbf{u}^0,\alpha^0,\psi^0)} = 1. \quad (4.168)$$

Note that the expression of $\Psi_{fl}^0(z)$ in Equation 4.70 is the same as in Equation 4.36. Therefore, Equation 4.166 is the same as Equation 4.90, Equation 4.167 is the same as Equation 4.91, and Equation 4.168 is clearly the same as Equation 4.92. Hence, the following multiple equalities hold for the following nonzero second-order sensitivities:

$$\frac{\partial^2 T_{fl}(z)}{\partial W \partial q} = \frac{\partial^2 T_s(z)}{\partial W \partial q} = \frac{\partial^2 T(r_0, z)}{\partial W \partial q} = -\frac{a^2 \ell}{\left(W^0\right)^2 c_p^0}\left(\sin\frac{\pi z}{\ell} + 1\right), -\frac{\ell}{2} \le z \le \frac{\ell}{2} \quad (4.169)$$

$$\frac{\partial^2 T_{fl}(z)}{\partial W^2}=\frac{\partial^2 T_s(z)}{\partial W^2}=\frac{\partial^2 T(r_0,z_0)}{\partial W^2}=\frac{2q^0a^2\ell}{\left(W^0\right)^3 c_p^0}\left(\sin\frac{\pi z}{\ell}+1\right),-\frac{\ell}{2}\le z\le\frac{\ell}{2}, \quad (4.170)$$

$$\frac{\partial^2 T_{fl}(z)}{\partial W\partial c_p}=\frac{\partial^2 T_s(z)}{\partial W\partial c_p}=\frac{\partial^2 T(r_0,z)}{\partial W\partial c_p}=\frac{q^0a^2\ell}{\left(W^0\right)^2\left(c_p^0\right)^2}\left(\sin\frac{\pi z}{\ell}+1\right),-\frac{\ell}{2}\le z\le\frac{\ell}{2}, \quad (4.171)$$

$$\frac{\partial^2 T_{fl}(z)}{\partial c_p\partial q}=\frac{\partial^2 T_s(z)}{\partial c_p\partial q}=\frac{\partial^2 T(r_0,z)}{\partial c_p\partial q}=-\frac{a^2\ell}{W^0\left(c_p^0\right)^2}\left(\sin\frac{\pi z}{\ell}+1\right).-\frac{\ell}{2}\le z\le\frac{\ell}{2}, \quad (4.172)$$

$$\frac{\partial^2 T_{fl}(z)}{\partial c_p^2}=\frac{\partial^2 T_s(z)}{\partial c_p^2}=\frac{\partial^2 T(r_0,z)}{\partial c_p^2}=\frac{2q^0a^2\ell}{W^0\left(c_p^0\right)^3}\left(\sin\frac{\pi z}{\ell}+1\right),-\frac{\ell}{2}\le z\le\frac{\ell}{2}. \quad (4.173)$$

Examining Equation 4.163 reveals that it has the same form as Equations 4.166 and 4.167, except for the factor multiplying the respective integral over $\Psi_{fl}^0(z)$, which does not contain the parameter q. Hence, it follows that the following consideration will hold for computing the G-differential of Equation 4.163: (i) the indirect-effect term will be identically zero, since $\Psi_{fl}^0(z)$ is independent of the model parameters; (ii) the only nonzero expressions in the direct-effect term will be the partial derivatives with respect to W and c_p, but these have already been computed in Equations 4.169 and 4.172. Hence, all of the nonzero second-order sensitivities of $T_{fl}(z_2)$(are obtained without needing to solve any additional differential equations, by simply examining the structures of the respective responses in terms of the corresponding 1st-LASS and 2nd-LASS.

As Equations 4.169 through 4.173 indicate, the nonzero second-order sensitivities of $T_{fl}(z)$ are identical to the corresponding ones for $T_s(z)$. However, the statistical characteristics of the distribution for $T_{fl}(z)$ will differ from the characteristics of the distribution for $T_s(z)$ because $\partial^2 T_{fl}(z)/\partial h^2\equiv 0$, but $\partial^2 T_s(z)/\partial h^2\neq 0$. The second-order statistics for $T_{fl}(z)$ are as follows:

1. The expression, $E_{SO}[T_{fl}(z_2)]$, comprising the contributions of the *second-order* effects in the expected value of $T_{fl}(z_2)$, is obtained as

$$E_{SO}\left[T_{fl}(z_2)\right]=\frac{1}{2}\left[\frac{\partial^2 T_{fl}(z_2)}{\partial W^2}\sigma_W^2+\frac{\partial^2 T_{fl}(z_2)}{\partial c_p^2}\sigma_{c_p}^2\right]. \quad (4.174)$$

2. The expression, $\mathrm{var}_{SO}[T_{fl}(z_2)]$, comprising the contributions of the *second-order effects* to the variance of the distribution of $T_{fl}(z_2)$ is obtained as

$$\mathrm{var}_{SO}\left[T_{fl}(z_2)\right]=\frac{1}{2}\left[\left(\frac{\partial^2 T_{fl}(z_2)}{\partial W^2}\right)^2\sigma_W^4+\left(\frac{\partial^2 T_{fl}(z_2)}{\partial c_p^2}\right)^2\sigma_{c_p}^4\right]. \tag{4.175}$$

3. The *third-order moment*, $\mu_3[T_{fl}(z_2)]$, of the distribution of $T_{fl}(z_2)$, is obtained as

$$\mu_3\left[T_{fl}(z_2)\right]=3\left[\left(\frac{\partial T_{fl}(z_2)}{\partial W}\right)^2\frac{\partial^2 T_{fl}(z_2)}{\partial W^2}\sigma_W^4+\left(\frac{\partial T_{fl}(z_2)}{\partial c_p}\right)^2\frac{\partial^2 T_{fl}(z_2)}{\partial c_p^2}\sigma_{c_p}^4\right]. \tag{4.176}$$

4. The *skewness* of the distribution of $T_{fl}(z_2)$ is obtained from its definition as

$$\gamma_1\left[T_{fl}(z_2)\right]=\frac{\mu_3\left[T_{fl}(z_2)\right]}{\left[var_{FO}T_{fl}(z_2)+var_{SO}T_{fl}(z_2)\right]^{3/2}}. \tag{4.177}$$

Figure 4.46 depicts the spatial variation of the second-order contributions, $E_{SO}[T_{fl}(z)]$, to the expected value of $T_{fl}(z_2)$, along with the spatial variation of the nominal value of $T_{fl}(z)$, showing that the contributions stemming from $E_{SO}[T_{fl}(z)]$ are less than 1% of the corresponding nominally computed values

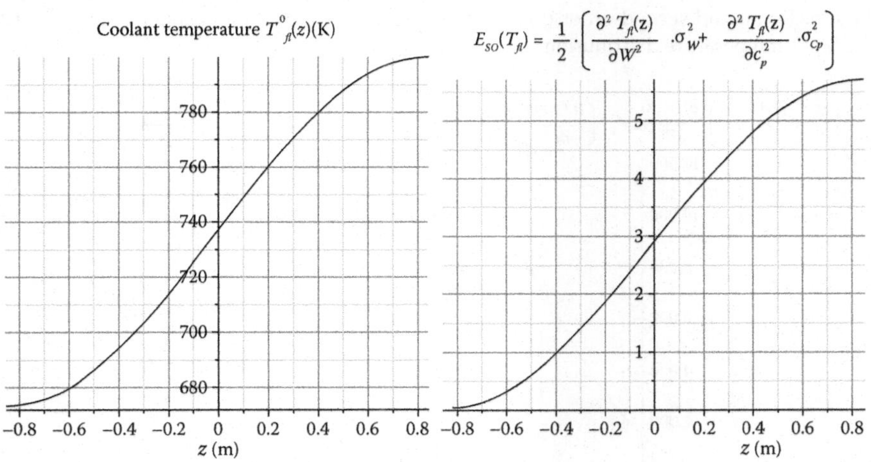

(a) (b)

FIGURE 4.46
Coolant's nominal temperature distribution (a) and 2nd-order contributions (b) to its spatially dependent expected value.

of $T_{fl}(z)$. Analogous to the feature of the rod's surface temperature, Figure 4.47 shows that the second-order contributions, $\text{var}_{SO}[T_{fl}(z)]$, to the variance of the coolant temperature distribution, $T_{fl}(z)$, are less than 0.6% of the corresponding values of the first-order contributions, $\text{var}_{FO}[T_{fl}(z)]$, thereby adding relatively little to the total standard deviation (uncertainty), $Std.Dev.[T_{fl}(z)] = \{\text{var}_{FO}[T_{fl}(z)] + \text{var}_{SO}[T_{fl}(z)]\}^{1/2}$, of the temperature distribution, $T_{fl}(z)$, within the coolant. Also analogous to the features of the rod's surface temperature,

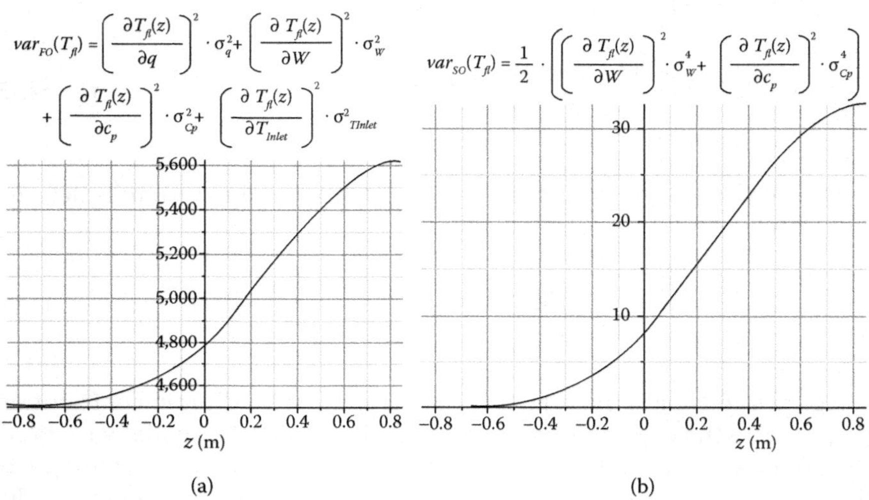

FIGURE 4.47
First-order (a) and second-order (b) contributions to the spatially dependent variance of the coolant's temperature distribution.

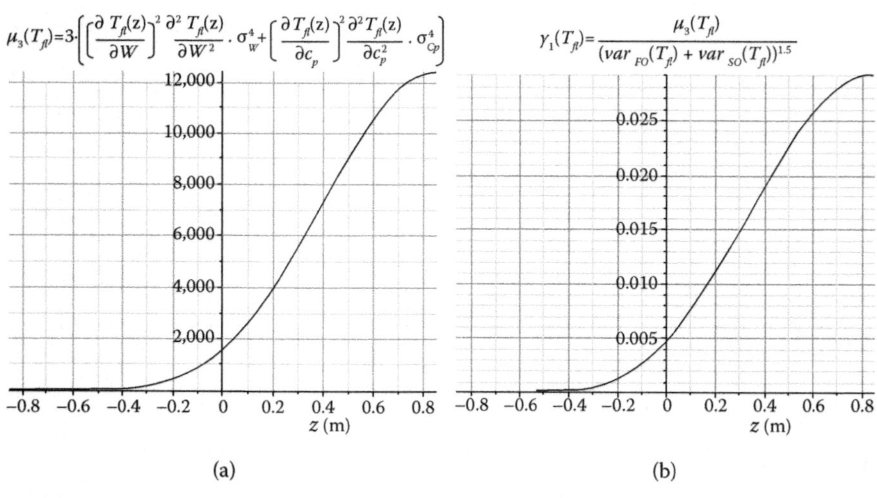

FIGURE 4.48
Third-order moment (a) and skewness (b) of the coolant's temperature distribution.

Figure 4.48 indicates that $\gamma_1[T_{fl}(z)] > 0$ everywhere, so that the distribution of $T_{fl}(z)$ is *not Gaussian* but is skewed towards temperatures higher than the mean temperature. However, this skewness is relatively small, so the departure from normality of the distribution of $T_{fl}(z)$ is not expected to be large.

4.4 Concluding Remarks

This chapter has presented a heat transport benchmark problem modeling the steady-state radial conduction in a heated rod coupled to the axial heat convection in a coolant surrounding the rod and flowing along it in a turbulent steady state. The development of this benchmark problem, which was presented in Section 4.2, was motivated by the need to verify the numerical results produced by the FLUENT Adjoint Solver (ANSYS, 2015), a code that was used by Cacuci et al. (2016a) to simulate thermal-hydraulics processes within the G4M reactor (Gen4Energy, Inc., 2012). The G4M reactor is envisaged as a small, modular, fast reactor cooled by LBE. The *2nd-ASAM* was applied to compute, in Section 4.2, the analytical expressions of the first-level adjoint state functions for this heat transport benchmark problem. In turn, these adjoint functions were used for computing the exact first-order sensitivities of the temperature distributions within the rod and coolant to the thermal-hydraulics parameters characteristic of both the G4M reactor and of an experimental facility ("test section") under conceptual design for studying thermal-hydraulics phenomena characteristic of this reactor. Section 4.3 presented the efficient derivation of the exact expressions of the benchmark's second-order sensitivities of the temperature distributions to the thermal-hydraulics parameters, by applying the general theory underlying the *2nd-ASAM for linear systems* as developed in Chapter 3. Second-order sensitivities are needed to quantify the impact of nonlinearities on the uncertainties in the temperature distributions within the heated rod and surrounding coolant. The paradigm benchmark problem has six thermal-hydraulics parameters. Therefore, for each functional type response, such as the temperature, $T(r_0, z_0)$, at a specific location (r_0, z_0) within the rod and/or on its surface, there are *six* first-order sensitivities and *twenty-one* second-order sensitivities. Furthermore, there are an additional *six* first-order sensitivities and additional *twenty-one* second-order sensitivities for the coolant temperature, $T_{fl}(z_2)$, at any location z_2. By applying the 2nd-ASAM, all of these first-order and second-order sensitivities were obtained using only *five independent adjoint computations*. For the heat transport benchmark problem analyzed in this chapter, the most important impact of the second-order sensitivities was to cause the temperature distributions within the rod, on the rod's surface, and in the coolant to become non-Gaussian, asymmetric, and skewed towards temperatures higher than the respective mean temperatures, as all three temperature distributions turned out to have positive skewness. For the temperature distributions on the rod's surface and within the coolant, the corresponding skewnesses are relatively small, so the departure from normality of these distributions is not large.

The derivations presented in Section 4.3 showed that the construction and solution of the 2nd-LASS require very little additional effort beyond the construction of the 1st-LASS needed for computing the first-order sensitivities. Very significantly, the computation of the second-level adjoint functions required modifications of only the sources for the solver used for computing the first-level adjoint functions; the solver of the respective differential equations remained unchanged. This very important property of the 2nd-ASAM opens up the possibility of using commercially available CFD codes, such as the FLUENT Adjoint Solver, to compute adjoint sensitivity functions, provided the user has access to modifying the "sources" for the respective "solver." This option is currently unavailable within the FLUENT Adjoint Solver. This fact underscores the importance of developing benchmarks, such as presented in this chapter, that enable the actual computation and quantification of first-order and second-order contributions of model sensitivities and uncertainties to the uncertainties in model responses.

5

Application of the 2nd-ASAM to a Linear Particle Diffusion Problem

Following the work by Cacuci (2015b), this chapter presents an illustrative application of the 2nd-ASAM developed in Chapter 3 to a paradigm particle diffusion problem that admits an exact solution. In this problem, all of the first-order relative response sensitivities to the model parameters will be shown to have significantly large values, of order unity. Also importantly, most of the second-order relative detector sensitivities to cross sections turn out to be actually larger than the corresponding first-order relative sensitivities, contrary to the tacit assumption that second-order sensitivities to cross sections are negligible in neutron diffusion problems, which was prevalent in sensitivity analysis works in the 1990s. This illustrative example will show that the second-order sensitivities have the following major impacts on the computed moments of the response distribution:

1. They cause the "expected" value of the response to differ from the "computed" nominal value of the response.
2. Their contributions to the response variances and covariances are relatively minor by comparison to the contributions stemming from the first-order response sensitivities.
3. They contribute decisively to causing asymmetries in the response distribution. Indeed, neglecting the second-order sensitivities would nullify the third-order response correlations and hence would nullify the *skewness* of the response. In the absence of second-order sensitivities, events occurring in a response's long and/or short tails, which are characteristic of rare but possibly decisive events, would likely be missed.

As in the applications of the 2nd-ASAM presented in the previous chapters, it will be shown in this chapter that the actual number of adjoint computations needed for computing all of the second-order response sensitivities is significantly less than the number of model parameters. Also as in previous chapters, the construction and solution of the 2nd-LASS requires very little additional effort beyond the construction of the adjoint sensitivity system needed for computing the first-order sensitivities. Only the *sources* on the right sides of the diffusion differential operator need to be modified, thus confirming the general results obtained in Chapter 3.

5.1 Problem Description

Consider the diffusion of neutrons in a pool of water containing spent-fuel rods discharged from a nuclear reactor. For simplicity, the pool is considered to have one dimension (say, the length) much larger than the other two, so that the pool could essentially be considered as a one-dimensional "slab" of water. The fuel elements are considered to be uniformly distributed within the pool and are considered to emit Q neutrons/cm^3. The neutrons are considered to be monoenergetic, all possessing the average thermal energy corresponding to the pool's water temperature. Mathematically, this simple physical system can be modeled by using the linear neutron diffusion equation

$$D\frac{d^2\varphi}{dx^2}-\Sigma_a\varphi+Q=0,\quad x\in(-a,a),\varphi(\pm a)=0, \tag{5.1}$$

where $\varphi(x)$ denotes the neutron flux, D denotes the diffusion coefficient, Σ_α denotes the macroscopic absorption cross section, Q denotes the distributed source term, and the pool is considered to be a slab of extrapolated thickness $2a$. In view of the problem's symmetry, the origin $x = 0$ can be conveniently chosen at the middle (center) of the slab. For simplicity, the boundary conditions chosen for Equation 5.1 are that the neutron flux must vanish at the extrapolated distance $x = a$.

The linear differential equation in Equation 5.1 can be solved readily to obtain the general solution

$$\varphi(x)=\frac{Q}{\Sigma_a}\left[1-\frac{\cosh(xk)}{\cosh(ak)}\right],\quad k\triangleq\sqrt{\Sigma_a/D},\quad x\in(-a,a), \tag{5.2}$$

where $k\triangleq\sqrt{\Sigma_a/D}$ denotes the reciprocal diffusion length. The above expression for the neutron flux $\varphi(x)$ holds for any physical values of D, Σ_a, and Q. For practical problems, the flux cannot be obtained analytically as above but would need to be computed numerically, using the nominal parameter values Σ_a^0, D^0, and Q^0. Such a numerical computation would yield the nominal value of the flux, $\varphi^0(x)$, namely

$$\varphi^0(x)=\frac{Q^0}{\Sigma_a^0}\left[1-\frac{\cosh\left(xk^0\right)}{\cosh\left(ak^0\right)}\right],\quad k^0\triangleq\sqrt{\Sigma_a^0/D^0},x\in(-a,a). \tag{5.3}$$

A generic response, $R(x)$, for the paradigm neutron diffusion problem modeled by Equation 5.1 is the reading of a detector within the slab, for example, at a distance b from the slab's midline at $x = 0$. The "reading" of such a detector response would indicate the neutrons' reaction rate within the detectors reactive material and would be mathematically represented in the form

$$R(\mathbf{e}) = \Sigma_d \varphi(b), \tag{5.4}$$

where Σ_d represents the detector's equivalent reaction cross section, assumed to be constant, for simplicity.

The system parameters for this problem are thus the positive scalars Σ_a, D, Q, and Σ_d. Referring to the general mathematical notation used in Chapter 3, we note the correspondences

$$\boldsymbol{\alpha} \triangleq \left(\Sigma_a, D, Q, \Sigma_d\right)^{\dagger}, \ \mathbf{u}(\mathbf{x}) \to \varphi(x), \ \mathbf{e} \triangleq \left(\mathrm{o}, \boldsymbol{\alpha}\right)^{\dagger}, \tag{5.5}$$

where the dagger (†) denotes transposition. For this illustrative problem, the nominal values of the state function (the neutron flux) and model parameters will be denoted as

$$\boldsymbol{\alpha}^0 \triangleq \left(\Sigma_a^0, D^0, Q^0, \Sigma_d^0\right)^{\dagger}, \ \mathbf{u}^0(\mathbf{x}) \to \varphi^0(x), \ \ \mathbf{e}^0 \triangleq \left(\varphi^0, \boldsymbol{\alpha}^0\right)^{\dagger}. \tag{5.6}$$

Using Equation 5.3 in Equation 5.4 gives the following form for the nominal response value, $R(\mathbf{e}^0)$:

$$R\left(\mathbf{e}^0\right) = \Sigma_d^0 \varphi^0(b) = \frac{Q^0 \Sigma_d^0}{\Sigma_a^0}\left[1 - \frac{\cosh\left(bk^0\right)}{\cosh\left(ak^0\right)}\right], \ \mathbf{e}^0 \equiv \left(\varphi^0, \boldsymbol{\alpha}^0\right)^{\dagger}, \ x \in (-a, a). \tag{5.7}$$

For a fixed value of b, the response $R(\mathbf{e})$ is a functional (i.e., a scalar-valued operator) that acts linearly on φ, as evidenced by Equation 5.7. On the contrary, even though Equation 5.1 is linear in φ, the solution $\varphi(x)$ depends nonlinearly on $\boldsymbol{\alpha}$, as highlighted by Equation 5.3. The same is true of the response $R(\mathbf{e})$. Even though $R(\mathbf{e})$ is linear separately in φ and in $\boldsymbol{\alpha}$, as shown in Equation 5.4, $R(\mathbf{e})$ *is not simultaneously linear in* φ *and* $\boldsymbol{\alpha}$; this fact leads to a nonlinear dependence of $R(\mathbf{e})$ on $\boldsymbol{\alpha}$. This fact is explicitly highlighted by the exact expression of $R(\mathbf{e})$ given in Equation 5.7.

5.2 Applying the 2nd-ASAM to Compute the First-Order Response Sensitivities to Model Parameters

The parameters in this problem are determined from experiments afflicted by uncertainties; in particular, the uncertainties for the basic (microscopic) neutron cross sections are provided in centrally deposited "covariance files."

Therefore, the system parameters $\boldsymbol{\alpha}$ can and do vary (because of uncertainties, external influences, etc.) from their nominal values $\boldsymbol{\alpha}^0$ by amounts $\delta\boldsymbol{\alpha}$, defined as

$$\delta\boldsymbol{\alpha} \triangleq \left(\delta\Sigma_a, \delta D, \delta Q, \delta\Sigma_d\right)^{\dagger}. \tag{5.8}$$

In practice, the variations $\delta\Sigma_a$, δD, δQ, and $\delta\Sigma_d$ usually correspond to the standard deviations quantifying the uncertainties in the respective model parameters. As shown in Chapter 3, the sensitivities of the response $R(\mathbf{e})$ to the variations $\delta\boldsymbol{\alpha}$ are given by the (first-order) G-differential $\delta R(\mathbf{e}^0; \mathbf{h})$ of $R(\mathbf{e})$ at $\mathbf{e}^0$, which is defined as

$$\delta R\left(\mathbf{e}^0;\mathbf{h}\right) \equiv \frac{d}{d\varepsilon}\left\{R\left(\mathbf{e}^0+\varepsilon\mathbf{h}\right)\right\}_{\varepsilon=0}, \text{ with } \mathbf{h} \triangleq \left(\delta\varphi, \delta\boldsymbol{\alpha}\right). \tag{5.9}$$

Applying the above definition to the response defined by Equation 5.4 gives

$$\delta R\left(\mathbf{e}^0;\mathbf{h}\right) \equiv \frac{d}{d\varepsilon}\left\{\left[\Sigma_d^0+\varepsilon\left(\delta\Sigma_d\right)\right]\left[\varphi^0(x)+\varepsilon\,\delta\varphi(x)\right]\right\}_{\varepsilon=0} = \frac{\partial R}{\partial\boldsymbol{\alpha}}\delta\boldsymbol{\alpha}+\frac{\partial R}{\partial\varphi}\delta\varphi, \tag{5.10}$$

where the direct-effect term $\frac{\partial R}{\partial\boldsymbol{\alpha}}\delta\boldsymbol{\alpha}$ is obtained as

$$\frac{\partial R}{\partial\boldsymbol{\alpha}}\delta\boldsymbol{\alpha} = \varphi^0(b)\left[(0,0,0,1)\bullet\ \left(\delta\Sigma_a, \delta D, \delta Q, \delta\Sigma_d\right)^{\dagger}\right] = \left(\delta\Sigma_d\right)\varphi^0(b), \tag{5.11}$$

while the indirect-effect term $\frac{\partial R}{\partial\varphi}\delta\varphi$ is obtained as

$$\frac{\partial R}{\partial\varphi}\delta\varphi = \Sigma_d^0\,\delta\varphi(x). \tag{5.12}$$

The direct-effect term $\frac{\partial R}{\partial\boldsymbol{\alpha}}\delta\boldsymbol{\alpha}$ can be evaluated immediately at this stage by substituting Equation 5.7 into Equation 5.11, to obtain

$$\frac{\partial R}{\partial\boldsymbol{\alpha}}\delta\boldsymbol{\alpha} = \left(\delta\Sigma_d\right)\frac{Q^0}{\Sigma_a^0}\left[1-\frac{\cosh\left(xk^0\right)}{\cosh\left(ak^0\right)}\right], \quad x\in(-a,a). \tag{5.13}$$

As Equation 5.10 indicates, the operator $\delta R(\mathbf{e}^0;\mathbf{h})$ is linear in $\mathbf{h}$; in particular, the indirect-effect term $\frac{\partial R}{\partial\varphi}\delta\varphi$ is a linear operator on $\delta\varphi(x)$. However, this indirect-effect term cannot be evaluated at this stage, since $\delta\varphi(x)$ is not yet available. As shown in Chapter 3, $\delta\varphi(x)$ is related to the parameter variations $\delta\boldsymbol{\alpha}$ through the 1st-LFSS, which is obtained by G-differentiating Equation 5.1. Performing this G-differentiation while using the general notation from Chapter 3 gives

$$L(\boldsymbol{\alpha}^0)\delta\varphi(x) = Q^{(1)}(\boldsymbol{\alpha}^0, \varphi^0; \delta\boldsymbol{\alpha}), \tag{5.14}$$

together with the boundary conditions

$$\delta\varphi(\pm a) = 0. \tag{5.15}$$

In Equation 5.14, the operator $L(\boldsymbol{\alpha}^0)$ is defined as

$$L(\boldsymbol{\alpha}^0) \triangleq D^0 \frac{d^2}{dx^2} - \Sigma_a^0, \tag{5.16}$$

while the "source term" $Q^{(1)}(\boldsymbol{\alpha}^0, \varphi^0; \delta\boldsymbol{\alpha})$ is defined below:

$$Q^{(1)}(\boldsymbol{\alpha}^0, \varphi^0; \delta\boldsymbol{\alpha}) \triangleq (\delta D)\frac{d^2\varphi^0}{dx^2} - (\delta\Sigma_a)\varphi^0 + (\delta Q). \tag{5.17}$$

Equation 5.14 together with the boundary conditions in Equation 5.15 constitutes the 1st-LFSS. For every possible parameter variation $\delta\varphi(x)$, the 1st-LFSS would need to be solved anew to obtain its corresponding solution, $\delta\varphi(x)$; each such computation represents a "large-scale" computation. Thus, solving repeatedly the 1st-LFSS would be advantageous only if the number of responses exceeded the number of model parameters, which is seldom the case in practice. Nevertheless, Equations 5.14 and 5.15 provide an independent way to verify the results that will be obtained later in this section using the 2nd-ASAM. Therefore, we provide below the expression of the solution of Equations 5.14 and 5.15:

$$\begin{aligned}\delta\varphi(x) &= C_1\left[\cosh(xk^0) - \cosh(ak^0)\right] \\ &+ C_2\left[x\sinh(xk^0)\cosh(ak^0) - a\sinh(ak^0)\cosh(xk^0)\right], \quad x \in (-a, a),\end{aligned} \tag{5.18}$$

where the constants C_1 and C_2 are defined, respectively, as

$$C_1 \equiv \frac{\left[(\delta\Sigma_a)Q^0/\Sigma_a^0 - (\delta Q)\right]}{\Sigma_a^0\left(\cosh ak^0\right)}, \tag{5.19}$$

and

$$C_2 \equiv \frac{\left[(\delta D)/D^0 - (\delta\Sigma_a)/\Sigma_a^0\right]Q^0}{2\sqrt{D^0\Sigma_a^0}\left(\cosh ak^0\right)^2}. \tag{5.20}$$

Hence, the expression of the indirect-effect term is obtained by replacing Equation 5.18 in Equation 5.12 to obtain:

$$\frac{\partial R}{\partial \varphi}\delta\varphi = \Sigma_d^0 C_1\left[\cosh\left(xk^0\right)-\cosh\left(ak^0\right)\right]$$
$$+\Sigma_d^0 C_2\left[x\sinh\left(xk^0\right)\cosh\left(ak^0\right)-a\sinh\left(ak^0\right)\cosh\left(xk^0\right)\right],\quad x\in(-a,a). \tag{5.21}$$

As was generally shown in Chapter 3, the fundamental prerequisite for applying the 2nd-ASAM is the introduction of a Hilbert space, $\mathcal{H}_u$, appropriate to the problem at hand. For our illustrative example, it is appropriate to choose $\mathcal{H}_u$ to be the real Hilbert space $\mathcal{H}_u \equiv L_2(\Omega)$, with $\Omega \equiv (-a, a)$, equipped with the inner product

$$\langle f(x), g(x)\rangle \triangleq \int_{-a}^{a} f(x)g(x)dx, \text{ for } f,\ g\in\mathcal{H}_u \equiv L_2(\Omega),\ \Omega\equiv(-a,a). \tag{5.22}$$

In view of Equation 5.22, the indirect-effect term $\frac{\partial R}{\partial \varphi}\delta\varphi$ can be expressed in inner product form as follows:

$$\frac{\partial R}{\partial \varphi}\delta\varphi \triangleq \Sigma_d^0\,\delta\varphi(b) = \int_{-a}^{a}\Sigma_d^0\,\delta\varphi(x)\,\delta(x-b)dx = \left\langle \Sigma_d^0\delta(x-b),\,\delta\varphi(x)\right\rangle. \tag{5.23}$$

The next step is the construction of the operator $L^*(\boldsymbol{\alpha}^0)$ that is formally adjoint to $L^*(\boldsymbol{\alpha}^0)$. For this purpose, Equation 5.14 is multiplied by a square-integrable, but at this stage otherwise still arbitrary, function $\psi(x) \in \mathcal{H}_Q = L_2(\Omega)$ and is subsequently integrated over x to obtain

$$\int_{-a}^{a}\psi(x)\left[D^0\frac{d^2}{dx^2}\delta\varphi(x)-\Sigma_a^0\delta\varphi(x)\right]dx = -\int_{-a}^{a}\psi(x)\left[(\delta D)\frac{d^2\varphi^0}{dx^2}-(\delta\Sigma_a)\varphi^0+(\delta Q)\right]dx. \tag{5.24}$$

Integrating the left side of Equation 5.24 by parts twice, to transfer all operations on $\delta\varphi(x)$ to operations on $\psi(x)$, leads to

$$\int_{-a}^{a}\psi(x)\left[D^0\frac{d^2}{dx^2}\delta\varphi(x)+\Sigma_a^0\delta\varphi(x)\right]dx$$
$$=\int_{-a}^{a}\left[D^0\frac{d^2\psi(x)}{dx^2}-\Sigma_a^0\psi(x)\right]\delta\varphi(x)\,dx + D^0\left[\psi\frac{d}{dx}\delta\varphi(x)-\delta\varphi(x)\frac{d\psi}{dx}\right]_{-a}^{a}. \tag{5.25}$$

The right side of Equation 5.25 shows that the formal adjoint of $L(\boldsymbol{\alpha}^0)$ is the operator

$$L^*\left(\alpha^0\right) \triangleq D^0 \frac{d^2}{dx^2} - \Sigma_a^0 . \tag{5.26}$$

Note that the function $\psi(x)$ is still arbitrary at this stage, except for the requirement that $\psi \in \mathcal{H}_Q = L_2(\Omega)$. The next step in the construction of the 1st-LASS is the identification of its source term, which is achieved by requiring the first term on the right side of Equation 5.23 and the first term on the right side of Equation 5.25, respectively, to represent the same functional. Imposing this requirement yields the equation

$$L^*\left(\alpha^0\right)\psi \triangleq D^0 \frac{d^2\psi}{dx^2} - \Sigma_a^0 \psi(x) = \Sigma_d^0 \delta(x-b), \tag{5.27}$$

The boundary conditions for $\psi(x)$ can now be selected by requiring that "unknown values of $\delta\varphi(x)$" be eliminated from Equation 5.25. Since the values of $\delta\varphi(x)$ at $x = \pm a$ are known from Equation 5.15, it follows that the "unknown values of $\delta\varphi(x)$" in Equation 5.25 are the derivatives $\left\{\frac{d}{dx}\delta\varphi(x)\right\}_{-a}^{a}$. The elimination from Equation 5.25 of these unknown values can be accomplished by choosing the following boundary conditions:

$$\psi(\pm a) = 0. \tag{5.28}$$

The above selection of the adjoint boundary conditions completes the construction of the 1st-LASS, which comprises Equations 5.27 and 5.28. Since the boundary conditions in Equation 5.28 for the adjoint function $\psi(x)$ are the same as the boundary conditions for $\delta\varphi(x)$ in Equation 5.15, it follows from this and from Equations 5.16 and 5.26 that the operators $L^*(\boldsymbol{\alpha}^0)$ and $L(\boldsymbol{\alpha}^0)$ are not just formally self-adjoint but bona fide self-adjoint.

Using Equations 5.24 and 5.27 in Equation 5.23 leads to the following expression for the indirect-effect term $R'_\varphi\left(\mathbf{e}^0\right)h_\varphi$ in terms of the adjoint function $\psi(x)$:

$$R'_\varphi\left(\mathbf{e}^0\right)h_\varphi = -\int_{-a}^{a} \psi(x)\left[(\delta D)\frac{d^2\varphi^0}{dx^2} - (\delta\Sigma_a)\varphi^0(x) + (\delta Q)\right]dx. \tag{5.29}$$

As expected, *the adjoint sensitivity system*, cf. Equations 5.27 and 5.28, *is independent of parameter variations* $\mathbf{h}_\alpha$, so it needs to be solved only once to obtain the adjoint function $\psi(x)$. Very important, too, is the fact—characteristic of linear systems such as the diffusion equation under consideration—that the adjoint system is independent of the original solution $\varphi^0(x)$ and can therefore be solved independently (and without any knowledge) of the forward neutron flux $\varphi(x)$. Of course, the adjoint system depends on the response, which provides the source term as shown in Equation 5.27. Solving the adjoint system for our illustrative example yields the following expression for the adjoint function $\psi(x)$:

$$\psi(x)=\frac{\Sigma_d^0}{\sqrt{\Sigma_a^0 D^0}}\left\{\frac{\sinh\left[(b-a)k^0\right]}{\sinh\left(2ak^0\right)}\sinh\left[(x+a)k^0\right]+H(x-b)\sinh\left[(x-b)k^0\right]\right\}, \tag{5.30}$$

where $H(x-b)$ is the Heaviside step functional defined as

$$H(x)=\begin{cases}0, & for\ x<0\\ 1, & for\ x\geq 0\end{cases}. \tag{5.31}$$

Replacing the adjoint function $\psi(x)$ from Equation 5.30 in Equation 5.29 and carrying out the respective integration over x yields the *first-order total sensitivity* $\delta R(\mathbf{e}^0;\mathbf{h})$ of $R(\mathbf{e})$ at $\mathbf{e}^0$ to variations $\delta\boldsymbol{\alpha}$ in the system parameters:

$$\delta R\left(\mathbf{e}^0;\mathbf{h}\right)=\frac{\partial R}{\partial \Sigma_a}(\delta\Sigma_a)+\frac{\partial R}{\partial D}(\delta D)+\frac{\partial R}{\partial Q}(\delta Q)+\frac{\partial R}{\partial \Sigma_d}(\delta\Sigma_d), \tag{5.32}$$

where

$$\begin{aligned} S_1\left(\boldsymbol{\alpha}^0\right)&\triangleq\frac{\partial R}{\partial \Sigma_a}=\int_{-a}^{a}\psi^0(x)\,\varphi^0(x)\,dx\\ &=Q^0\Sigma_d^0\left[-\left(\Sigma_a^0\right)^{-2}A\left(k^0\right)+\frac{1}{2\Sigma_a^0\sqrt{D^0\Sigma_a^0}}B\left(k^0\right)\right], \end{aligned} \tag{5.33}$$

$$\begin{aligned} S_2\left(\boldsymbol{\alpha}^0\right)&\triangleq\frac{\partial R}{\partial D}=\int_{-a}^{a}\psi^0(x)\frac{d^2\varphi^0}{dx^2}dx=-\frac{\Sigma_a^0}{D^0}\int_{-a}^{a}\psi^0(x)\varphi^0(x)\,dx+\frac{Q^0}{D^0}\int_{-a}^{a}\psi^0(x)dx\\ &=-\frac{Q^0\Sigma_d^0}{2D^0\Sigma_a^0}k^0B\left(k^0\right), \end{aligned} \tag{5.34}$$

$$S_3\left(\boldsymbol{\alpha}^0\right)\triangleq\frac{\partial R}{\partial Q}=-\int_{-a}^{a}\psi^0(x)dx=\Sigma_d^0\left(\Sigma_a^0\right)^{-1}A\left(k^0\right), \tag{5.35}$$

$$S_4\left(\boldsymbol{\alpha}^0\right) \triangleq \frac{\partial R}{\partial \Sigma_d} = \int_{-a}^{a} \varphi^0(x)\,\delta(x-b)\,dx = Q^0\left(\Sigma_a^0\right)^{-1} A\left(k^0\right), \tag{5.36}$$

and where the quantities $A(k^0)$ and $B(k^0)$ are defined, respectively, as

$$A\left(k^0\right) \triangleq 1 - \frac{\cosh\left(bk^0\right)}{\cosh\left(ak^0\right)}, \tag{5.37a}$$

$$B\left(k^0\right) \triangleq \left.\frac{dA}{dk}\right|_{k^0} = \left[\cosh\left(ak^0\right)\right]^{-2}\left[a\sinh\left(ak^0\right)\cosh\left(bk^0\right) - b\sinh\left(bk^0\right)\cosh\left(ak^0\right)\right]. \tag{5.37b}$$

One of the main uses of sensitivities is for ranking the relative importance of parameter variations in influencing variations in responses. Relative sensitivities are used for this purpose, since they are dimensionless numbers; the relative sensitivity of a response $R(\mathbf{e})$ to the ith parameter, α_i, is defined as $S_i^{rel} \triangleq \left(\partial R/\partial \alpha_i\right)_{\mathbf{e}^0}\left[\alpha_i^0/R\left(\mathbf{e}^0\right)\right]$. It is evident from Equations 5.7, 5.35, and 5.36 that the relative sensitivities of the detector response $R(\mathbf{e})$ to the source ($\alpha_3 \triangleq Q$) and to the detector's reaction cross section ($\alpha_4 \triangleq \Sigma_d$) are unity, that is, $S_3^{rel} \triangleq \left(\partial R/\partial Q\right)_{\mathbf{e}^0}\left[Q^0/R\left(\mathbf{e}^0\right)\right] = 1$ and $S_4^{rel} \triangleq \left(\partial R/\partial \Sigma_d\right)_{\mathbf{e}^0}\left[\Sigma_d^0/R\left(\mathbf{e}^0\right)\right] = 1$, regardless of the position ($x = b$) of the location of the detector or the material properties of the medium. This means that a 1% increase/decrease in either Q or Σ_d will induce a 1% increase/decrease in nominal value of the response $R(\mathbf{e})$. These are rather large sensitivities, indeed! The relative influence of a response's sensitivities to the medium's macroscopic absorption cross section, Σ_a, and to the medium's diffusion coefficient, D, can best be illustrated by considering that the medium in the pool is water, for which the absorption cross section and diffusion coefficient for thermal neutrons have the following nominal values: $\Sigma_a^0 = 0.0197\,\text{cm}^{-1}$, $D^0 = 0.16$ cm. Furthermore, consider that the slab pool's half-thickness is $a = 50$ cm, and the distributed neutron sources emit nominally $S^0 = 10^7$ neutrons·cm^{-3}·s^{-1}. Consider also that measurements are performed with an idealized, infinitely thin detector having an indium-like detector cross section with nominal value $\Sigma_d^0 = 7.438\,\text{cm}^{-1}$.

Consider now six responses, at the following detector locations, which are pair-wise symmetric: (i) $b = \pm 10$ cm, located close to the water pool's center; (ii) $b = \pm 40$ cm, located close to the pool's boundaries; and (iii) $b = \pm 49.5$ cm, located very close to the edge, where the neutron flux's gradient becomes very steep and the diffusion coefficient markedly influences the detector's response. The detector's responses at these locations are designated as follows: R_1 (10 cm), R_2 (40 cm), R_3 (49.5 cm), R_4 (−10 cm), R_5 (−40 cm), and R_6 (−49.5 cm). These symmetric locations, at which Equation 5.7 clearly indicates that $R(x) = R(-x)$, were deliberately chosen in order to verify the preservation of symmetry during and after the computation of the respective responses' sensitivities to the model parameters. In units of neutrons·cm^{-3}·s^{-1}, the nominal computed response values obtained using Equation 5.7 for these detector locations were as follows: $R_1(10\text{ cm}) = R_4(-10\text{ cm}) = 3.77 \times 10^9$ $R_2(-40\text{ cm}) = 3.66 \times 10^9$ and $R_3(49.5\text{ cm}) = R_6(-49.5\text{ cm}) = 6.076 \times 10^8$.

Note also that $R(x) = R(-x)$, so that only *three* (rather than six) *"large-scale" adjoint sensitivity computations* (i.e., solving Equations 5.27 and 5.28) will suffice to obtain all of the results presented below in Table 5.1 (absolute sensitivities, in the respective units, which are omitted from the table for brevity) and Table 5.2 (relative sensitivities).

The absolute sensitivities presented in Table 5.1 are difficult to use for ranking the importance of parameters in influencing the respective responses. On the contrary, the relative sensitivities presented in Table 5.2 readily indicate the importance of the model parameters in influencing the detector responses. Of course, these "parameter importance(s)" reflect the relative importance of the physical processes associated with the respective parameters in influencing the detector's response. As already mentioned, $S_3^{rel} = S_4^{rel} = 1$, and these sensitivities turn out to be the most important ones for the detector response, regardless of the detector's position. This result is justified on physical considerations since the neutron source strength and the detector's interaction cross section are expected to be very important. Note, in particular, that the sensitivity S_4^{rel} describes a direct-effect term specific to

TABLE 5.1

Absolute Sensitivities for Six Detector Responses

First-Order Absolute Sensitivities	R_1 (10 cm) R_4 (−10 cm)	R_2 (40 cm) R_5 (−40 cm)	R_3 (49.5 cm) R_6 (−49.5 cm)
S_1 (Σ_a)	-1.917×10^{11}	-1758×10^{11}	-1.673×10^{10}
S_2 (D)	-1.331×10^{5}	-1.239×10^{9}	-1.737×10^{9}
S_3 (Q)	3.776×10^{2}	3.663×10^{2}	6.076×10^{1}
S_4 (Σ_d)	5.076×10^{8}	4.924×10^{8}	8.168×10^{7}

TABLE 5.2

Relative Sensitivities for Six Detector Responses

First-Order Relative Sensitivities	R_1 (10 cm) R_4 (−10 cm)	R_2 (40 cm) R_5 (−40 cm)	R_3 (49.5 cm) R_6 (−49.5 cm)
S_1 (Σ_a) *relative*	−1.00	−0.95	−0.54
S_2 (D) *relative*	-5.64×10^{-6}	−0.05	−0.46
$S_3(Q)$ *relative*	1.00	1.00	1.00
S_4 (Σ_d) *relative*	1.00	1.00	1.00

the detector's properties (in this illustrative paradigm example, S_4^{rel} is actually independent of the medium's material properties).

The other two relative sensitivities, namely $S_1^{rel} \triangleq (\partial R/\partial \Sigma_a)_{\mathbf{e}^0} \left[\Sigma_a^0 / R(\mathbf{e}^0) \right]$ and $S_2^{rel} \triangleq (\partial R/\partial D)_{\mathbf{e}^0} \left[D^0 / R(\mathbf{e}^0) \right]$, describe opposing physical processes and hence opposite influences on the detector responses. Thus, S_1^{rel} is very important in the center of the slab, where it has a large negative value of comparable absolute magnitude to S_3^{rel}. Towards the edge of the slab, S_1^{rel} decreases in absolute magnitude but remains negative throughout the medium. Physically, the behavior of S_1^{rel} indicates that the absorption of neutron is a loss mechanism (hence the negative sign of S_1^{rel} throughout the medium), and its importance is almost as large as the production by the neutron sources (as indicated by the positive sign and magnitude of S_3^{rel}, throughout most of the medium). Very close to the physical boundary of the medium, however, where the neutron flux must vanish because of the chosen boundary conditions, the absorption mechanism diminishes rapidly in importance. The flux gradient is very steep over a small distance towards the boundary, indicating a boundary-layer-like behavior of the absorption process, which obviously must cease at the medium's extrapolated boundary, where the flux must mathematically vanish due to the imposed boundary conditions.

On the contrary, the behavior of $S_2^{rel} \triangleq (\partial R/\partial D)_{\mathbf{e}^0} \left[D^0 / R(\mathbf{e}^0) \right]$ indicates that the diffusion coefficient, and consequently the details of the diffusion process, plays a very minor role in influencing the detector response, except close to the boundaries of the slab. The sign of S_2^{rel} is negative, indicating a "loss" of neutrons in contributing to the detector's response (i.e., increasing the diffusion coefficient, and hence the importance of the diffusion process, causes fewer neutrons to interact with the detector and hence contributes to its response). Towards the extrapolated boundary of the medium, namely at 0.5 cm from the respective extrapolated boundaries, the magnitude of S_2^{rel} becomes comparable to that of S_1^{rel}, both being negative, indicating neutron losses from the detector's response. This result is expected since the neutron flux was (mathematically) required to vanish at the extrapolated boundaries, and the physical mechanisms that reduce the neutron flux can only stem from absorption combined with diffusion.

5.3 Applying the 2nd-ASAM to Compute the Second-Order Response Sensitivities to Model Parameters

Applying the general theory presented in Chapter 3, the starting point for computing the second-order response sensitivities $S_{ij}(\boldsymbol{\alpha}^0) \triangleq \partial^2 R / \partial \alpha_i \partial \alpha_j = \partial^2 R / \partial \alpha_j \partial \alpha_i$ is the consideration of the first-order sensitivities to be functionals of the form $S_i(\varphi, \Psi, \boldsymbol{\alpha})$, $i = 1, 2, 3, 4$. Based on this fundamental consideration, the 2nd-ASAM proceeds by computing the first-order G-differential, $\delta S_i(\mathbf{e}^0, \boldsymbol{\psi}^0, \mathbf{g})$, of the various functionals $S_i(\mathbf{u}, \boldsymbol{\alpha}, \boldsymbol{\psi})$, at the point $(\mathbf{e}^0, \boldsymbol{\psi}^0)$, from the definition

$$\delta S_i\left(\varphi^0,\psi^0,\alpha^0;\delta\varphi,\delta\psi,\delta\alpha\right)\triangleq\left\{\frac{d}{d\varepsilon}\left[S_i\left(\varphi^0+\varepsilon\delta\varphi,\psi^0+\varepsilon\delta\psi,\alpha^0+\varepsilon\delta\alpha\right)\right]\right\}_{\varepsilon=0} \quad (5.38)$$

for an arbitrary scalar $\varepsilon \in \mathcal{F}$, and all (i.e., arbitrary) vectors $(\delta\varphi, \delta\Psi, \delta\boldsymbol{\alpha}) \in \mathcal{H}_\varphi(\Omega_x) \times \mathcal{H}_\psi(\Omega_x) \times \mathcal{H}_{\boldsymbol{\alpha}}$. For the illustrative diffusion problem, there are $N_{\boldsymbol{\alpha}}(N_{\boldsymbol{\alpha}} + 1)/2 = 10$ distinct second-order derivatives, due to the symmetry property $\partial^2 R/\partial\alpha_i\partial\alpha_j = \partial^2 R/\partial\alpha_j\partial\alpha_i$. It is convenient to compute the sensitivities $S_{ij}(\boldsymbol{\alpha}^0)$ by starting with the simplest ones and progressing to the more difficult ones. An examination of Equations 5.33 through 5.36 readily reveals that $S_4(\boldsymbol{\alpha}^0)$ and $S_3(\boldsymbol{\alpha}^0)$ have the simplest expressions, while $S_2(\boldsymbol{\alpha}^0)$ has the most "complicated" expression. Therefore, it is computationally advantageous to compute first the four second-order sensitivities $S_{4i}(\alpha^0) \triangleq \partial^2 R/\partial Q\partial\alpha_i$, $(i = 1, 2, 3, 4)$ followed by the three second-order sensitivities $S_{3i}(\alpha^0) \triangleq \partial^2 R/\partial\Sigma_d\partial\alpha_i$, $(i = 1, 2, 3)$ since $S_{34}(\boldsymbol{\alpha}^0)$ would already be available. The computations would then proceed by determining $S_{1i}(\alpha^0) \triangleq \partial^2 R/\partial\Sigma_a \partial\alpha_i$, $(i = 1, 2)$ and finally the sensitivity $S_{22}(\boldsymbol{\alpha}^0) \equiv \partial^2 R/\partial D^2$. Applying the general theory presented in Chapter 3, Section 3.3, to the illustrative diffusion problem under consideration indicates that every G-differential $\delta S_i(\varphi^0, \psi^0, \boldsymbol{\alpha}^0; \delta\varphi, \delta\Psi, \delta\boldsymbol{\alpha})$, $i = 1, 2, 3, 4$, will have a corresponding 2nd-LASS whose solution will be a second-level adjoint function of the form $\psi_i^{(2)}(x)\triangleq\left[\lambda_i(x),\theta_i(x)\right]^{\dagger}$, $i=1, 2, 3, 4$, with $\lambda_i(x)$ corresponding to the variation $\delta\varphi(x)$ and $\theta_i(x)$ corresponding to the variation $\delta\psi(x)$.

5.3.1 Applying the 2nd-ASAM to Compute the Second-Order Response Sensitivities $S_{4i} \triangleq \partial^2 R/(\partial\sum_d\partial\alpha_i)$

Applying the definition shown in Equation 5.38 to Equation 5.36 yields the corresponding G-differential, $DS_4(\boldsymbol{\alpha}^0)$, of the first-order sensitivity $S_4(\boldsymbol{\alpha})$, as

$$DS_4\left(\boldsymbol{\alpha}^0\right)=\frac{d}{d\varepsilon}\left\{\int_{-a}^{a}\left(\varphi^0+\varepsilon\delta\varphi\right)\delta(x-b)dx\right\}_{\varepsilon=0}=\int_{-a}^{a}\delta\varphi(x)\,\delta(x-b)dx, \quad (5.39)$$

where the function $\delta\varphi(x)$ is the solution of Equations 5.14 and 5.15. Note that the entire contribution to $DS_4(\boldsymbol{\alpha}^0)$ comes from the indirect-effect term; there is no direct-effect term contribution to $DS_4(\boldsymbol{\alpha}^0)$. Furthermore, the quantity $DS_4(\boldsymbol{\alpha}^0)$ does not depend on $\delta\psi(x)$. Applying the general theory presented in Chapter 3, Section 3.3, yields the following 2nd-LASS for computing the second-level adjoint function $\psi_4^{(2)}(x) \triangleq [\lambda_4(x), \theta_4(x)]^\dagger$ for computing the G-differential $DS_4(\boldsymbol{\alpha})$:

$$D^0 \frac{d^2\lambda_4}{dx^2} - \Sigma_a^0 \lambda_4(x) = \delta(x-b), \tag{5.40}$$

$$\lambda_4(\pm a) = 0. \tag{5.41}$$

Since $DS_4(\boldsymbol{\alpha}^0)$ does not depend on $\delta\psi(x)$, the second adjoint function, $\theta_4(x)$, which would have corresponded to $\delta\psi(x)$, is identically zero, since the adjoint system for $\theta_4(x)$ would have been a linear homogeneous equation with zero source and zero boundary conditions. In terms of the second-level adjoint function $\psi_4^{(2)}(x) \triangleq [\lambda_4(x), 0]^\dagger$, the expression of the differential $DS_4(\boldsymbol{\alpha}^0)$ becomes

$$\begin{aligned} DS_4(\boldsymbol{\alpha}^0) &= \int_{-a}^{a} \lambda_4(x) \left[-(\delta D)\frac{d^2\varphi^0}{dx^2} + (\delta\Sigma_a)\varphi^0(x) - (\delta Q) \right] dx \\ &= S_{41}(\delta\Sigma_a) + S_{42}(\delta D) + S_{43}(\delta Q) + S_{44}(\delta\Sigma_d), \end{aligned} \tag{5.42}$$

where

$$S_{41} \triangleq \frac{\partial^2 R}{\partial\Sigma_d\, \partial\Sigma_a} = \int_{-a}^{a} \lambda_4(x)\varphi^0(x)dx, \tag{5.43}$$

$$S_{42} \triangleq \frac{\partial^2 R}{\partial\Sigma_d\, \partial D} = -\int_{-a}^{a} \lambda_4(x)\frac{d^2\varphi^0}{dx^2}dx, \tag{5.44}$$

$$S_{43} \triangleq \frac{\partial^2 R}{\partial\Sigma_d\, \partial Q} = -\int_{-a}^{a} \lambda_4(x)dx, \tag{5.45}$$

$$S_{44} \triangleq \frac{\partial^2 R}{\partial\Sigma_d\, \partial\Sigma_d} \equiv 0. \tag{5.46}$$

Note that $S_{44} \equiv 0$, since the expression of $DS_4(\boldsymbol{\alpha}^0)$, namely Equation 5.39, does not contain the variation $(\delta\Sigma_d)$. Note also that the adjoint system satisfied by $\lambda_4(x)$, comprising Equations 5.40 and 5.41, is the same as would be obtained by setting $\Sigma_d^0 \equiv 1$ into the adjoint system satisfied by the adjoint function $\psi(x)$, cf. Equations 5.27 and 5.28. Therefore, $\lambda_4(x)$ is obtained by simply dividing the result in Equation 5.30 by Σ_d^0 or, equivalently, setting $\Sigma_d^0 \equiv 1$ in Equation 5.30 to obtain

$$\lambda_4(x) = \{\psi(x)\}_{\Sigma_d^0 \equiv 1}$$

$$= \frac{1}{\sqrt{\Sigma_a^0 D^0}} \left\{ \frac{\sinh\left[(b-a)k^0\right]}{\sinh\left(2ak^0\right)} \sinh\left[(x+a)k^0\right] + H(x-b)\sinh\left[(x-b)k^0\right] \right\}. \tag{5.47}$$

Even more, since $\lambda_4(x) = \{\psi(x)\}_{\Sigma_d^0 \equiv 1}$, it is not even necessary to solve Equations 5.40 and 5.41 to compute $\lambda_4(x)$, and it is not necessary to introduce Equation 5.47 into Equations 5.43 through 5.45 and to perform the respective integrations. In fact, *one simply replaces* $\Sigma_d^0 \equiv 1$ *in Equations 5.33 through 5.35 or, equivalently, divides these equations through* Σ_d^0 *to obtain*

$$S_{41}(\boldsymbol{\alpha}^0) \triangleq \frac{\partial^2 R}{\partial \Sigma_d \, \partial \Sigma_a} = \frac{S_1(\boldsymbol{\alpha}^0)}{\Sigma_d^0} = Q^0 \left[-\left(\Sigma_a^0\right)^{-2} A\left(k^0\right) + \frac{1}{2\Sigma_a^0 \sqrt{D^0 \Sigma_a^0}} B\left(k^0\right) \right], \tag{5.48}$$

$$S_{42}(\boldsymbol{\alpha}^0) \triangleq \frac{\partial^2 R}{\partial \Sigma_d \, \partial D} = \frac{S_2(\boldsymbol{\alpha}^0)}{\Sigma_d^0} = -\frac{Q^0}{2D^0 \Sigma_a^0} k^0 B\left(k^0\right), \tag{5.49}$$

$$S_{43}(\boldsymbol{\alpha}^0) \triangleq \frac{\partial^2 R}{\partial \Sigma_d \, \partial Q} = \frac{S_3(\boldsymbol{\alpha}^0)}{\Sigma_d^0} = \left(\Sigma_a^0\right)^{-1} A\left(k^0\right). \tag{5.50}$$

5.3.2 Applying the 2nd-ASAM to Compute the Second-Order Response Sensitivities $S_{3i} \triangleq \partial^2 R / (\partial Q \partial \alpha_i)$

The G-differential, $DS_3(\boldsymbol{\alpha}^0)$, of the first-order sensitivity $S_3(\boldsymbol{\alpha})$ is computed by applying the definition given in Equation 5.38 to Equation 5.35 to obtain

$$DS_3\left(\boldsymbol{\alpha}^0\right) = -\frac{d}{d\varepsilon} \left\{ \int_{-a}^{a} \left(\psi^0 + \varepsilon \delta\psi\right) dx \right\}_{\varepsilon=0} = -\int_{-a}^{a} \delta\psi(x)\, dx, \tag{5.51}$$

where the function $\delta\psi(x)$ is the solution of the G-differentiated 1st-LASS, cf. Equations 5.27 and 5.28. Applying the definition of the G-differential to Equations 5.27 and 5.28 yields

$$D^0 \frac{d^2 h_\psi}{dx^2} - \Sigma_a^0 h_\psi(x) = -(\delta D)\frac{d^2 \psi^0}{dx^2} + (\delta \Sigma_a)\psi^0 + \left(\delta \Sigma_d^0\right)\delta(x-b), \tag{5.52}$$

$$h_\psi(\pm a) = 0. \tag{5.53}$$

As Equation 5.51 indicates, the entire contribution to $DS_3(\boldsymbol{\alpha}^0)$ comes from the indirect effect term; there is no direct-effect term contribution to $DS_3(\boldsymbol{\alpha}^0)$. Applying the general theory presented in Chapter 3, Section 3.3, yields the following 2nd-LASS for computing the second-level adjoint function $\boldsymbol{\psi}_3^{(2)}(x) \triangleq [\lambda_3(x), \theta_3(x)]^\dagger$:

$$D^0 \frac{d^2\theta_3}{dx^2} - \Sigma_a^0 \theta_3(x) = 1, \tag{5.54}$$

$$\theta_3(\pm a) = 0. \tag{5.55}$$

Since $DS_3(\boldsymbol{\alpha}^0)$ does not depend on $\delta\varphi(x)$, the adjoint function $\lambda_3(x)$, which would have corresponded to $\delta\varphi(x)$, is identically zero, since the adjoint system for $\lambda_3(x)$ would have been a linear homogeneous equation with zero source and zero boundary conditions. In terms of the second-level adjoint function $\boldsymbol{\psi}_3^{(2)}(x) \triangleq [0, \theta_3(x)]^\dagger$, the expression of the differential $DS_3(\boldsymbol{\alpha}^0)$ becomes

$$\begin{aligned} DS_3(\boldsymbol{\alpha}^0) &= \int_{-a}^{a} \theta_3(x) \left[(\delta D) \frac{d^2\psi^0}{dx^2} - (\delta\Sigma_a)\psi^0(x) - (\delta\Sigma_d)\delta(x-b) \right] dx \\ &= S_{31}(\delta\Sigma_a) + S_{32}(\delta D) + S_{33}(\delta Q) + S_{34}(\delta\Sigma_d), \end{aligned} \tag{5.56}$$

where

$$S_{31} \triangleq \frac{\partial^2 R}{\partial Q \partial \Sigma_a} = -\int_{-a}^{a} \theta_3(x)\psi^0(x)dx, \tag{5.57}$$

$$\begin{aligned} S_{32} &\triangleq \frac{\partial^2 R}{\partial Q \partial D} = \int_{-a}^{a} \theta_3(x) \frac{d^2\psi^0}{dx^2} dx \\ &= \left(\Sigma_a^0/D^0\right) \int_{-a}^{a} \theta_3(x)\psi^0(x)dx + \left(\Sigma_d^0/D^0\right) \int_{-a}^{a} \theta_3(x)\, \delta(x-b)dx, \end{aligned} \tag{5.58}$$

$$S_{33} \triangleq \frac{\partial^2 R}{\partial Q \partial Q} = 0, \tag{5.59}$$

$$S_{34} \triangleq \frac{\partial^2 R}{\partial Q \partial \Sigma_d} = -\int_{-a}^{a} \theta_3(x)\delta(x-b)dx = -\theta_3(b). \tag{5.60}$$

Note that *the adjoint system comprising Equations 5.54 and 5.55 does not even need to be solved,* since a simple comparison of this system to the original diffusion Equation 5.1 for the neutron flux $\varphi(x)$ shows that $\theta_3(x)=\varphi(x)/Q$. This fact can be readily verified by actually solving Equations 5.54 and 5.55 to obtain

$$\theta_3(x)=\frac{1}{\Sigma_a^0}\left[\frac{\cosh(xk^0)}{\cosh(ak^0)}-1\right]. \tag{5.61}$$

Introducing the above expression for $\theta_3(x)$ into Equations 5.57 through 5.60 and performing the respective integrations leads to the following results for the second-order sensitivities S_{3i}:

$$S_{31}(\boldsymbol{\alpha}^0)\triangleq\frac{\partial^2 R}{\partial Q\,\partial\Sigma_a}=-\frac{\Sigma_d^0}{\left(\Sigma_a^0\right)^2}A(k^0)+\frac{\Sigma_d^0}{2\Sigma_a^0\sqrt{D^0\Sigma_a^0}}B(k^0), \tag{5.62}$$

$$S_{32}(\boldsymbol{\alpha}^0)\triangleq\frac{\partial^2 R}{\partial Q\,\partial D}=-\frac{\Sigma_d^0 k^0}{2D^0\Sigma_a^0}B(k^0), \tag{5.63}$$

$$S_{34}(\boldsymbol{\alpha}^0)\triangleq\frac{\partial^2 R}{\partial Q\,\partial\Sigma_d}=-\theta_3(b)=\frac{A(k^0)}{\Sigma_a^0}. \tag{5.64}$$

Note that the computation of S_{34} via Equation 5.60 gives the same result as has already been obtained from the computation of S_{43} via Equation 5.50. This important consideration provides an *independent verification* that the adjoint functions $\theta_3(x)$ and, respectively, $\lambda_4(x)$, have indeed been computed correctly.

5.3.3 Applying the 2nd-ASAM to Compute the Second-Order Response Sensitivities $S_{1i}\triangleq\partial^2R/(\partial\Sigma_a\partial\alpha_i)$

The G-differential, $DS_1(\boldsymbol{\alpha}^0)$, of the first-order sensitivity $S_1(\boldsymbol{\alpha})$ is computed by applying the definition given in Equation 5.38 to Equation 5.33, to obtain

$$DS_1(\boldsymbol{\alpha}^0)=\frac{d}{d\varepsilon}\left\{\int_{-a}^{a}(\psi^0+\varepsilon\delta\psi)(\varphi^0+\varepsilon\delta\varphi)\,dx\right\}_{\varepsilon=0}=\int_{-a}^{a}(\psi^0\delta\varphi+\varphi^0\delta\psi)\,dx. \tag{5.65}$$

The G-differential $DS_1(\boldsymbol{\alpha}^0)$ depends on both $\delta\varphi$ and $\delta\psi$, where $\delta\varphi$ is the solution of Equations 5.14 and 5.15, while $\delta\psi$ is the solution of Equations 5.52 and 5.53. The direct-effect term contribution in Equation 5.65 is identically zero. The entire contribution to $DS_1(\boldsymbol{\alpha}^0)$ comes from the indirect-effect term. Since $DS_1(\boldsymbol{\alpha}^0)$ depends on both $\delta\varphi$ and $\delta\psi$, the second-level adjoint function

$\psi_1^{(2)}(x) \triangleq \left[\lambda_1(x), \theta_1(x)\right]^\dagger$, which will be used to compute $DS_1(\boldsymbol{\alpha}^0)$, will have two nonzero components, labeled as $\lambda_1(x)$ and $\theta_1(x)$, which will correspond to the variations $\delta\varphi$ and $\delta\psi$, respectively. Thus, the adjoint system for $\lambda_1(x)$, corresponding to Equations 5.14 and 5.15, as well as to the response $DS_1(\boldsymbol{\alpha}^0)$, is obtained as

$$D^0 \frac{d^2\lambda_1}{dx^2} - \Sigma_a^0 \lambda_1(x) = \psi^0, \tag{5.66}$$

$$\lambda_1(\pm a) = 0, \tag{5.67}$$

while the adjoint system for $\theta_1(x)$, corresponding to Equations 5.52 and 5.53, as well as to the response $DS_1(\boldsymbol{\alpha}^0)$, is obtained as

$$D^0 \frac{d^2\theta_1}{dx^2} - \Sigma_a^0 \theta_1(x) = \varphi^0, \tag{5.68}$$

$$\theta_1(\pm a) = 0. \tag{5.69}$$

In view of Equations 5.65 through 5.69, the G-differential $DS_1(\boldsymbol{\alpha}^0)$ can be expressed in terms of the adjoint functions $\lambda_1(x)$ and $\theta_1(x)$ in the form

$$\begin{aligned} DS_1(\boldsymbol{\alpha}^0) &= \int_{-a}^{a} \lambda_1(x)\left[-(\delta D)\frac{d^2\varphi^0}{dx^2} + (\delta\Sigma_a)\varphi^0(x) - (\delta Q)\right]dx \\ &+ \int_{-a}^{a} \theta_1(x)\left[-(\delta D)\frac{d^2\psi^0}{dx^2} + (\delta\Sigma_a)\psi^0(x) + (\delta\Sigma_d)\delta(x-b)\right]dx \\ &= S_{11}(\delta\Sigma_a) + S_{12}(\delta D) + S_{13}(\delta Q) + S_{14}(\delta\Sigma_d), \end{aligned} \tag{5.70}$$

where

$$S_{11}(\boldsymbol{\alpha}^0) \triangleq \frac{\partial^2 R}{\partial\Sigma_a \partial\Sigma_a} = \int_{-a}^{a} \left[\lambda_1(x)\varphi^0(x) + \theta_1(x)\psi^0(x)\right]dx, \tag{5.71}$$

$$\begin{aligned} S_{12}(\boldsymbol{\alpha}^0) \triangleq \frac{\partial^2 R}{\partial D \partial\Sigma_a} &= -\int_{-a}^{a}\left[\lambda_1(x)\frac{d^2\varphi^0}{dx^2} + \theta_1(x)\frac{d^2\psi^0}{dx^2}\right]dx \\ &= -\left(\Sigma_a^0/D\right)\int_{-a}^{a}\left[\lambda_1(x)\varphi^0(x) + \theta_1(x)\psi^0(x)\right]dx + \left(Q^0/D\right)\int_{-a}^{a}\lambda_1(x)dx \\ &-\left(\Sigma_d^0/D\right)\int_{-a}^{a}\theta_1(x)\delta(x-b)dx, \end{aligned} \tag{5.72}$$

$$S_{13}(\boldsymbol{\alpha}^0) \triangleq \frac{\partial^2 R}{\partial Q \partial \Sigma_a} = -\int_{-a}^{a} \lambda_1(x)dx, \tag{5.73}$$

$$S_{14}(\boldsymbol{\alpha}^0) \triangleq \frac{\partial^2 R}{\partial \Sigma_d \partial \Sigma_a} = \int_{-a}^{a} \theta_1(x)\,\delta(x-b)dx = \theta_1(b). \tag{5.74}$$

The solution $\lambda_1(x)$ of the adjoint system shown in Equations 5.66 and 5.67 is

$$\lambda_1(x) = \lambda_{1p}(x) - \sinh(x)\frac{\lambda_{1p}(a)-\lambda_{1p}(-a)}{2\sinh\left(ak^0\right)} - \cosh(x)\frac{\lambda_{1p}(a)+\lambda_{1p}(-a)}{2\cosh\left(ak^0\right)}, \tag{5.75}$$

with

$$\begin{aligned}
\lambda_{1p}(x) = {} & \frac{\Sigma_d^0}{D^0\Sigma_a^0}\frac{\sinh\left(bk^0-ak^0\right)}{\cosh\left(ak^0\right)} \\
& \times\left\{\frac{x}{2}\cosh\left(xk^0+ak^0\right)+\frac{1}{4k^0}\cosh\left(2xk^0+ak^0\right)\left[\sinh\left(xk^0\right)-\cosh\left(xk^0\right)\right]\right\} \\
& -\frac{\Sigma_d^0}{D^0\Sigma_a^0}H(x-b) \\
& \times\left\{\frac{x}{2}\cosh\left(xk^0-bk^0\right)+\frac{1}{4k^0}\cosh\left(2xk^0-bk^0\right)\left[\sinh\left(xk^0\right)-\cosh\left(xk^0\right)\right]\right\}.
\end{aligned} \tag{5.76}$$

Similarly, solving the adjoint system consisting of Equations 5.68 and 5.69 yields

$$\begin{aligned}
\theta_1(x) = {} & \frac{Q^0}{2\Sigma_a^0\sqrt{D^0\Sigma_a^0}}\frac{a\sinh\left(ak^0\right)\cosh\left(xk^0\right)-x\sinh\left(xk^0\right)\cosh\left(ak^0\right)}{\cosh^2\left(ak^0\right)} \\
& +\frac{Q^0}{\left(\Sigma_a^0\right)^2}\left[\frac{\cosh\left(xk^0\right)}{\cosh\left(ak^0\right)}-1\right].
\end{aligned} \tag{5.77}$$

The second-order sensitivity $S_{14}(\boldsymbol{\alpha}^0)$ can now be computed from Equations 5.77 and 5.74 to obtain the same expression as was obtained in Equation 5.48 for $S_{41}(\boldsymbol{\alpha}^0)$, namely

$$S_{14}(\boldsymbol{\alpha}^0) = \theta_1(b) = S_{41}(\boldsymbol{\alpha}^0) = Q^0\left[-\left(\Sigma_a^0\right)^{-2} A\left(k^0\right) + \frac{1}{2\Sigma_a^0\sqrt{D^0\Sigma_a^0}} B\left(k^0\right)\right]. \quad (5.48^*)$$

The above result provides an independent verification that the adjoint functions $\theta_1(x)$ and $\lambda_4(x)$ have been computed correctly. Next, inserting the expressions of $\lambda_1(x)$ and $\theta_1(x)$ into Equations 5.71 through 5.73, and performing the respective integrations, yields the following expressions:

$$S_{11}(\boldsymbol{\alpha}^0) \triangleq \frac{\partial^2 R}{\partial \Sigma_a^2} = \frac{Q^0\Sigma_d^0}{\left(\Sigma_a^0\right)^3}\left[2A\left(k^0\right) - \frac{5}{4}k^0 B\left(k^0\right) + \frac{1}{4}\left(k^0\right)^2 C\left(k^0\right)\right], \quad (5.78)$$

$$S_{21}(\boldsymbol{\alpha}^0) \triangleq \frac{\partial^2 R}{\partial D \partial \Sigma_a} = \frac{Q^0\Sigma_d^0}{4D^0\left(\Sigma_a^0\right)^2} k^0\left[B\left(k^0\right) - k^0 C\left(k^0\right)\right], \quad (5.79)$$

$$S_{31}(\boldsymbol{\alpha}^0) \triangleq \frac{\partial^2 R}{\partial Q \partial \Sigma_a} = S_{13}(\boldsymbol{\alpha}^0) = \frac{\Sigma_d^0}{\Sigma_a^0}\left[-\frac{A\left(k^0\right)}{\Sigma_a^0} + \frac{B\left(k^0\right)}{2\sqrt{D^0\Sigma_a^0}}\right]. \quad (5.62^*)$$

where

$$C\left(k^0\right) \triangleq \left.\frac{dB}{dk}\right|_{k^0} = \left.\frac{d^2 A}{dk^2}\right|_{k^0} = 2a^2\cosh\left(bk^0\right)\left[\cosh\left(ak^0\right)^{-3}\right]$$

$$+2ab\sinh\left(ak^0\right)\sinh\left(bk^0\right)\left[\cosh\left(ak^0\right)\right]^{-2} - \left(a^2+b^2\right)\cosh\left(bk^0\right)\left[\cosh\left(ak^0\right)\right]^{-1}. \quad (5.80)$$

Of course, the expression of $S_{31}(\boldsymbol{\alpha}^0) = S_{13}(\boldsymbol{\alpha}^0)$ has already been computed in Equation 5.62, which is the reason why it was accordingly labeled above, so its computation from Equation 5.73 serves only as an additional independent verification that the adjoint function $\lambda_1(x)$ has been correctly computed. This verification is in the same spirit as the verification that was performed on the adjoint functions $\theta_1(x)$ when computing $S_{14}(\boldsymbol{\alpha}^0)$ from Equation 5.74 and then ensuring that the resulting expression was identical to that already obtained in Equation 5.48 for $S_{41}(\boldsymbol{\alpha}^0)$. On the contrary, the expressions of $S_{11}(\boldsymbol{\alpha}^0)$ and $S_{21}(\boldsymbol{\alpha}^0)$ obtained in Equations 5.78 and 5.79, respectively, are in addition to the expressions already obtained for the second-order sensitivities in Sections 5.3.1 and 5.3.2.

5.3.4 Applying the 2nd-ASAM to Compute the Second-Order Response Sensitivities $S_{2i} \triangleq \partial^2 R/(\partial D \partial \alpha_i)$

The G-differential, $DS_2(\boldsymbol{\alpha}^0)$, of the first-order sensitivity $S_2(\boldsymbol{\alpha})$ is computed by applying the definition given in Equation 5.38 to Equation 5.34, to obtain

$$
\begin{aligned}
&DS_2(\boldsymbol{\alpha}^0) \\
&= \frac{d}{d\varepsilon}\left\{-\frac{\left[\Sigma_a^0+\varepsilon(\delta\Sigma_a)\right]}{D^0+\varepsilon(\delta D)}\int_{-a}^{a}\left(\psi^0+\varepsilon h_\psi\right)\left(\varphi^0+\varepsilon h_\varphi\right)dx+\left[Q^0+\varepsilon(\delta Q)\right]\int_{-a}^{a}\left(\psi^0+\varepsilon h_\psi\right)dx\right\}_{\varepsilon=0} \\
&= \left\{DS_2(\boldsymbol{\alpha}^0)\right\}_{direct}+\left\{DS_2(\boldsymbol{\alpha}^0)\right\}_{indirect},
\end{aligned}
\tag{5.81}
$$

where

$$
\begin{aligned}
\left\{DS_2(\boldsymbol{\alpha}^0)\right\}_{direct} &\triangleq \frac{(\delta D)\Sigma_a^0-(\delta\Sigma_a)D^0}{\left(D^0\right)^2}\int_{-a}^{a}\psi^0(x)\varphi^0(x)dx \\
&+\frac{D^0(\delta Q)-(\delta D)Q^0}{\left(D^0\right)^2}\int_{-a}^{a}\psi^0(x)dx,
\end{aligned}
\tag{5.82}
$$

and

$$
\left\{DS_2(\boldsymbol{\alpha}^0)\right\}_{indirect} \triangleq \int_{-a}^{a} h_\varphi(x)\left[-\frac{\Sigma_a^0}{D^0}\psi^0(x)\right]dx+\int_{-a}^{a} h_\psi(x)\left[\frac{Q^0}{D^0}-\frac{\Sigma_a^0}{D^0}\varphi^0(x)\right]dx. \tag{5.83}
$$

In Equations 5.81 through 5.83, the function $\delta\varphi$ is the solution of Equations 5.14 and 5.15, while the function $\delta\psi$ is the solution of Equations 5.52 and 5.53, just as was the case in the previous sections. Note that, in contradistinction to the G-differentials $DS_{i\neq 2}(\boldsymbol{\alpha}^0)$, the G-differential $DS_2(\boldsymbol{\alpha}^0)$ comprises not only an indirect-effect term but also a (nonzero) direct-effect term. The direct-effect term can be evaluated immediately since all of the quantities in Equation 5.82 are already available. On the contrary, the indirect-effect term $\{DS_2(\boldsymbol{\alpha}^0)\}_{indirect}$ depends on both $\delta\varphi$ and $\delta\psi$, so that both components of the second-level adjoint function $\psi_2^{(2)}(x) \triangleq \left[\lambda_2(x), \theta_2(x)\right]^\dagger$, which will be used to compute the G-differential $DS_2(\boldsymbol{\alpha}^0)$, will be nonzero. The components $\lambda_2(x)$ and $\theta_2(x)$ will be solutions corresponding to the systems adjoint to those satisfied by $\delta\varphi$ and $\delta\psi$, respectively, but with sources derived from the representation of $\{DS_2(\boldsymbol{\alpha}^0)\}_{indirect}$ in Equation 5.83.

Applying by now the general theory presented in Chapter 3 yields the following *adjoint system for* $\lambda_2(x)$ (which corresponds to Equations 5.14 and 5.15, as well as to the response in Equation 5.83):

$$D^0 \frac{d^2\lambda_2}{dx^2} - \Sigma_a^0 \lambda_2(x) = -\frac{\Sigma_a^0}{D^0}\psi^0, \tag{5.84}$$

$$\lambda_2(\pm a) = 0, \tag{5.85}$$

and the following *adjoint system for* $\theta_2(x)$:

$$D^0 \frac{d^2\theta_2}{dx^2} - \Sigma_a^0 \theta_2(x) = \frac{Q^0}{D^0} - \frac{\Sigma_a^0}{D^0}\varphi^0, \tag{5.86}$$

$$\theta_2(\pm a) = 0. \tag{5.87}$$

Note that Equations 5.86 and 5.87 correspond to Equations 5.52 and 5.53.

The G-differential $\{DS_2(\boldsymbol{\alpha}^0)\}_{indirect}$ can now be expressed in terms of the adjoint functions $\lambda_2(x)$ and $\theta_2(x)$ as

$$\begin{aligned} \left\{DS_2(\boldsymbol{\alpha}^0)\right\}_{indirect} &= \int_{-a}^{a} \lambda_2(x)\left[-(\delta D)\frac{d^2\varphi^0}{dx^2} + (\delta\Sigma_a)\varphi^0(x) - (\delta Q)\right]dx \\ &+ \int_{-a}^{a} \theta_2(x)\left[-(\delta D)\frac{d^2\psi^0}{dx^2} + (\delta\Sigma_a)\psi^0(x) + (\delta\Sigma_d)\delta(x-b)\right]dx. \end{aligned} \tag{5.88}$$

Combining the above result with the direct-effect term, $\left\{DS_2(\boldsymbol{\alpha}^0)\right\}_{direct}$, defined in Equation 5.82, yields

$$DS_2(\boldsymbol{\alpha}^0) = S_{21}(\delta\Sigma_a) + S_{22}(\delta D) + S_{23}(\delta Q) + S_{24}(\delta\Sigma_d), \tag{5.89}$$

where

$$\begin{aligned} S_{21}(\boldsymbol{\alpha}^0) &\triangleq \frac{\partial^2 R}{\partial D \partial \Sigma_a} = (-1/D^0)\int_{-a}^{a} \varphi^0(x)\psi^0(x)dx + \int_{-a}^{a} \lambda_2(x)\varphi^0(x)dx \\ &+ \int_{-a}^{a} \theta_2(x)\psi^0(x)dx, \end{aligned} \tag{5.90}$$

$$S_{22}(\boldsymbol{\alpha}^0) \triangleq \frac{\partial^2 R}{\partial D^2} = \frac{\Sigma_a^0}{\left(D^0\right)^2}\int_{-a}^{a}\varphi^0(x)\psi^0(x)dx - \frac{Q^0}{\left(D^0\right)^2}\int_{-a}^{a}\varphi^0(x)\psi^0(x)dx$$
$$\int_{-a}^{a}\left[\lambda_2(x)\frac{d^2\varphi^0}{dx^2} + \theta_2(x)\frac{d^2\psi^0}{dx^2}\right]dx, \tag{5.91}$$

$$S_{23}(\boldsymbol{\alpha}^0) \triangleq \frac{\partial^2 R}{\partial D\partial Q} = \left(1/D^0\right)\int_{-a}^{a}\psi^0(x)dx - \int_{-a}^{a}\lambda_2(x)dx, \tag{5.92}$$

$$S_{24}(\boldsymbol{\alpha}^0) \triangleq \frac{\partial^2 R}{\partial D\partial \Sigma_d} = \int_{-a}^{a}\theta_2(x)\delta(x-b)dx = \theta_2(b). \tag{5.93}$$

To compute the last integral on the right side of Equation 5.91, we use Equations 5.1 and 5.27 to obtain

$$\lambda_2(x)\frac{d^2\varphi^0}{dx^2} + \theta_2(x)\frac{d^2\psi^0}{dx^2} = \lambda_2(x)\left[\frac{\Sigma_a^0}{D^0}\varphi^0(x) - \frac{Q^0}{D^0}\right] + \theta_2(x)\left[\frac{\Sigma_a^0}{D^0}\psi^0(x) + \frac{\Sigma_d^0}{D^0}\delta(x-b)\right],$$

and subsequently replace the right side of the above expression in the last integral on the right side of Equation 5.91.

Comparing Equations 5.84 and 5.85 with Equations 5.66 and 5.67 readily reveals that the sources of these two linear systems are proportional to each other by the factor $\left(-\Sigma_a^0/D^0\right)$. Hence, the solutions of these two systems will be proportional to each other by the same factor. Consequently, *Equations 5.84 and 5.85 need not even be solved* since the solution $\lambda_2(x)$ can be immediately written down in terms of the $\lambda_1(x)$ as

$$\lambda_2(x) = \left(-\Sigma_a^0/D^0\right)\lambda_1(x), \tag{5.94}$$

with the expression of $\lambda_1(x)$ provided in Equation 5.75. On the contrary, the solution $\theta_2(x)$ of Equations 5.86 and 5.87 is obtained by conventional methods as

$$\theta_2(x) = \frac{Q^0}{2k^0\left(D^0\right)^2}\frac{x\sinh\left(xk^0\right)\cosh\left(ak^0\right) - a\sinh\left(ak^0\right)\cosh\left(xk^0\right)}{\cosh^2\left(ak^0\right)}. \tag{5.95}$$

Since the expression of the second-order sensitivity $S_{42}(\boldsymbol{\alpha}^0)$ is already available from Equation 5.49, it follows that Equation 5.93 provides an *exact verification* of the adjoint function $\theta_2(x)$. This is indeed the case, as demonstrated by the fact that

$$S_{24}(\boldsymbol{\alpha}^0) = S_{42}(\boldsymbol{\alpha}^0) = \theta_2(b). \tag{5.96}$$

The expression of $S_{21}(\boldsymbol{\alpha}^0)$ has already been obtained in Equation 5.79, and the expression of $S_{32}(\boldsymbol{\alpha}^0)$ has already been obtained in Equation 5.63. Therefore, Equations 5.90 through 5.92 can serve as multiple verifications of the correctness of the various computations. This is indeed the case: inserting the expressions given in Equations 5.94 and 5.95 [for $\lambda_2(x)$ and $\theta_2(x)$, respectively] together with the Equations 5.3 and 5.30 [i.e., the expressions of $\varphi^0(x)$ and $\psi^0(x)$] into Equations 5.90 through 5.92 and performing the respective integrations leads to the following results:

$$S_{21}\left(\boldsymbol{\alpha}^0\right) = S_{12}\left(\boldsymbol{\alpha}^0\right) = \frac{Q^0\,\Sigma_d^0}{4D^0\left(\Sigma_a^0\right)^2} k^0\left[B\left(k^0\right) - k^0 C\left(k^0\right)\right], \tag{5.97}$$

and

$$S_{23}(\boldsymbol{\alpha}^0) = S_{32}(\boldsymbol{\alpha}^0) = -\frac{\Sigma_d^0\,k^0}{2D^0\Sigma_a^0} B\left(k^0\right). \tag{5.98}$$

The only new result that is actually obtained when computing $DS_2(\boldsymbol{\alpha}^0)$, after all of the other differentials $DS_{i\neq 2}(\boldsymbol{\alpha}^0)$ have already been computed, is the expression for the second-order sensitivity $S_{22}(\boldsymbol{\alpha}^0)$. Thus, performing the integrations in Equation 5.91 yields

$$S_{22}(\boldsymbol{\alpha}^0) \triangleq \frac{\partial^2 R}{\partial D^2} = \frac{Q^0\Sigma_d^0}{4\left(D^0\right)^2}\left[3\frac{k^0}{\Sigma_a^0} B\left(k^0\right) + \frac{1}{D^0} C\left(k^0\right)\right]. \tag{5.99}$$

Having now finished the computation of all of the second-order response sensitivities, it is instructive to count the number of adjoint systems that needed to be solved, since these computations would be the equivalent of the "large-scale" computations performed when the 2nd-ASAM is used in practice. For one functional-type response, as for this illustrative example, the count is as follows:

1. *One adjoint computation* to determine the function $\psi^0(x)$, which suffices to compute, using just quadratures, all of the first-order sensitivities $S_i \triangleq \partial R/\partial\alpha_i$, $i = 1, 2, 3, 4$ and for subsequently determining (through a trivial division) all of the second-order sensitivities $S_{4i} \triangleq \partial^2 R/\partial Q\,\partial\alpha_i$, $i = 1, 2, 3, 4$
2. *Two adjoint computations* for determining $S_{1i} \triangleq \partial^2 R/\partial\Sigma_a\partial\alpha_i$, $i = 1, 2$
3. *One adjoint computation* for determining $S_{22} \equiv \partial^2 R/\partial D^2$

In summary, for each response, *four large-scale adjoint computations sufficed for the complete and exact computations of all (4) first- and (10) distinct second-order derivatives, including the various verifications of the adjoint functions and derivative-symmetry properties. Also, for all of the adjoint computations, only the sources on the right side of Equation 5.1* [i.e., the initial diffusion equation for the neutron flux $\varphi^0(x)$] *needed modifications*. The actual solver for the diffusion equation has remained unchanged for all of the adjoint computations. By comparison, forward methods, for example, would require 14 large-scale forward computations (i.e., solutions of the diffusion equation) for computing all of the first- and (distinct) second-order sensitivities.

It is instructive to compare the second-order sensitivities to the first-order response sensitivities, using the same data as used for producing the results shown in Tables 5.1 and 5.2. The numerical results for the second-order absolute sensitivities are presented in Table 5.3 (omitting the corresponding units), while the numerical results for the second-order relative sensitivities are presented in Table 5.4.

The values of the absolute second-order sensitivities presented in Table 5.3 will be used in the following section (Section 5.4) to illustrate their essential role for quantifying non-Gaussian features (e.g., asymmetries) of the various response distributions. To quantify asymmetries in a distribution, the third-order ("skewness") response correlations need to be computed (at the very least), which require the exact computation of (at least) the first- and second-order response sensitivities to model parameters. Furthermore, the results shown in Table 5.4 indicate that "conventional folklore," which holds that second-order sensitivities are sufficiently small to be neglected, potentially ignores significant information about the system under study. On the contrary, the results presented in Table 5.4 clearly indicate that second-order relative sensitivities can be just as large as, or even significantly larger than, the first-order sensitivities.

TABLE 5.3

Second-Order Absolute Sensitivities for Six Detector Responses

Second-Order Absolute Sensitivities	**R_1 (10 cm) R_4 (–10 cm)**	**R_2 (40 cm) R_5 (–40 cm)**	**R_3 (49.5 cm) R_6 (–49.5 cm)**
S_{11}	1.95×10^{13}	1.67×10^{13}	1.28×10^{12}
S_{12}	5.08×10^{7}	1.42×10^{11}	5.18×10^{10}
S_{13}	-1.92×10^{4}	-1.76×10^{4}	-1.67×10^{3}
S_{14}	-2.58×10^{10}	-2.36×10^{10}	-2.25×10^{9}
S_{22}	-4.59×10^{6}	-1.97×10^{9}	1.53×10^{10}
S_{23}	-1.33×10^{-2}	-1.24×10^{2}	-1.74×10^{2}
S_{24}	-1.79×10^{4}	-1.67×10^{8}	-2.33×10^{8}
S_{33}	0	0	0
S_{34}	5.08×10^{1}	4.92×10^{1}	8.16
S_{44}	0	0	0

Note: Units omitted.

TABLE 5.4

Second-Order Relative Sensitivities for Six Detector Responses

Second-Order Relative Sensitivities	R_1 (10 cm) R_4 (−10 cm)	R_2 (40 cm) R_5 (−40 cm)	R_3 (49.5 cm) R_6 (−49.5 cm)
S_{11}_rel	2.00	1.77	0.82
S_{12}_rel	4.24×10^{-5}	0.12	0.27
S_{13}_rel	−1.00	−0.95	−0.54
S_{14}_rel	−1.00	−0.95	−0.54
S_{22}_rel	-3.11×10^{-5}	−0.014	0.65
S_{23}_rel	-5.64×10^{-6}	−0.054	−0.46
S_{24}_rel	-5.64×10^{-6}	−0.054	−0.46
S_{33}_rel	0	0	0
S_{34}_rel	1.00	1.00	1.00
S_{44}_rel	0	0	0

5.4 Role of Second-Order Response Sensitivities for Quantifying Non-Gaussian Features of the Response Uncertainty Distribution

The effects of second-order response sensitivities for the paradigm neutron diffusion problem considered in this chapter can be readily illustrated by considering a simple case where (i) the model parameters are uncorrelated and normally distributed and (ii) only the first- and second-order response sensitivities are available. For these particular conditions, the response correlations presented in Chapter 1 reduce to the following expressions for the first three response moments:

1. The expected value of a response r_k, denoted as $[E(r_k)]^{UG}$, which arises due to uncertainties in uncorrelated normally distributed model parameters (the superscript *UG* indicates *uncorrelated Gaussian parameters*), is given by the following simplified form of Equation 1.18 from Chapter 1:

$$\left[E(r_k)\right]^{UG} = r_k\left(\boldsymbol{\alpha}^0\right) + \frac{1}{2}\sum_{i=1}^{N_\alpha} \frac{\partial^2 r_k}{\partial \alpha_i^2}\sigma_i^2, \qquad (5.100)$$

   where $r_k(\boldsymbol{\alpha}^0)$ denotes the computed nominal value of the response.

2. The covariance, $cov(r_k, r_l)$, between two responses r_k and r_l arising from normally distributed uncorrelated parameters is given by the following simplified form of Equation 1.19 from Chapter 1:

$$\left[cov(r_k, r_\ell)\right]^{UG} = \sum_{i=1}^{N_\alpha}\left(\frac{\partial r_k}{\partial \alpha_i}\frac{\partial r_\ell}{\partial \alpha_i}\right)\sigma_i^2 + \frac{1}{2}\sum_{i=1}^{N_\alpha}\left(\frac{\partial^2 r_k}{\partial \alpha_i^2}\right)\left(\frac{\partial^2 r_\ell}{\partial \alpha_i^2}\right)\sigma_i^4. \qquad (5.101)$$

The variance, $\text{var}(r_k)$, of a response r_k is obtained by setting $r_k \equiv r_l$ in the above expression to obtain

$$\left[\text{var}(r_k)\right]^{UG} = \sum_{i=1}^{N_\alpha}\left(\frac{\partial r_k}{\partial \alpha_i}\right)^2 \sigma_i^2 + \frac{1}{2}\sum_{i=1}^{N_\alpha}\left(\frac{\partial^2 r_k}{\partial \alpha_i^2}\right)^2 \sigma_i^4. \tag{5.102}$$

3. The third-order response correlation, $\mu_3(r_k, r_l, r_m)$, among three responses (r_k, r_l and r_m) is given by the following simplified form of Equation 1.20 from Chapter 1:

$$\left[\mu_3(r_k, r_l, r_m)\right]^{UG} = \sum_{i=1}^{N_\alpha}\left(\frac{\partial^2 r_k}{\partial \alpha_i^2}\frac{\partial r_l}{\partial \alpha_i}\frac{\partial r_m}{\partial \alpha_i} + \frac{\partial r_k}{\partial \alpha_i}\frac{\partial^2 r_l}{\partial \alpha_i^2}\frac{\partial r_m}{\partial \alpha_i} + \frac{\partial r_k}{\partial \alpha_i}\frac{\partial r_l}{\partial \alpha_i}\frac{\partial^2 r_m}{\partial \alpha_i^2}\right)\sigma_i^4. \tag{5.103}$$

In particular, the third-order central moment, $\mu_3(r_k)$, of the response r_k is determined by setting $k = l = m$ in Equation 5.103 to obtain

$$\left[\mu_3(r_k)\right]^{UG} = 3\sum_{i=1}^{N_\alpha}\left(\frac{\partial r_k}{\partial \alpha_i}\right)^2 \frac{\partial^2 r_k}{\partial \alpha_i^2}\sigma_i^4. \tag{5.104}$$

4. The *skewness*, $\gamma_1(r_k)$, of a response r_k can be computed using the customary definition

$$\gamma_1(r_k) \equiv \mu_3(r_k) / \left[\text{var}(r_k)\right]^{3/2}. \tag{5.105}$$

Recall from Chapter 1 that the *skewness* of a distribution quantifies the departure of the subject distribution from symmetry. Symmetric univariate distributions (e.g., the Gaussian) are characterized by $\gamma_1(r_k) = 0$. A distribution with a long right tail would have a positive skewness, while a distribution with a long left tail would have a negative skewness. In other words, if $\gamma_1(r_k) < 0$, then the respective distribution is skewed towards the left of the mean $[E(r_k)]^{UG}$, favoring lower values of r_k relative to $[E(r_k)]^{UG}$. On the contrary, if $\gamma_1(r_k) > 0$, then the respective distribution is skewed towards the right of the mean $[E(r_k)]^{UG}$, favoring higher values of r_k relative to $[E(r_k)]^{UG}$.

As discussed in Chapter 1, and as indicated by the expressions in Equations 5.100 through 5.105 for the particular case of uncorrelated and normally distributed parameters, the second-order sensitivities have the following impacts on the response moments:

1. They cause the expected value of the response, $[E(r_k)]^{UG}$, to differ from the computed nominal value of the response, $r_k(\boldsymbol{\alpha}^0)$.
2. They contribute to the response variances and covariances; however, since the contributions involving the second-order sensitivities are

multiplied by the fourth power of the parameters' standard deviations, the total of these contributions is expected to be relatively smaller than the contributions stemming from the first-order response sensitivities.

3. They *contribute decisively to causing asymmetries in the response distribution.* As Equation 5.103 indicates, neglecting the second-order sensitivities would nullify the third-order response correlations and hence would nullify the *skewness*, $\gamma_1(r_k)$, of a response r_k, cf. Equation 5.105. Consequently, any events occurring in a response's long and/or short tails, which are characteristic of rare but decisive events, would likely be missed.

To illustrate the above points numerically for our paradigm neutron diffusion problem, it is convenient to recall here that $a = 50$ cm, $S^0 = 10^7$ neutrons$\cdot$cm$^{-3}\cdot$s^{-1}, $\Sigma_a^0 = 0.0197\,\text{cm}^{-1}$, $D^0 = 0.16$ cm, and $\Sigma_d^0 = 7.438\,\text{cm}^{-1}$. Recall also that the computed responses had the following values in units of neutrons$\cdot$cm$^{-3}\cdot$s^{-1}: $R_1(10\text{ cm}) = R_4(-10\text{ cm}) = 3.77 \times 10^9$, $R_2(40\text{ cm}) = R_5(-40\text{ cm}) = 3.66 \times 10^9$ and $R_3(49.5\text{ cm}) = R_6(-49.5\text{ cm}) = 6.076 \times 10^8$.

The effects of response sensitivities and parameter uncertainties on the resulting uncertainties induced in the respected responses will be quantified by using Equations 5.100 through 5.105. Several combinations of parameters' relative standard deviations, denoted as $\delta\Sigma_A(\%)$, $\delta D(\%)$, $\delta Q\ (\%)$, and $\delta\Sigma_D(\%)$, will be considered for this purpose; the various cases and corresponding relative standard deviations are listed in Table 5.5.

The expression given in Equation 5.102 has been used to compute the standard deviations induced in responses by the parameter uncertainties listed in Table 5.5. The results of these computations are shown in Table 5.6.

TABLE 5.5

Model Parameters' Relative Standard Deviations

Case	$\delta\Sigma_a(\%)$	$\delta D(\%)$	$\delta Q\ (\%)$	$\delta\Sigma_d(\%)$
1	0	0	15	0
2	0	0	0	15
3	15	0	0	0
4	0	15	0	0
5	10	10	10	10

TABLE 5.6

Response Relative Standard Deviations

Case	$\sigma_{rel}(R_1) = \sigma_{rel}(R_4)$	$\sigma_{rel}(R_2) = \sigma_{rel}(R_5)$	$\sigma_{rel}(R_3) = \sigma_{rel}(R_6)$
1	0.150	0.150	0.150
2	0.150	0.150	0.150
3	0.150	0.145	0.082
4	9.8×10^{-7}	0.008	0.069
5	0.174	0.171	0.158

Examination of Cases 1 and 2 indicates that uncertainties arising solely in Q and Σ_d, respectively, propagate "one to one," inducing corresponding uncertainties in all of the responses; this fact is expected in view of their first-order relative sensitivities of unity, for all responses (as shown in Table 5.2). The uncertainties in Σ_a propagate almost one to one to produce the response uncertainties in the interior of the slab but cause diminishingly smaller uncertainties in the detector response towards the slab's boundaries; again, this behavior is expected in view of the values for the corresponding first-order sensitivities shown in Table 5.2. The impact of the uncertainties in D on the various responses is also dictated by the behavior of the corresponding first-order sensitivities shown in Table 5.2: the uncertainties in D have a negligible impact on the responses in the interior of the slab, but their impact increases as the detector responses approach the medium's boundaries. Finally, when all of the parameters are uncertain, their combined contribution to the respective response uncertainty is larger than any one parameter's uncertainty but is smaller than "the square-root of the sums of squares," because the response sensitivities to Σ_a and D are not unity. We also mention that the results (which are not shown, for brevity) of using Equation 5.101 to compute the response correlations (value ranges 0.998 to 1.000) indicate that the responses are fully (or almost fully) correlated. In all of these computations, *the second-order sensitivities had a negligible impact on the responses' standard deviations and correlations.*

As indicated by the results presented in Table 5.7 for the skewness of the various responses, the second-order sensitivities have a decisive impact on the asymmetries of the induced response distributions. Since the second-order sensitivities of the responses to both Q and Σ_d are zero, cf. Table 5.3, it follows that uncertainties arising solely in these quantities would not affect the symmetry of the resulting response distribution in our illustrative paradigm problem. This expectation is indeed confirmed by the results presented under Cases 1 and 2 in Table 5.7. On the contrary, Case 3 in Table 5.7 indicates that uncertainties stemming solely from the absorption cross section Σ_a cause the resulting response distributions to be significantly skewed to the right of their corresponding expected values, throughout the medium; this asymmetry diminishes somewhat

TABLE 5.7

Response Skewness

Case	$\gamma_1(R_1) = \gamma_1(R_5)$	$\gamma_1(R_2) = \gamma_1(R_6)$	$\gamma_1(R_3) = \gamma_1(R_7)$
1	0	0	0
2	0	0	0
3	0.8425	0.7945	0.6519
4	−1.5965	−0.1143	0.6147
5	0.1143	0.0955	0.0283

towards the medium's boundaries. Case 4 indicates a somewhat counter-intuitive behavior of the response distributions, in that uncertainties arising solely from the diffusion coefficient *D* cause the response distribution to undergo a complex behavior, as follows:

1. The response distribution is skewed significantly to the left of the response's expected value in the medium's interior.
2. The above-mentioned asymmetry diminishes towards the medium's boundary.
3. The response distribution becomes symmetric close to the boundary and then dramatically "shifts directions" by becoming markedly skewed to the right in the "boundary layer" close to the medium's physical boundary.

Finally, Case 5 in Table 5.7 indicates the "compensatory effects" that are induced by similar uncertainties in Σ_a and D, respectively. Similar standard deviations in Σ_a and D induce opposite asymmetries in the response distributions, but collectively they tend to diminish these opposing effects. Overall, the distribution remains somewhat skewed to the right of the responses' respective mean values throughout the medium.

Of course, none of the effects implied by the results shown in Table 5.7 would have been observed in the absence of the second-order response sensitivities.

The general theory underlying 2nd-ASAM indicates that, for a physical system comprising N_α parameters, the computation of all of the first- and second-order response sensitivities requires at most N_α large-scale computations involving 2nd-LASS. However, the illustrative application presented in this work has shown that the actual number of adjoint computations needed for computing all of the first- and second-order response sensitivities may be far less than N_α per response. In particular, for this illustrative problem, four large-scale adjoint computations sufficed for the complete and exact computations of all 4 first- and 10 distinct second-order derivatives. We have also shown that the construction and solution of the 2nd-LASS requires very little additional effort beyond the construction of the adjoint sensitivity system needed for computing the first-order sensitivities. Very significantly, only the sources on the right sides of the solver for the diffusion equation needed to be modified, while the left side of the differential equations remained unchanged.

All of the first-order response sensitivities to the model parameters had significant relative values of (or approaching) unity. After discussing the physical significance of the first-order sensitivities, it has been shown that most of the second-order relative sensitivities have magnitudes comparable to (i.e., as large as or larger than) those of the first-order relative sensitivities.

It has been also shown that the second-order sensitivities play the following important roles:

1. They cause the expected value of the response to differ from the computed nominal value of the response.
2. They contribute to increasing the response variances and modifying the response covariances, but their contribution is smaller than that stemming from the first-order response sensitivities.
3. They contribute decisively to causing asymmetries in the response distribution. Indeed, neglecting the second-order sensitivities would nullify the third-order response correlations and hence would nullify the *skewness* of the response.

5.5 Illustrative Application of First-Order Response Sensitivities for Predictive Modeling

This section will illustrate the use and influence of sensitivities for performing "predictive modeling," which, as discussed in Chapter 1, combines computational and experimental information to obtain optimally predicted "best-estimate" nominal values for the model responses and parameters while reducing their respective predicted uncertainties. For uncorrelated and normally distributed model parameters, the expressions provided in Equations 1.27, 1.37, 1.39, 1.42, 1.45, and 1.47 in Chapter 1 take on the following simplified forms:

$$\boldsymbol{\alpha}^{pred} = \boldsymbol{\alpha}^0 - \mathbf{C}_\alpha \mathbf{S}^\dagger \left(\mathbf{C}_{rc} + \mathbf{C}_m\right)^{-1} \left[\mathbf{R}\left(\boldsymbol{\alpha}^0\right) - \mathbf{r}_m\right], \tag{5.106}$$

$$\mathbf{r}^{pred} = \mathbf{r}^m + \mathbf{C}_m \left(\mathbf{C}_{rc} + \mathbf{C}_m\right)^{-1} \left[\mathbf{R}\left(\boldsymbol{\alpha}^0\right) - \mathbf{r}_m\right], \tag{5.107}$$

$$\mathbf{C}_\alpha^{pred} = \mathbf{C}_\alpha - \mathbf{C}_\alpha \mathbf{S}^\dagger \left(\mathbf{C}_{rc} + \mathbf{C}_m\right)^{-1} \left(\mathbf{C}_\alpha \mathbf{S}^\dagger\right)^\dagger, \tag{5.108}$$

$$\mathbf{C}_r^{pred} = \mathbf{C}_m - \mathbf{C}_m \left(\mathbf{C}_{rc} + \mathbf{C}_m\right)^{-1} \mathbf{C}_m^\dagger, \tag{5.109}$$

$$\mathbf{C}_{\alpha r}^{pred} = \mathbf{C}_\alpha \mathbf{S}^\dagger \left(\mathbf{C}_{rc} + \mathbf{C}_m\right)^{-1} \mathbf{C}_m^\dagger, \tag{5.110}$$

$$\chi^2 = \left[\mathbf{R}\left(\boldsymbol{\alpha}^0\right) - \mathbf{r}_m\right]^\dagger \left(\mathbf{C}_{rc} + \mathbf{C}_m\right)^{-1} \left[\mathbf{R}\left(\boldsymbol{\alpha}^0\right) - \mathbf{r}_m\right]. \tag{5.111}$$

The covariance matrix of the computed responses, $\mathbf{C}_{rc}$, which appears in Equations 5.106 through 5.111, arises from the covariance matrix of the parameters and the response sensitivities. Using just the first-order sensitivities, the matrix $\mathbf{C}_{rc}$ has the following "sandwich" form, which is a particular form of Equation 1.19 in Chapter 1:

$$\mathbf{C}_{rc} \triangleq \left\langle \left[\mathbf{r}-\mathbf{r}^{c}\left(\alpha^{0}\right)\right]\left[\mathbf{r}-\mathbf{r}^{c}\left(\alpha^{0}\right)\right]^{\dagger} \right\rangle = \mathbf{S}\mathbf{C}_{\alpha}\mathbf{S}^{\dagger}. \tag{5.112}$$

5.5.1 Assimilating an Imprecise but Consistent Measurement

Consider, on the one hand, that a measurement is performed at the location $b = 10$ cm in the positive direction (i.e., to the right of the origin at $x = 0$), yielding a nominal value of $r_1^m = 3.40 \times 10^9$ neutrons $\cdot$ cm$^{-3} \cdot$ s^{-1}, with a relative standard deviation of 25%, that is, $\Delta r_m / r_m = 8.50 \times 10^8$ neutrons$\cdot$cm$^{-3}\cdot$s^{-1}, which corresponds to a variance $C_{rr}^{m} = 7.22 \times 10^{17} \left(\text{neutrons} \cdot \text{cm}^{-3} \cdot \text{s}^{-1}\right)^2$. On the other hand, the nominal value of the computed response at $b = 10$ cm is obtained by using Equation 5.7 as $R(10\text{ cm}) \triangleq r_1^c = 3.77 \times 10^9$ neutrons $\cdot$ cm$^{-3} \cdot$ s^{-1}.

Omitting, for simplicity, the various units, the absolute sensitivities $\mathbf{S}_{r\alpha}^{1\times 4}$ of the response R(10 cm) to the model parameters are computed from Equations 5.33 through 5.36 to obtain the row vector:

$$\mathbf{S}_{r\alpha}^{1\times 4} \equiv \left[\partial R(10\text{ cm})/\partial \Sigma_a,\ \partial R(10\text{ cm})/\partial D,\ \partial R(10\text{ cm})/\partial S,\ \partial R(10\text{ cm})/\partial \Sigma_d\right]$$
$$= \left[-1.92 \times 10^{11},\ -1.33 \times 10^{5},\ 3.77 \times 10^{2},\ 5.08 \times 10^{8}\right]. \tag{5.113}$$

In particular, the relative sensitivities of R(10 cm) with respect to Σ_a and D are, respectively,

$$\left[\partial R(10\text{ cm})/\partial \Sigma_a\right]\left[\Sigma_a^0 / R(10\text{ cm})\right] = -0.99999, \tag{5.114}$$

$$\left[\partial R(10\text{ cm})/\partial D\right]\left[D^0 / R(10\text{ cm})\right] = -5.64 \times 10^{-6}. \tag{5.115}$$

Consider, for this illustrative example, that all of the uncorrelated model parameters have the following relative standard deviations: $\Delta\Sigma_a^0 / \Sigma_a^0 = 5\%$, $\Delta D^0/D^0 = 5\%$, $\Delta S^0/S^0 = 15\%$, so that the covariance matrix for the model parameters is

$$\mathbf{C}_{\alpha\alpha} = \begin{pmatrix} \left(9.85 \times 10^{-5}\right)^2 & 0 & 0 & 0 \\ 0 & \left(8.0 \times 10^{-3}\right)^2 & 0 & 0 \\ 0 & 0 & \left(1.5 \times 10^{6}\right)^2 & 0 \\ 0 & 0 & 0 & \left(7.44 \times 10^{-1}\right)^2 \end{pmatrix}. \tag{5.116}$$

Since there is only one response in this example, Equation 5.112 reduces to a scalar value for the covariance, C_{rr}^{comp}, of r_1^c. Thus, using the sensitivities provided in Equation 5.113 together with the covariance matrix given in Equation 5.116 yields the following value for the variance of R(10 cm):

$$C_{rr}^{comp}\left(r_1^c\right)=\left[\mathbf{S}_{r\alpha}^{1\times 4}\right]\mathbf{C}_{\alpha\alpha}\left[\mathbf{S}_{r\alpha}^{1\times 4}\right]^{\dagger}=4.98\times 10^{17}\left(\mathrm{n}\cdot\mathrm{cm}^{-3}\cdot\mathrm{s}^{-1}\right)^2\equiv\left[\Delta C_{rr}^{comp}\left(r_1^c\right)\right]^2, \quad (5.117)$$

where $\Delta C_{rr}^{comp}\left(r_1^c\right)$ denotes the *absolute standard deviation* of r_1^c; the *relative standard deviation*, $\Delta C_{rr}^{comp}\left(r_1^c\right)/r_1^c$, is readily obtained to be $\Delta C_{rr}^{comp}\left(r_1^c\right)/r_1^c=18.72\%$. Note that this standard deviation is smaller than would be naively expected from the popular recipe of taking the square root of the sum of the squared errors, which would in this case amount to 19.37%. The smaller actual computed standard deviation is due to the fact that the sensitivity of R(10 cm) with respect to D is vanishingly small, so that the error in D does not practically contribute to the computed error in R(10 cm).

Note also that the measurement and computation can be considered as being consistent with one another, since

$$(std)_{m,1}+(std)_{comp,1}=1.56\times 10^9\ \mathrm{n}\cdot\mathrm{cm}^{-3}\cdot\mathrm{s}^{-1}>\left|d_1\right|\triangleq\left|r_1^c-r_1^m\right|=3.7\times 10^8\ \mathrm{n}\cdot\mathrm{cm}^{-3}\cdot\mathrm{s}^{-1}.$$

Using the information from Equations 5.116, 5.113, and 5.117 in Equations 5.106 through 5.110 leads to the following predicted best-estimate nominal values for the calibrated parameters and their accompanying predicted covariance matrices:

$$\begin{aligned}&\Sigma_a^{pred}=0.0197\ \mathrm{cm}^{-1},\quad D^{pred}=0.16\ \mathrm{cm},\\&S^{pred}=9.7372\times 10^6\ \mathrm{neutrons}\cdot\mathrm{cm}^{-3}\cdot\mathrm{s}^{-1},\quad \Sigma_d^{pred}=7.351\ \mathrm{cm}^{-1},\end{aligned} \quad (5.118)$$

$$\begin{aligned}C_{\alpha\alpha}^{pred}=&\begin{pmatrix}9.71\times 10^{-4} & 0 & 0 & 0\\ 0 & 0.80\times 10^{-2} & 0 & 0\\ 0 & 0 & 1.28\times 10^{6} & 0\\ 0 & 0 & 0 & 6.99\times 10^{-1}\end{pmatrix}\\ &\times\begin{pmatrix}1.0 & -1.67\times 10^{-7} & 1.03\times 10^{-1} & 0.63\times 10^{-1}\\ -1.67\times 10^{-7} & 1.0 & 5.75\times 10^{-7} & 3.50\times 10^{-7}\\ 1.03\times 10^{-1} & 5.75\times 10^{-7} & 1.0 & -2.17\times 10^{-1}\\ 0.63\times 10^{-1} & 3.50\times 10^{-7} & -2.17\times 10^{-1} & 1.0\end{pmatrix}\\ &\times\begin{pmatrix}9.71\times 10^{-4} & 0 & 0 & 0\\ 0 & 0.80\times 10^{-2} & 0 & 0\\ 0 & 0 & 1.28\times 10^{6} & 0\\ 0 & 0 & 0 & 6.99\times 10^{-1}\end{pmatrix},\end{aligned} \quad (5.119)$$

$$r_1^{pred} = 3.62 \times 10^9 \, \text{n} \cdot \text{cm}^{-3} \cdot \text{s}^{-1}, \; C_{rr}^{pred} = 2.95 \times 10^{17} \left(\text{n} \cdot \text{cm}^{-3} \cdot \text{s}^{-1}\right)^2, \quad (5.120)$$

$$C_{\alpha r}^{pred} = \left(5.43 \times 10^8\right) \times \begin{pmatrix} -2.08 \times 10^{-1} & -1.15 \times 10^{-6} & 7.18 \times 10^{-1} & 4.37 \times 10^{-1} \end{pmatrix} \times \begin{pmatrix} 9.71 \times 10^{-4} & 0 & 0 & 0 \\ 0 & 0.80 \times 10^{-2} & 0 & 0 \\ 0 & 0 & 1.28 \times 10^6 & 0 \\ 0 & 0 & 0 & 6.99 \times 10^{-1} \end{pmatrix}. \quad (5.121)$$

Note from Equation 5.120 that the predicted absolute and, respectively, relative standard deviations are

$$\Delta C_{rr}^{pred} = 5.43 \times 10^8 \, \text{neutrons} \cdot \text{cm}^{-3} \cdot \text{s}^{-1}, \; \Delta C_{rr}^{pred} / r_1^{pred} = 14.99\%. \quad (5.122)$$

Figure 5.1 shows the spatial variation of the original nominal computed values and standard deviations (depicted using dotted lines), together with the best-estimate values and corresponding standard deviations

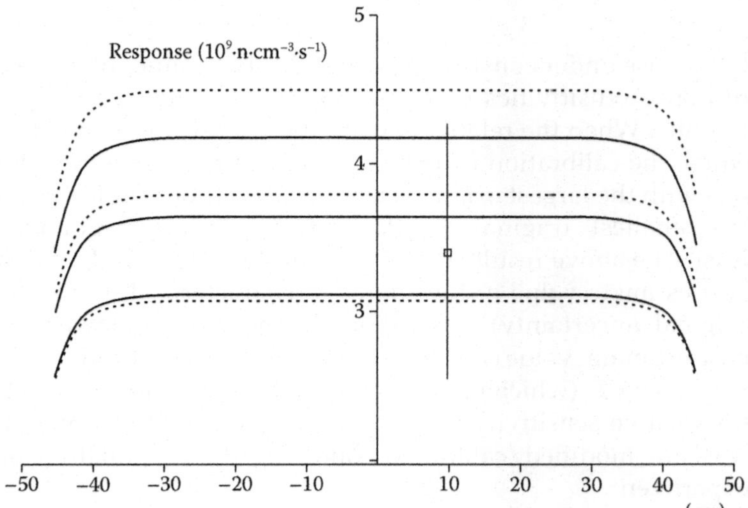

FIGURE 5.1
Original nominal computed values and standard deviations (depicted using dotted lines) together with the best-estimate values and corresponding standard deviations (depicted using solid lines), after assimilating an imprecise (relative standard deviation = 25%) but consistent measurement.

(depicted using solid lines). The value of the consistency indicator is computed from Equation 5.109 to obtain $\chi^2 = 0.12$, which is a reasonable value for a single measurement.

All together, the information obtained in the foregoing leads to the following conclusions:

1. As would be expected intuitively, the best-estimate response value, r_1^{pred}, is predicted by the predicted modeling formalism to fall in between the experimentally measured and the originally computed values. The predicted value, r_1^{pred}, is somewhat closer to the computed values, because the relative standard deviation of the computed response, $\Delta C_{rr}^{comp}/R(10\,\text{cm}) = 18.72\%$, is smaller than the relative standard deviation of the measured response (25%), that is,

$$r_1^m = 3.40 < r_1^{pred} = 3.62 < r_1^c = 3.77\left(\times 10^9\ \text{n}\cdot\text{cm}^{-3}\cdot\text{s}^{-1}\right). \tag{5.123}$$

2. Again as expected, the assimilation of a consistent experiment reduces the best-estimate variance, C_{rr}^{pred}, of r_1^{pred} to a value that is smaller than the variances of both the computed and measured responses, that is,

$$\begin{aligned} C_{rr}^{pred} &= 2.95\times 10^{17} < C_{rr}^{comp} = 4.98\times 10^{17} < C_{rr}^{m} \\ &= 7.22\times 10^{17}\left[\left(\text{neutrons}\cdot\text{cm}^{-3}\cdot\text{s}^{-1}\right)^2\right]. \end{aligned} \tag{5.124}$$

3. In the case under consideration, the absolute values of the relative response sensitivities with respect to S, Σ_d, and Σ_a are all equal to unity. When the relative sensitivities to several parameters are equal, the calibration usually affects most markedly the parameters with the largest original standard deviation, while those with the smallest original standard deviation are calibrated the least. The above results underscore this general trend, in that the best-estimate value of the source S (characterized by the largest original uncertainty) was calibrated the most, followed by the best-estimate value for Σ_d (the second largest original uncertainty) and Σ_a (which had the smallest original uncertainty). Since the relative sensitivity to D is vanishingly small, this parameter was not modified/calibrated following the assimilation of the experiment.
4. Comparing the matrices in Equations 5.116 and 5.119 indicates that the predicted best-estimate standard deviations for the parameters are also reduced by comparison to their original values. Due to the same reasons as discussed above (i.e., equal absolute values for the relative sensitivities of S, Σ_d, and Σ_a and the values of the respective

original standard deviations), the best-estimate standard deviations for S and Σ_d are reduced the most, and the standard deviation for Σ_a is reduced the least, while that for D is not reduced (due to the extremely small accompanying sensitivity).

5. Although the parameters were originally considered as uncorrelated, cf. Equation 5.116, the assimilation of the experimental response induces correlations among the best-estimate parameters, as indicated by the correlation matrix in Equation 5.119, as follows:
   a. Σ_a^{pred} becomes positively correlated with both S^{pred} and Σ_d^{pred}.
   b. S^{pred} and Σ_d^{pred} become anticorrelated.
   c. D^{pred} remains uncorrelated to the other best-estimate parameters, due to its vanishingly small relative sensitivity.
6. As indicated by the correlation matrix $C_{\alpha r}^{pred}$ in Equation 5.121, the best-estimate response becomes anticorrelated to Σ_a^{pred}, remains uncorrelated to D^{pred}, and becomes positively correlated to both S^{pred} and Σ_d^{pred}, even though originally the response was considered to have been uncorrelated to any of the original model parameters.
7. Due to its vanishingly small relative sensitivity, the diffusion parameter is the sole quantity that has remained essentially unaffected by the assimilation of the experiment and subsequent calibration procedure, that is, $D^0 = D^{pred}$ and $\Delta D^0/D^0 = \Delta D^{pred}/D^{pred} = 5\%$.

5.5.2 Assimilating a Precise and Consistent Measurement

Consider now that the measurement described in the previous section 5.5.1, with a nominally measured value of $r_1^m = 3.40\times10^9$ neutrons $\cdot\,\mathrm{cm}^{-3}\cdot\mathrm{s}^{-1}$, is performed more precisely, with a relative standard deviation $\Delta r_1^m/r_1^m = 5\%$ (rather than 25%) or corresponding variance $C_{rr}^m = 2.89\times10^{15}\left(\text{neutrons}\cdot\mathrm{cm}^{-3}\cdot\mathrm{s}^{-1}\right)^2$. In this case, therefore, the experimental variance is considerably smaller than the computed response variance $C_{rr}^{comp} = 4.98\times10^{17}\left(\text{neutrons}\cdot\mathrm{cm}^{-3}\cdot\mathrm{s}^{-1}\right)^2$. The measurement and computation can be considered as being consistent with one another, since

$$(std)_{m,1}+(std)_{comp,1}=1.41\times10^9\,\mathrm{n}\cdot\mathrm{cm}^{-3}\cdot\mathrm{s}^{-1}>|d_1|\equiv\left|r_1^c-r_1^m\right|=3.7\times10^8\,\mathrm{n}\cdot\mathrm{cm}^{-3}\cdot\mathrm{s}^{-1}.$$

Figure 5.2 shows the spatial variation of the original nominal computed values and standard deviations (depicted using dotted lines), together with the best-estimate response values and corresponding standard deviations (depicted using solid lines). The value of the consistency indicator is $\chi^2 = 0.27$, which quantitatively highlights the improvement over the imprecise measurement considered in the previous section.

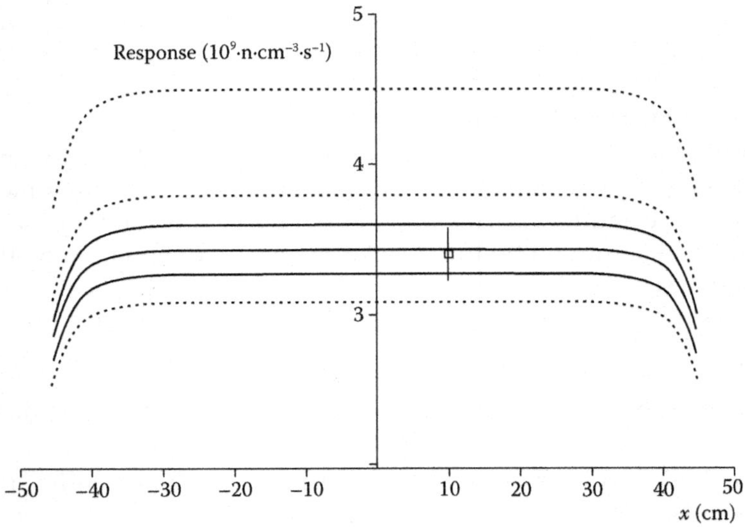

FIGURE 5.2
Original nominal computed values and standard deviations (depicted using dotted lines), together with the best-estimate values and corresponding standard deviations (depicted using solid lines), after assimilating a precise (relative standard deviation = 5%) and consistent measurement.

After assimilating the "precise" experiment, the best-estimate predicted values are obtained as follows:

$$\Sigma_a^{pred} = 0.0198\ \text{cm}^{-1};\ D^{pred} = 0.16\ \text{cm};$$
$$S^{pred} = 9.393\times 10^6\ \text{neutrons}\cdot\text{cm}^{-3}\cdot\text{s}^{-1};\ \Sigma_d^{pred} = 7.2371\ \text{cm}^{-1}; \tag{5.125}$$

$$C_{\alpha\alpha}^{pred} = \begin{pmatrix} 9.51\times10^{-4} & 0 & 0 & 0 \\ 0 & 0.80\times10^{-2} & 0 & 0 \\ 0 & 0 & 9.39\times10^{5} & 0 \\ 0 & 0 & 0 & 6.35\times10^{-1} \end{pmatrix}$$
$$\times \begin{pmatrix} 1.0 & -3.94\times10^{-7} & 3.34\times10^{-1} & 1.63\times10^{-1} \\ 3.94\times10^{-7} & 1.0 & 1.82\times10^{-6} & 8.91\times10^{-7} \\ 3.34\times10^{-1} & 1.82\times10^{-6} & 1.0 & -7.57\times10^{-1} \\ 1.63\times10^{-1} & 8.91\times10^{-7} & -7.57\times10^{-1} & 1.0 \end{pmatrix} \tag{5.126}$$
$$\times \begin{pmatrix} 9.51\times10^{-4} & 0 & 0 & 0 \\ 0 & 0.80\times10^{-2} & 0 & 0 \\ 0 & 0 & 9.39\times10^{5} & 0 \\ 0 & 0 & 0 & 6.35\times10^{-1} \end{pmatrix}.$$

The adjusted best-estimate response value, the corresponding best-estimate response variance, and the best-estimate response-parameter correlation matrix are obtained, respectively, as

$$r_1^{pred} = 3.42 \times 10^9 \mathrm{n} \cdot \mathrm{cm}^{-3} \cdot \mathrm{s}^{-1}, C_{rr}^{pred} = 2.73 \times 10^{16} \left(\mathrm{n} \cdot \mathrm{cm}^{-3} \cdot \mathrm{s}^{-1}\right)^2, \quad (5.127)$$

$$C_{\alpha r}^{pred} = \left(1.65 \times 10^8\right)$$
$$\times \begin{pmatrix} -6.47 \times 10^{-2} & -3.52 \times 10^{-7} & 2.99 \times 10^{-1} & 1.46 \times 10^{-1} \end{pmatrix} \quad (5.128)$$
$$\times \begin{pmatrix} 9.51 \times 10^{-4} & 0 & 0 & 0 \\ 0 & 0.80 \times 10^{-2} & 0 & 0 \\ 0 & 0 & 9.39 \times 10^5 & 0 \\ 0 & 0 & 0 & 6.35 \times 10^{-1} \end{pmatrix}.$$

The following conclusions can be drawn from the information shown in Equations 5.125 through 5.128:

1. The more precise measurement considered in this section has a considerably stronger influence on the best-estimate predicted values than the less precise measurement considered in the previous section, pulling the predicted values strongly towards the experimentally measured value, as depicted in Figure 5.2. Thus, the predicted best-estimate response value $r_1^{pred} = 3.42 \times 10^9 \,\mathrm{neutrons} \cdot \mathrm{cm}^{-3} \cdot \mathrm{s}^{-1}$ is much closer to the experimentally measured value than to the originally computed value, as shown below:

$$r_1^m = 3.40 < r_1^{pred} = 3.42 < r_1^c = 3.77 \left(\times 10^9 \,\mathrm{n} \cdot \mathrm{cm}^{-3} \cdot \mathrm{s}^{-1}\right). \quad (5.129)$$

2. The best-estimate response variance C_{rr}^{pred} is also reduced more significantly in this case than in the case of the less precise measurement considered in the Section 5.5.1, the best-estimate predicted relative standard deviation is 4.83%. As before, the predicted best-estimate response variance is smaller than both the measured and the computed response variances, that is,

$$\begin{aligned} C_{rr}^{pred} &= 2.73 \times 10^{16} < C_{rr}^m = 2.89 \times 10^{16} < C_{rr}^{comp} \\ &= 4.98 \times 10^{17} \left[\left(\mathrm{neutrons} \cdot \mathrm{cm}^{-3} \cdot \mathrm{s}^{-1}\right)^2\right]. \end{aligned} \quad (5.130)$$

3. The response sensitivities to model parameters are obviously unaffected by measurements, so they have remained the same as

in the Section 5.5.1. Thus, the amount of parameter calibration would also be expected to mirror the kind of calibrations noticed in the Section 5.5.1. This is indeed the case, as indicated by the results in Equation 5.125: the largest adjustments occurred for the source S (characterized by the largest original uncertainty), followed by Σ_d (the second largest original uncertainty); Σ_a underwent a minute adjustment, while D was not adjusted (just as in the previous section). However, since the experimental response variance is much smaller than in the Section 5.5.1, the corresponding parameter adjustments are also larger in the present case than in the Section 5.5.1.

4. Comparing the best-estimate parameter covariance matrix $C_{\alpha\alpha}^{pred}$ in Equation 5.126 with the original parameter covariance matrix in Equation 5.116 shows that the best-estimate standard deviations for the parameters are reduced more by the more precise experiment considered in this case (by comparison to the uncertainty reductions achieved in Section 5.5.1). The largest reductions of the standard deviations have occurred in the following order: for S (best-estimate relative standard deviation reduced to 10%), for Σ_d (best-estimate relative standard deviation reduced to 8.78%), and for Σ_a (best-estimate relative standard deviation reduced to 4.79%). Due to its small sensitivity, the uncertainty in D was hardly reduced (to 4.99%).
5. Just as in Section 5.5.1, the originally uncorrelated parameters became correlated after calibration, as indicated by the correlation matrix $C_{\alpha\alpha}^{pred}$ in Equation 5.126. The nature of the induced correlations has retained the same character as in Section 5.5.1 (i.e., Σ_a^{pred} has become positively correlated with both S^{pred} and Σ_d^{pred}; S^{pred} and Σ_d^{pred} have become anticorrelated; and D^{pred} remained uncorrelated to the other best-estimate parameters), but the induced best-estimate correlations have become stronger (i.e., larger in absolute values) than in the case presented in Section 5.5.1, by a factor of about three.
6. Just as in the Section 5.5.1, the best-estimate response has become anticorrelated to Σ_a^{pred}, remained uncorrelated to D^{pred}, and became positively correlated to both S^{pred} and Σ_d^{pred}. However, as indicated by the correlation matrix $C_{\alpha r}^{pred}$ in Equation 5.128, these correlations have become weaker, also by a factor of about three, by comparison to the assimilation of the imprecise measurement considered in Section 5.5.1.
7. As in Section 5.5.1, because of its vanishingly small relative sensitivity, the diffusion coefficient has remained essentially unaffected by the assimilation of the experiment and subsequent calibration procedure.

5.5.3 Assimilating Two Consistent Measurements

Consider now that two measurements are performed at the locations $b = \pm 10$ cm, symmetric with respect to the slab's mid-plane at $x = 0$. Consider that the measured nominal values of the detector response and the corresponding absolute standard deviations (*std*) and relative standard deviations (*rsd*) are as follows:

$$\begin{aligned} &r_1^m \equiv R(\textit{measured at } 10 \text{ cm}) = 3.40\times 10^9 \,\text{n}\cdot\text{cm}^{-3}\cdot\text{s}^{-1}, \\ &(std)_{m,1} = 1.7\times 10^8 \,\text{n}\cdot\text{cm}^{-3}\cdot\text{s}^{-1}, \quad rsd(r_1^m) = 5\%, \end{aligned} \tag{5.131}$$

$$\begin{aligned} &r_2^m \equiv R(\textit{measured at } -10 \text{ cm}) = 3.59\times 10^9 \,\text{n}\cdot\text{cm}^{-3}\cdot\text{s}^{-1}, \\ &(std)_{m,2} = 2.15\times 10^8 \,\text{n}\cdot\text{cm}^{-3}\cdot\text{s}^{-1}, \quad rsd(r_2^m) = 6\%. \end{aligned} \tag{5.132}$$

Consider further that the above measurements are uncorrelated, so that the (2×2) covariance matrix of the measured responses is simply

$$\mathbf{C}_{rr}^{(2\times 2)m} = \begin{pmatrix} (1.7\times 10^8)^2 & 0 \\ 0 & (2.15\times 10^8)^2 \end{pmatrix}. \tag{5.133}$$

The nominal computed response values at the above locations are obtained from Equation 5.7. In view of the problem's symmetry, that is,

$$R(x) = R(-x) \tag{5.134}$$

a single computation of the response suffices to obtain

$$R(\textit{computed at } 10 \text{ cm}) \equiv r_1^c = 3.77\times 10^9 \text{n}\cdot\text{cm}^{-3}\cdot\text{s}^{-1}, \tag{5.135}$$

$$R(\textit{computed at } -10 \text{ cm}) \equiv r_2^c = 3.77\times 10^9 \text{n}\cdot\text{cm}^{-3}\cdot\text{s}^{-1}. \tag{5.136}$$

Dealing with two responses means that two distinct adjoint computations (one for each of the responses) would be needed to compute the matrix of response sensitivities to the four model parameters. Because of the symmetric responses considered in this illustrative example, however, a single computation of the adjoint function $\psi(x)$ suffices to obtain all eight sensitivities. Omitting the details of these straightforward computations, the results for both the absolute values of the (2×4) matrix of sensitivities (omitting the obvious units, for simplicity) and for the relative sensitivities are as follows:

$$\mathbf{S}_{r\alpha}^{(2\times 4)} \equiv \left(\frac{\partial R_i}{\partial \alpha_j}\right)_{i=1,2;j=1,\dots,4} = \begin{pmatrix} -1.92\times 10^{11} & -1.33\times 10^5 & 3.78\times 10^2 & 5.08\times 10^8 \\ -1.92\times 10^{11} & -1.33\times 10^5 & 3.78\times 10^2 & 5.08\times 10^8 \end{pmatrix}, \tag{5.137}$$

$$\mathbf{S}_{r\alpha}^{(2\times 4)rel} \equiv \left(\frac{\partial R_i}{\partial \alpha_j}\frac{\Delta\alpha_j}{R_i}\right)_{i=1,2;j=1,\ldots,4} = \begin{pmatrix} -1.00 & -5.64\times 10^{-6} & 1.00 & 1.00 \\ -1.00 & -5.64\times 10^{-6} & 1.00 & 1.00 \end{pmatrix}. \quad (5.138)$$

As indicated by the above sensitivities, the responses are equally (and strongly) sensitive to the model parameters Σ_a, S, and Σ_d. The responses are insensitive to D, indicating that the neutron diffusion process in the region around the two measurements is practically irrelevant to the detectors' responses.

Omitting again (for easy notation) the explicit writing of the various units, replacing the absolute sensitivities from Equation 5.137 and the parameter covariance matrix from Equation 5.116 into Equation 5.112 yields the following covariance matrix (i.e., uncertainties in computed responses due to the uncertainties in the model parameters) for the computed responses:

$$\mathbf{C}_{rr}^{(2\times 2)comp} = \mathbf{S}_{r\alpha}^{(2\times 4)}\mathbf{C}_{\alpha\alpha}\left(\mathbf{S}_{r\alpha}^{(2\times 4)}\right)^{\dagger} = \begin{pmatrix} 4.99\times 10^{17} & 4.99\times 10^{17} \\ 4.99\times 10^{17} & 4.99\times 10^{17} \end{pmatrix}. \quad (5.139)$$

Note that $\mathbf{C}_{rr}^{(2\times 2)comp}$ has a rank of unity (rather than rank two), since the two *computed* (as opposed to measured!) responses are actually identical due to the problem's symmetry. As the result in Equation 5.139 indicates, the absolute standard deviations of the two computed responses are identical, both having the value

$$(std)_{comp,1} = (std)_{comp,2} = 7.06\times 10^{8}\ \mathrm{n}\cdot\mathrm{cm}^{-3}\cdot\mathrm{s}^{-1}. \quad (5.140)$$

Note also that the sums of the respective measured and computed response standard deviations (in units of $\mathrm{n}\cdot\mathrm{cm}^{-3}\cdot\mathrm{s}$) are larger than the "deviations" between the nominally measured and computed response values, namely:

$$(std)_{m,1} + (std)_{comp,1} = 8.76\times 10^{8} > d_1 \equiv r_1^c - r_1^m = 3.78\times 10^{8}, \quad (5.141)$$

$$(std)_{m,2} + (std)_{comp,2} = 9.22\times 10^{8} > d_2 = r_2^c - r_2^m = 1.89\times 10^{8}. \quad (5.142)$$

The above relations indicate that both measurements are "consistent" with the computations. This fact is confirmed by computing the value of the "consistency indicator" χ^2 using Equation 5.111 to obtain $\chi^2 = 0.327$, which indicates good consistency of the measured and computed data for the problem's two responses (i.e., two degrees of freedom).

The predicted optimal values for responses, parameters, and their predicted reduced uncertainties are computed using the results from Equations 5.116, and 5.131 through 5.139, into Equations 5.106 through 5.110. Performing these computations leads to the following predicted value for the best-estimate parameters and (reduced) accompanying predicted standard deviations:

$$\Sigma_a^{pred} = 0.0198 \text{ cm}^{-1}, \quad rsd\left(\Sigma_a^{pred}\right) = 4.8\%, \tag{5.143}$$

$$D^{pred} = 0.160 \text{ cm}, \quad rsd\left(D^{pred}\right) = 5.00\%, \tag{5.144}$$

$$S^{pred} = 9.498 \times 10^6 \text{n} \cdot \text{cm}^{-3} \cdot \text{s}^{-1}, \quad rsd\left(S^{pred}\right) = 9.72\%, \tag{5.145}$$

$$\Sigma_d^{pred} = 7.272 \text{cm}^{-1}, \quad rsd\left(\Sigma_d^{pred}\right) = 8.7\%. \tag{5.146}$$

The predicted parameter covariance matrix is

$$\mathbf{C}_{\alpha\alpha}^{pred} = \begin{pmatrix} 9.03\times10^{-7} & -3.06\times10^{-12} & 3.06\times10^{2} & 1.01\times10^{-4} \\ -3.06\times10^{-12} & 6.40\times10^{-5} & 1.40\times10^{-2} & 4.63\times10^{-9} \\ 3.06\times10^{2} & 1.40\times10^{-2} & 8.53\times10^{11} & -4.62\times10^{5} \\ 1.01\times10^{-4} & 4.63\times10^{-9} & -4.62\times10^{5} & 4.01\times10^{-1} \end{pmatrix}, \tag{5.147}$$

indicating that the initially uncorrelated parameters have become correlated after assimilation of the experimental information.

The best-estimate predicted response values together with their predicted relative standard deviations are obtained as follows:

$$at\ (10 \text{ cm}):\ r_1^{pred} = 3.48\times10^9 \text{n}\cdot\text{cm}^{-3}\cdot\text{s}^{-1}, \quad rsd\left(r_1^{pred}\right) = 3.8\%, \tag{5.148}$$

$$at\ (-10 \text{ cm}):\ r_2^{pred} = 3.48\times10^9 \text{n}\cdot\text{cm}^{-3}\cdot\text{s}^{-1}, \quad rsd\left(r_2^{pred}\right) = 3.8\%. \tag{5.149}$$

The complete predicted response covariance matrix is

$$\mathbf{C}_{rr}^{(2\times2)pred} = \begin{pmatrix} 1.72\times10^{16} & 1.72\times10^{16} \\ 1.72\times10^{16} & 1.72\times10^{16} \end{pmatrix}, \tag{5.150}$$

indicating that *the predicted best-estimate responses have become fully correlated after the assimilation of the experimental information even though the measurements were uncorrelated. This full correlation is actually introduced* not by the measurements but *by the symmetry of the problem, via the respective response sensitivities* (cf. Equation 5.139).

Finally, the predicted response–parameter correlation matrix is obtained from Equation 5.141 as

$$\mathbf{C}_{\alpha r}^{pred} = \begin{pmatrix} -6.40\times10^{3} & -2.93\times10^{-1} & 2.92\times10^{13} & 9.67\times10^{6} \\ -6.40\times10^{3} & -2.93\times10^{-1} & 2.92\times10^{13} & 9.67\times10^{6} \end{pmatrix}, \quad (5.151)$$

indicating that, after assimilating experimental information, the predicted parameters and responses have become correlated. Note that $\mathbf{C}_{\alpha r}^{pred}$ is also of rank one (rather than rank two), just as $\mathbf{C}_{rr}^{pred}$, even though the two measurements differed from each other. These facts are consequences of the problem's symmetry (i.e., the fact that the location of the measurements was symmetrical with respect to the problem's mid-plane, even though the nominal/mean values and standard deviations of the two measurements differed from each other).

Figure 5.3 depicts the computed responses with their computed standard deviations (depicted with dotted lines) and also the predicted best-estimate responses together with their predicted standard deviations (depicted with solid lines).

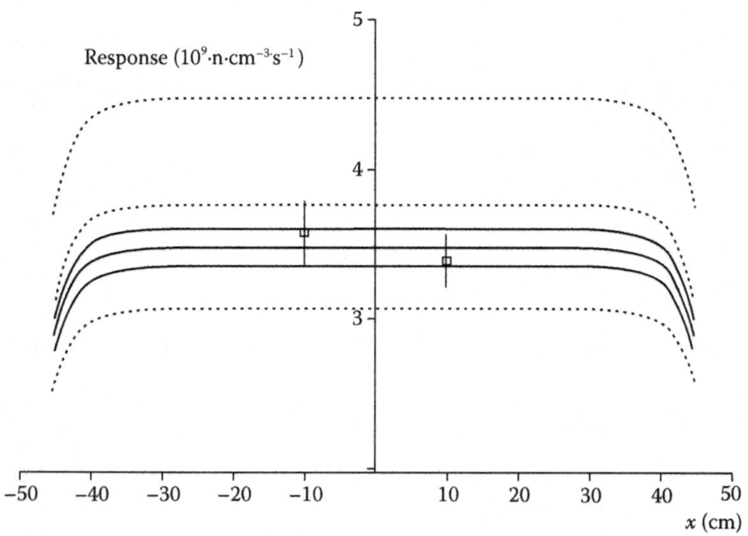

FIGURE 5.3
Computed and predicted responses and their corresponding standard deviations after assimilating two consistent measurements for predictive modeling.

The following conclusions can be drawn from the results obtained in this subsection:

1. As depicted in Figure 5.3, the best-estimate response value $r_1^{pred} = r_2^{pred} = 3.42 \times 10^9 \text{neutrons} \cdot \text{cm}^{-3} \cdot \text{s}^{-1}$ is much closer to the experimentally measured value than to the originally computed value.

$$r_1^m = 3.40 < r_1^{pred} = r_2^{pred} = 3.48 < r_2^m = 3.59 < r_1^c = r_2^c \\ = 3.77\left(\times 10^9 \,\text{n} \cdot \text{cm}^{-3} \cdot \text{s}^{-1}\right). \tag{5.152}$$

2. The predicted relative (and also the predicted absolute) response standard deviation is smaller than both the measured and the computed response variances:

$$rsd\left(r_1^{pred}\right) = rsd\left(r_2^{pred}\right) = 3.8\% < rsd\left(r_1^m\right) = 5\% < rsd\left(r_2^m\right) = 6\% \\ < rsd\left(r_1^c\right) = rsd\left(r_2^c\right) = 18.72\%; \tag{5.153}$$

$$(std)_{pred,1} = (std)_{pred,2} = 1.32 < (std)_{m,1} = 1.7 < (std)_{m,2} = 2.15 \\ < (std)_{comp,1} = (std)_{comp,2} = 7.06 \left(\times 10^8 \text{n} \cdot \text{cm}^{-3} \cdot \text{s}\right). \tag{5.154}$$

3. The largest adjustments occurred for the source S (characterized by the largest original uncertainty and largest sensitivity), followed by Σ_d (the second largest original uncertainty and also largest sensitivity); Σ_a underwent a minute adjustment (because its original uncertainty was small, although its sensitivity was comparable to those of S and Σ_d), while D was not adjusted (because of its vanishingly small sensitivity).
4. Comparing the best-estimate parameter covariance matrix $C_{\alpha\alpha}^{pred}$ in Equation 5.147 with the original parameter covariance matrix in Equation 5.116 shows that the best-estimate predicted standard deviations for the parameters are reduced. The largest reductions of the standard deviations occurred, in order, for S (best-estimate relative standard deviation reduced to 10%), Σ_d (best-estimate relative standard deviation reduced to 8.78%), and Σ_a (best-estimate relative standard deviation reduced to 4.79%), while that for D was hardly reduced (4.99%).
5. The originally uncorrelated parameters became correlated after calibration, as indicated by the correlation matrix $C_{\alpha\alpha}^{pred}$ in Equation 5.147: Σ_a^{pred} has become positively correlated with both S^{pred} and Σ_d^{pred}; S^{pred} and Σ_d^{pred} have become anticorrelated; and D^{pred} has remained uncorrelated to the other predicted best-estimate parameters.

6. As indicated by the correlation matrix $C_{\alpha r}^{pred}$ in Equation 5.151, the best-estimate response has become anticorrelated to Σ_a^{pred}, has remained essentially uncorrelated to D^{pred}, and has become positively correlated to both S^{pred} and Σ_d^{pred}.
7. Because of its vanishingly small relative sensitivity, the diffusion parameter has remained essentially unaffected by the assimilation of the experiments and subsequent calibration within the predictive modeling procedure.

5.5.4 Assimilating Four Consistent Measurements

Consider now that, in addition to the two measurements r_1^m and r_2^m performed at the locations $b = \pm 10\,\text{cm}$ (and previously analyzed in Section 5.5.3), two further measurements, denoted as r_3^m and r_4^m, are performed at the locations $b = \pm 40$ cm, also symmetrically with respect to the slab's mid-plane at $x = 0$. Consider, for simplicity, that all four measurements are uncorrelated and have the following mean values and standard deviations:

$$r_1^m \triangleq R(\textit{measured at } 10 \text{ cm}) = 3.40 \times 10^9 \left(\text{n} \cdot \text{cm}^{-3} \cdot \text{s}^{-1}\right), \quad rsd\left(r_1^m\right) = 5\%, \tag{5.155}$$

$$r_2^m \triangleq R(\textit{measured at } -10 \text{ cm}) = 3.59 \times 10^9 \left(\text{n} \cdot \text{cm}^{-3} \cdot \text{s}^{-1}\right), \quad rsd\left(r_2^m\right) = 6\%, \tag{5.156}$$

$$r_3^m \triangleq R(\textit{measured at } -40 \text{ cm}) = 3.77 \times 10^9 \left(\text{n} \cdot \text{cm}^{-3} \cdot \text{s}^{-1}\right), \quad rsd\left(r_3^m\right) = 5\%, \tag{5.157}$$

$$r_4^m \triangleq R(\textit{measured at } 40 \text{ cm}) = 3.74 \times 10^9 \left(\text{n} \cdot \text{cm}^{-3} \cdot \text{s}^{-1}\right), \quad rsd\left(r_4^m\right) = 5\%. \tag{5.158}$$

It follows from the data given in Equations 5.155 through 5.158 that the covariance matrix of these four uncorrelated measured responses is

$$\mathbf{C}_{rr}^{(4\times 4)m} = \begin{pmatrix} \left(1.7 \times 10^8\right)^2 & 0 & 0 & 0 \\ 0 & \left(2.15 \times 10^8\right)^2 & 0 & 0 \\ 0 & 0 & \left(1.89 \times 10^8\right)^2 & 0 \\ 0 & 0 & 0 & \left(1.87 \times 10^8\right)^2 \end{pmatrix}. \tag{5.159}$$

Using Equation 5.7 yields the following nominal values for the computed responses at the four locations:

$$r_1^c \triangleq R(\textit{computed at } 10 \text{ cm}) = 3.77 \times 10^9 \left(\text{n} \cdot \text{cm}^{-3} \cdot \text{s}^{-1}\right), \tag{5.160}$$

$$r_2^c \triangleq R(\textit{computed at } -10 \text{ cm}) = 3.77 \times 10^9 \left(\text{n} \cdot \text{cm}^{-3} \cdot \text{s}^{-1}\right), \quad (5.161)$$

$$r_3^c \triangleq R(\textit{computed at } -40 \text{ cm}) = 3.66 \times 10^9 \left(\text{n} \cdot \text{cm}^{-3} \cdot \text{s}^{-1}\right), \quad (5.162)$$

$$r_4^c \triangleq R(\textit{computed at } 40 \text{ cm}) = 3.66 \times 10^9 \left(\text{n} \cdot \text{cm}^{-3} \cdot \text{s}^{-1}\right). \quad (5.163)$$

Because of the problem's symmetry, that is, $R(x) = R(-x)$, the sensitivities and the responses to the model parameters are computed using only two (rather than four) adjoint computations: one adjoint computation for obtaining the sensitivities of R (at ±10 cm) and the second adjoint computation for obtaining the sensitivities of R (at ±10 cm). Omitting again the units, the matrix of absolute sensitivities thus obtained is presented below:

$$\mathbf{S}_{r\alpha}^{(4\times4)} \triangleq \left(\frac{\partial R_i}{\partial \alpha_j}\right)_{i,j=1,\ldots,4} = \begin{pmatrix} -1.92\times10^{11} & -1.33\times10^{5} & 3.78\times10^{2} & 5.08\times10^{8} \\ -1.92\times10^{11} & -1.33\times10^{5} & 3.78\times10^{2} & 5.08\times10^{8} \\ -1.76\times10^{11} & -1.24\times10^{9} & 3.56\times10^{2} & 4.92\times10^{8} \\ -1.76\times10^{11} & -1.24\times10^{9} & 3.56\times10^{2} & 4.92\times10^{8} \end{pmatrix}. \quad (5.164)$$

The matrix of relative sensitivities, which indicate the importance of model parameters in influencing the responses considered, can also be readily computed as

$$\mathbf{S}_{r\alpha}^{(4\times4)rel} \equiv \left(\frac{\partial R_i}{\partial \alpha_j}\frac{\Delta\alpha_j}{R_i}\right) = \begin{pmatrix} -1.00 & -5.64\times10^{-6} & 1.00 & 1.00 \\ -1.00 & -5.64\times10^{-6} & 1.00 & 1.00 \\ -9.46\times10^{-1} & -5.41\times10^{-2} & 1.00 & 1.00 \\ -9.46\times10^{-1} & -5.41\times10^{-2} & 1.00 & 1.00 \end{pmatrix}. \quad (5.165)$$

Replacing the absolute sensitivities from Equation 5.164 and the parameter covariance matrix from Equation 5.116 into Equation 5.112 yields the following covariance matrix (i.e., uncertainties in computed responses due to the uncertainties in the model parameters) for the computed responses (omitting the various units):

$$\mathbf{C}_{rr}^{(4\times4)comp} = \mathbf{S}_{r\alpha}^{4\times4}\mathbf{C}_{\alpha\alpha}\left(\mathbf{S}_{r\alpha}^{4\times4}\right)^{\dagger} = \begin{pmatrix} 4.99\times10^{17} & 4.99\times10^{17} & 4.82\times10^{17} & 4.82\times10^{17} \\ 4.99\times10^{17} & 4.99\times10^{17} & 4.82\times10^{17} & 4.82\times10^{17} \\ 4.82\times10^{17} & 4.82\times10^{17} & 4.66\times10^{17} & 4.66\times10^{17} \\ 4.82\times10^{17} & 4.82\times10^{17} & 4.66\times10^{17} & 4.66\times10^{17} \end{pmatrix}. \quad (5.166)$$

Note that the rank of $\mathbf{C}_{rr}^{(4\times4)comp}$ is *two* (rather than four), since only two of the four *computed* responses are actually distinct from one another: the problem's symmetry imposes $r_1^c \equiv r_2^c$ and $r_3^c \equiv r_4^c$, so that only the values of r_1^c and r_3^c are distinct from the other two values. This is in contrast to the *measured* responses: all four measured responses differ from one another.

Computing the square roots of the elements on the diagonal of the matrix obtained in Equation 5.166 yields the following values for the absolute standard deviations of the computed responses:

$$\begin{aligned} std\left(r_1^c\right) = std\left(r_2^c\right) = 7.06\times10^8\left(\mathrm{n}\cdot\mathrm{cm}^{-3}\cdot\mathrm{s}^{-1}\right), \\ std\left(r_3^c\right) = std\left(r_4^c\right) = 6.83\times10^8\left(\mathrm{n}\cdot\mathrm{cm}^{-3}\cdot\mathrm{s}^{-1}\right). \end{aligned} \tag{5.167}$$

Note also that the sums of the respective measured and computed response standard deviations (in units of $[\mathrm{n}\cdot\mathrm{cm}^{-3}\cdot\mathrm{s}^{-1}]$) are larger than the "deviations" between the nominally measured and computed response values, namely:

$$std\left(r_1^c\right)+std\left(r_1^m\right)=8.76\times10^8>\left|r_1^c-r_1^m\right|=3.78\times10^8, \tag{5.168}$$

$$std\left(r_2^c\right)+std\left(r_2^m\right)=9.22\times10^8>\left|r_2^c-r_2^m\right|=1.89\times10^8, \tag{5.169}$$

$$std\left(r_3^c\right)+std\left(r_3^m\right)=8.71\times10^8>\left|r_3^c-r_3^m\right|=1.1\times10^8, \tag{5.170}$$

$$std\left(r_4^c\right)+std\left(r_4^m\right)=8.69\times10^8>\left|r_4^c-r_4^m\right|=7.32\times10^7. \tag{5.171}$$

The above inequalities indicate that the measured and computed responses are consistent with each other. This fact is confirmed by using Equation 5.111 to compute the "consistency indicator" χ^2, which yields χ^2 (per 4 degrees of freedom) = 1.21; this value indicates very good consistency among the measured and computed data for the problem's four responses.

The predicted optimal values for responses, parameters, and their predicted reduced uncertainties are computed using the results from Equations 5.116 and 5.155 through 5.166 in Equations 5.106 through 5.110. Performing these computations leads to the following predicted value for the best-estimate parameters and (reduced) accompanying predicted standard deviations:

1. Predicted nominal values for the model parameters:

$$\Sigma_a^{pred}=0.0198\ \mathrm{cm}^{-1},\quad rsd\left(\Sigma_a^{pred}\right)=4.79\%, \tag{5.172}$$

$$D^{pred}=0.1591\ \mathrm{cm},\quad rsd\left(D^{pred}\right)=5.00\%, \tag{5.173}$$

$$S^{pred} = 9.85 \times 10^{6}\,\mathrm{n \cdot cm^{-3} \cdot s^{-1}}, \quad rsd\left(S^{pred}\right) = 9.21\%, \tag{5.174}$$

$$\Sigma_d^{pred} = 7.388\ \mathrm{cm^{-1}}, \quad rsd\left(\Sigma_d^{pred}\right) = 8.53\%. \tag{5.175}$$

2. Predicted covariance matrices for the model parameters:

$$\mathbf{C}_{\alpha\alpha}^{pred} = \begin{pmatrix} 9.03\times10^{-7} & 6.75\times10^{-9} & 3.03\times10^{2} & 1.00\times10^{-4} \\ 6.75\times10^{-9} & 6.38\times10^{-5} & 7.37\times10^{1} & 2.44\times10^{-5} \\ 3.03\times10^{2} & 7.37\times10^{1} & 8.24\times10^{11} & -4.71\times10^{5} \\ 1.00\times10^{-4} & 2.44\times10^{-5} & -4.71\times10^{5} & 3.97\times10^{-1} \end{pmatrix}$$

$$= \begin{pmatrix} 9.50\times10^{-4} & 0 & 0 & 0 \\ 0 & 7.99\times10^{-3} & 0 & 0 \\ 0 & 0 & 9.08\times10^{5} & 0 \\ 0 & 0 & 0 & 6.30\times10^{-1} \end{pmatrix}$$

$$\times \begin{pmatrix} 1.0 & -8.89\times10^{-4} & 3.51\times10^{-1} & 1.67\times10^{-1} \\ -8.89\times10^{-4} & 1.0 & 1.02\times10^{-2} & 4.84\times10^{-3} \\ 3.51\times10^{-1} & 1.02\times10^{-2} & 1.0 & -8.24\times10^{-1} \\ 1.67\times10^{-1} & 4.84\times10^{-3} & -8.24\times10^{-1} & 1.0 \end{pmatrix} \tag{5.176}$$

$$\times \begin{pmatrix} 9.50\times10^{-4} & 0 & 0 & 0 \\ 0 & 7.99\times10^{-3} & 0 & 0 \\ 0 & 0 & 9.08\times10^{5} & 0 \\ 0 & 0 & 0 & 6.30\times10^{-1} \end{pmatrix}.$$

3. Predicted nominal values for the model responses:

$$at\ (10\ \mathrm{cm}): \ r_1^{pred} = 3.66\times10^{9}\,\mathrm{n \cdot cm^{-3} \cdot s^{-1}}, \quad rsd\left(r_1^{pred}\right) = 2.58\%, \tag{5.177}$$

$$at\ (-10\ \mathrm{cm}): \ r_2^{pred} = 3.66\times10^{9}\,\mathrm{n \cdot cm^{-3} \cdot s^{-1}}, \quad rsd\left(r_2^{pred}\right) = 2.59\%, \tag{5.178}$$

$$at\ (-40\ \mathrm{cm}): \ r_3^{pred} = 3.56\times10^{9}\,\mathrm{n \cdot cm^{-3} \cdot s^{-1}}, \quad rsd\left(r_3^{pred}\right) = 2.59\%, \tag{5.179}$$

$$at\ (40\ \mathrm{cm}): \ r_4^{pred} = 3.56\times10^{9}\,\mathrm{n \cdot cm^{-3} \cdot s^{-1}}, \quad rsd\left(r_4^{pred}\right) = 2.58\%. \tag{5.180}$$

4. Predicted covariance matrix for the model responses:

$$\mathbf{C}_{rr}^{pred} = \begin{pmatrix} 9.04\times10^{15} & 9.04\times10^{15} & 8.64\times10^{15} & 8.64\times10^{15} \\ 9.04\times10^{15} & 9.04\times10^{15} & 8.64\times10^{15} & 8.64\times10^{15} \\ 8.64\times10^{15} & 8.64\times10^{15} & 8.45\times10^{15} & 8.45\times10^{15} \\ 8.64\times10^{15} & 8.64\times10^{15} & 8.45\times10^{15} & 8.45\times10^{15} \end{pmatrix}. \quad (5.181)$$

5. Predicted parameter–response correlation matrix:

$$\mathbf{C}_{\alpha r}^{pred} = \begin{pmatrix} 9.51\times10^{7} & 0 & 0 & 0 \\ 0 & 9.51\times10^{7} & 0 & 0 \\ 0 & 0 & 9.19\times10^{7} & 0 \\ 0 & 0 & 0 & 9.19\times10^{7} \end{pmatrix}$$

$$\times \begin{pmatrix} -8.65\times10^{-2} & 5.12\times10^{-2} & 1.60\times10^{-1} & 7.62\times10^{-2} \\ -8.65\times10^{-2} & 5.12\times10^{-2} & 1.60\times10^{-1} & 7.62\times10^{-2} \\ 1.72\times10^{-2} & -5.62\times10^{-2} & 1.96\times10^{-1} & 9.34\times10^{-2} \\ 1.72\times10^{-2} & -5.62\times10^{-2} & 1.96\times10^{-1} & 9.34\times10^{-2} \end{pmatrix} \quad (5.182)$$

$$\times \begin{pmatrix} 9.50\times10^{-4} & 0 & 0 & 0 \\ 0 & 7.99\times10^{-3} & 0 & 0 \\ 0 & 0 & 9.08\times10^{5} & 0 \\ 0 & 0 & 0 & 6.30\times10^{-1} \end{pmatrix}.$$

Figure 5.4 depicts the computed responses with their computed standard deviations (depicted using dotted lines), as well as the predicted best-estimate responses together with their predicted standard deviations (depicted using solid lines).

The following important physical conclusions can be drawn from the results presented in this section:

1. The predictive modeling procedure preserves the problem's symmetry, so that $r_1^{pred} = r_2^{pred}$ and $q_1^{pred} = q_2^{pred}$, while all four predicted responses remain consistent with one another within their "one-standard deviation" ranges.
2. As in the cases of one and, respectively, two assimilated measurements, the predicted relative (and also the predicted absolute) response standard deviation after the assimilation of four consistent measurements is smaller than both the measured and the computed

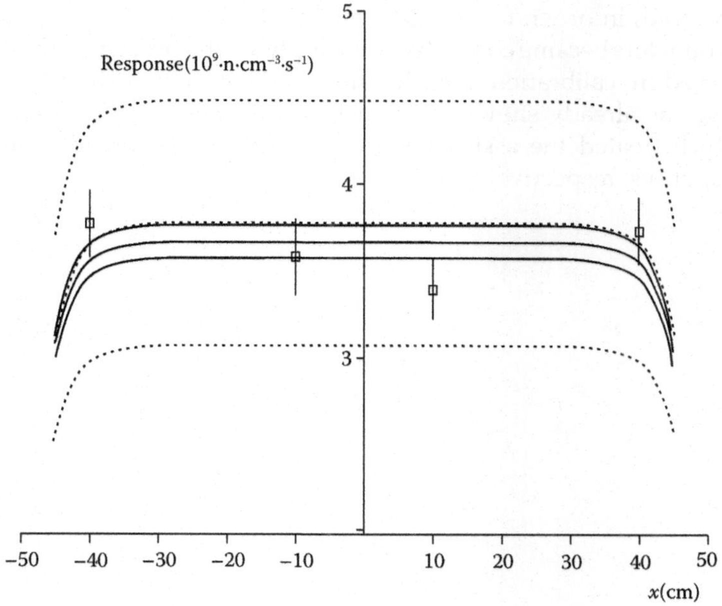

FIGURE 5.4
Computed and predicted responses and their corresponding standard deviations after assimilating four consistent measurements for predictive modeling.

response variances. The results clearly highlight the fact that the assimilation of additional consistent measurements further reduces the uncertainty in the predicted response. In all cases, the assimilation of measurements has reduced the uncertainty of the predicted responses considerably below the uncertainties (which arise from uncertainties in the model parameters) in the computed responses. The following sequence of inequalities holds:

$$rsd\left(r_i^{pred}\,\middle|\,with\,4\,measurements\right) \cong 2.59\% < rsd\left(r_i^{pred}\,\middle|\,with\,2\,measurements\right) \cong 3.8\%$$

$$< rsd\left(r_i^{pred}\,\middle|\,with\,1\,measurements\right) \cong 4.58\% < rsd\left(r_i^{m}\,\middle|\,all\,cases\right) \cong 5\%$$

$$< rsd\left(r_i^{c}\,\middle|\,all\,cases\right) \cong 18.7\%.$$

3. As the results in this subsection show, all of the conclusions reached previously in Sections 5.5.1 through 5.5.3 (i.e., after the assimilation of one and two measurements, respectively) also remain valid when four consistent measurements are assimilated. The largest adjustments have occurred for the source S, followed by the adjustments for Σ_d and Σ_a, while D was not adjusted because of its vanishingly small sensitivity. The largest reductions of the standard deviations

occurred, in order, for S, Σ_d, Σ_a, and D. The originally uncorrelated parameters became correlated after applying the predictive modeling procedure calibration, and the predicted correlations had the same trend as already shown previously in Sections 5.5.1 through 5.5.3, which treated the assimilation of measurements for one and two responses, respectively.

6

Application of the 2nd-ASAM for Computing Sensitivities of Detector Responses to Uncollided Radiation Transport

This chapter will present the application of the 2nd-ASAM to a problem of monoenergetic radiation transport through a heterogeneous nonmultiplying three-dimensional medium that does not scatter the respective radiation more than once. Such radiation transport phenomena are modeled mathematically by a simplified form of the Boltzmann transport equation, customarily called the "ray-tracing form," in which the seven-dimensional integro-differential Boltzmann transport equation is simplified to a partial differential equation in three spatial independent variables. This paradigm illustrative example has been chosen because—contrary to the illustrative examples presented in previous chapters—it is three-dimensional and does not admit an analytic solution, yet it is generic and, in particular, it is connected to the surrogate dissolver model analyzed in Chapter 2. The ray-tracing form is presented in Section 6.1. Albeit much simplified, the ray-tracing form of the transport equation nevertheless comprises hundreds of model parameters (number densities, cross sections, and sources characterizing the medium) and can only be solved numerically. Section 6.2 presents the computation of the first-order sensitivities of a radiation detector response to the model parameters, while Section 6.3 presents the computation of the second-order response sensitivities. Section 6.4 summarizes the salient conclusions for this application of the 2nd-ASAM and also highlights the connection between an idealized ray-tracing surrogate model and the idealized surrogate dissolver model analyzed in Chapter 2.

6.1 The Ray-Tracing Form of the Forward and Adjoint Boltzmann Transport Equations

The angular flux $\varphi(\mathbf{r},\mathbf{\Omega})$ of uncollided ("ray-trace") particles (neutrons or photons) in a medium satisfies the following simplified transport (Boltzmann) equation:

$$\mathbf{\Omega}\bullet\nabla\varphi(\mathbf{r},\mathbf{\Omega})+\Sigma_t(\mathbf{r};\boldsymbol{\alpha})\varphi(\mathbf{r},\mathbf{\Omega})=q(\mathbf{r};\boldsymbol{\alpha}), \tag{6.1}$$

subject to the "vacuum" boundary condition

$$\varphi(\mathbf{r}_s,\boldsymbol{\Omega})=0,\mathbf{r}_s\in\partial V,\boldsymbol{\Omega}\cdot\mathbf{n}<0, \tag{6.2}$$

which specifies that there is no incoming flux of particles. The quantities appearing in Equations 6.1 and 6.2 are defined as follows: $\boldsymbol{\Omega}$ denotes the unit vector in the direction of the particle's motion; $\mathbf{r}$ denotes the particle's position; $\Sigma_t(\mathbf{r};\boldsymbol{\alpha})$ denotes the total interaction cross section; $q(\mathbf{r};\boldsymbol{\alpha})$ denotes the particle source density (particles/cm^{3}sec) assumed to be isotropic; V denotes the body's volume; and $\mathbf{n}$ denotes the unit outer normal vector at any point $\mathbf{r}_s\in\partial V$ on the body's outer surface ∂V.

The quantity of interest in ray-tracing problems is a detector response, $R(\varphi,\boldsymbol{\alpha})$, of the form

$$R(\varphi,\alpha)=\int dV\int_{4\pi} d\Omega\,\Sigma_d(\mathbf{r},\Omega;\alpha)\varphi(\mathbf{r},\Omega), \tag{6.3}$$

where $\Sigma_d(\mathbf{r},\boldsymbol{\Omega})$ models the interaction cross section of the detector with the incident particles. Detector responses of particular interest are those that measure (i) the scalar flux at a point, in which case the detector-interaction function has the form

$$\Sigma_d(\mathbf{r},\Omega;\alpha)=\Sigma_d(\alpha)\delta(\mathbf{r}-\mathbf{r}_d), \tag{6.4}$$

where $\mathbf{r}_d$ represents the detector's location, and (ii) the current across a surface $\mathbf{r}_s\in\partial V$ in which case the detector-interaction function has the form

$$\Sigma_d(\mathbf{r},\Omega)=\Sigma_d(\alpha)\boldsymbol{\Omega}\cdot\mathrm{n}\delta\left(\min_{\mathbf{r}_s\in\partial V}|\mathbf{r}-\mathbf{r}_d|\right). \tag{6.5}$$

The quantities $\Sigma_t(\mathbf{r};\boldsymbol{\alpha})$, $q(\mathbf{r};\boldsymbol{\alpha})$, and $\Sigma_d(\mathbf{r},\boldsymbol{\Omega};\boldsymbol{\alpha})$ depend not only on the spatial variable, $\mathbf{r}$, but also on model parameters such as atomic number densities, microscopic cross sections, and weighting functions. Therefore, it is convenient to denote generically the model parameters as α_j and to consider that these model parameters are ordered as the components of a (column) "vector of model parameters" denoted as $\boldsymbol{\alpha}$ and defined as

$$\alpha\triangleq\left[\alpha_1,\ldots,\alpha_{N_\alpha}\right]^\dagger. \tag{6.6}$$

The model parameters are not perfectly well known. As in previous chapters, the parameter's nominal values, which will be denoted as $\boldsymbol{\alpha}^0\triangleq\left[\alpha_1^0,\ldots,\alpha_{N_\alpha}^0\right]^\dagger$, and their corresponding standard deviations, which will be denoted as $\boldsymbol{\sigma}\triangleq\left[\sigma_1,\ldots,\sigma_{N_\alpha}\right]^\dagger$, are considered to be available. The dagger (†) will be used to denote transposition and the superscript zero will be used to denote nominal values.

Although the medium in which the neutrons and/or gamma rays propagate is heterogeneous, the material properties $\Sigma_t(\mathbf{r};\boldsymbol{\alpha})$, $q(\mathbf{r};\boldsymbol{\alpha})$, and $\Sigma_d(\mathbf{r},\boldsymbol{\Omega};\boldsymbol{\alpha})$ are usually constant piecewise within the various material regions that make up the respective medium. Usually, the heterogeneous medium under consideration comprises N_m layers of materials having piecewise constant properties within each layer. In such cases, the quantities $\Sigma_t(\mathbf{r};\boldsymbol{\alpha})$, $q(\mathbf{r};\boldsymbol{\alpha})$, and $\Sigma_d(\mathbf{r},\boldsymbol{\Omega};\boldsymbol{\alpha})$ can be represented as follows:

$$\Sigma_t(\mathbf{r};\alpha) = \sum_{j=1}^{N_m} C_j(\alpha) f_j(\mathbf{r}), \tag{6.7}$$

$$q(\mathbf{r};\alpha) = \sum_{j=1}^{N_m} Q_j(\alpha) g_j(\mathbf{r}), \tag{6.8}$$

$$f_j(\mathbf{r}) = g_j(\mathbf{r}) = H(\mathbf{r}-\mathbf{r}_j) - H(\mathbf{r}-\mathbf{r}_{j+1}), \; j = 1,\ldots,N_m, \tag{6.9}$$

where $H(\mathbf{r} - \mathbf{r}_j)$ denotes the customary Heaviside unit functional and $\mathbf{r}_1$ denotes the coordinates of the innermost material boundary. When the detector consists of N_d layers of materials having piecewise constant properties, represented by interaction coefficients $\mu_k(\boldsymbol{\alpha})$ within each layer, the effective detector cross section may be represented in the form

$$\Sigma_d(\mathbf{r},\Omega;\alpha) = \sum_{k=1}^{N_d} \mu_k(\alpha) h_k(\mathbf{r},\Omega). \tag{6.10}$$

For this illustrative problem, the space-dependent functions $f_j(\mathbf{r})$, $g_j(\mathbf{r})$, and $h_k(\mathbf{r},\boldsymbol{\Omega})$, which describe internal boundaries, will be considered to be perfectly well known, although situations may arise in which the internal and/or external boundaries may also be affected by uncertainties. Linear and/or nonlinear spatial dependence of material properties can also be accommodated by suitable definitions of the functions $f_j(\mathbf{r})$, $g_j(\mathbf{r})$, and $h_k(\mathbf{r},\boldsymbol{\Omega})$.

6.2 Application of the 2nd-ASAM to Compute the First-Order Response Sensitivities to Variations in Model Parameters

The nominal value of the angular flux, denoted as $\varphi^0(\mathbf{r},\boldsymbol{\Omega})$, is obtained by solving Equations 6.1 and 6.2 using the nominal parameter values, that is, $\varphi^0(\mathbf{r},\boldsymbol{\Omega})$ is the solution of

$$\Omega\cdot\nabla\varphi^0(\mathbf{r},\Omega) + \Sigma_t^0(\mathbf{r};\alpha)\varphi^0(\mathbf{r},\Omega) = q^0(\mathbf{r};\alpha), \tag{6.11}$$

subject to the "vacuum" boundary condition, which specifies that there is no incoming flux of particles, that is,

$$\varphi^0(\mathbf{r}_s, \mathbf{\Omega}) = 0, \mathbf{r}_s \in \partial V, \mathbf{\Omega} \cdot \mathbf{n} < 0. \tag{6.12}$$

The nominal value of the detector response, denoted as R^0, is obtained by evaluating Equation 6.3 at the nominal flux and parameter values:

$$R^0 \triangleq R(\varphi^0, \alpha^0) = \int dV \int_{4\pi} d\Omega \, \Sigma_d^0(\mathbf{r}, \Omega; \alpha^0) \varphi^0(\mathbf{r}, \Omega). \tag{6.13}$$

The total sensitivity, $\delta R(\psi^0, \boldsymbol{\alpha}^0; \delta\varphi, \boldsymbol{\delta\alpha})$, of the detector response defined in Equation 6.3 to variations $\boldsymbol{\delta\alpha} \triangleq [\delta\alpha_1, ..., \delta\alpha_{N_\alpha}]^\dagger$ in the model parameters, around the nominal values $\boldsymbol{\alpha}^0$, is obtained by computing the G-differential of Equation 6.3 at the nominal parameter and flux values. The G-differential is obtained from its definition, as follows:

$$\begin{aligned} &\delta R(\varphi^0, \alpha^0; \delta\varphi, \delta\alpha) \\ &\triangleq \frac{d}{d\varepsilon} \left\{ \int dV \int_{4\pi} d\Omega \left[\Sigma_d^0(\mathbf{r}, \Omega) + \varepsilon \delta\Sigma_d(\mathbf{r}, \Omega) \right] \left[\varphi^0(\mathbf{r}, \Omega) + \varepsilon \delta\varphi(\mathbf{r}, \Omega) \right] \right\}_{\varepsilon=0} \\ &= \left\{ \delta R(\varphi^0, \alpha^0; \delta\alpha) \right\}_{dir} + \left\{ \delta R(\varphi^0, \alpha^0; \delta\varphi, \delta\alpha) \right\}_{ind}, \end{aligned} \tag{6.14}$$

where the direct-effect term is defined as

$$\left\{ \delta R(\varphi, \alpha; \delta\alpha) \right\}_{dir} = \int dV \int_{4\pi} \delta\Sigma_d(\mathbf{r}, \Omega; \alpha) \varphi(\mathbf{r}, \Omega) d\Omega, \tag{6.15}$$

and where the indirect-effect term is defined as

$$\left\{ \delta R(\varphi, \alpha; \delta\varphi, \delta\alpha) \right\}_{ind} \triangleq \int dV \int_{4\pi} d\Omega \, \Sigma_d(\mathbf{r}, \Omega; \alpha) \delta\varphi(\mathbf{r}, \Omega). \tag{6.16}$$

The variation $\delta\varphi(\mathbf{r}, \mathbf{\Omega})$, which appears in Equation 6.16, is the solution of the 1st-LFSS, which is obtained by G-differentiating Equations 6.1 and 6.2 to obtain

$$\Omega \bullet \nabla \delta\varphi(\mathbf{r}, \Omega) + \Sigma_t^0(\mathbf{r}; \alpha) \delta\varphi(\mathbf{r}, \Omega) = \delta q(\mathbf{r}) - \delta\Sigma_t(\mathbf{r}; \alpha) \varphi^0(\mathbf{r}, \Omega), \tag{6.17}$$

$$\delta\varphi(\mathbf{r}_s, \Omega) = 0, \mathbf{r}_s \in \partial V, \Omega\alpha \cdot \mathbf{n} < 0. \tag{6.18}$$

In view of Equations 6.7 through 6.9, the variations $\delta\Sigma_t(\mathbf{r};\boldsymbol{\alpha})$, $\delta q(\mathbf{r};\boldsymbol{\alpha})$, and $\delta\Sigma_d(\mathbf{r},\boldsymbol{\Omega};\boldsymbol{\alpha})$, which appear in Equations 6.15 and 6.17, are defined as follows:

$$\delta\Sigma_t(\mathbf{r}) = \sum_{j=1}^{N_m}\left[\sum_{k=1}^{N_\alpha}\frac{\partial C_j(\boldsymbol{\alpha})}{\partial\alpha_k}\delta\alpha_k\right]f_j(\mathbf{r}), \tag{6.19}$$

$$\delta q(\mathbf{r}) = \sum_{j=1}^{N_m}\left[\sum_{k=1}^{N_\alpha}\frac{\partial Q_j(\boldsymbol{\alpha})}{\partial\alpha_k}\delta\alpha_k\right]g_j(\mathbf{r}), \tag{6.20}$$

$$\delta\Sigma_d(\mathbf{r},\boldsymbol{\Omega}) = \sum_{j=1}^{N_d}\left[\sum_{k=1}^{N_\alpha}\frac{\partial \mu_j(\boldsymbol{\alpha})}{\partial\alpha_k}\delta\alpha_k\right]h_j(\mathbf{r},\boldsymbol{\Omega}). \tag{6.21}$$

The indirect-effect term defined in Equation 6.16 can be computed only after solving the 1st-LFSS, which is computationally expensive, since the 1st-LFSS would need to be solved anew for every variation in the model parameters. As is well known, the computationally expensive evaluation of the indirect-effect term by using Equation 6.16 can be avoided by expressing this indirect-effect term in terms of the solution of the 1st-LASS, which is constructed by implementing the following sequence of steps:

1. Define the inner product $\langle u(\mathbf{r},\boldsymbol{\Omega}), w(\mathbf{r},\boldsymbol{\Omega})\rangle$ of two functions $u(\mathbf{r},\boldsymbol{\Omega}) \in L_2(V \times \Omega)$ and $w(\mathbf{r},\boldsymbol{\Omega}) \in L_2(V \times \Omega)$ in the Hilbert space $L_2(V \times \Omega)$ of square-integrable functions, as follows:

$$\langle u(\mathbf{r},\boldsymbol{\Omega}), w(\mathbf{r},\boldsymbol{\Omega})\rangle \triangleq \int dV \int_{4\pi} d\Omega \; u(\mathbf{r},\boldsymbol{\Omega}) w(\mathbf{r},\boldsymbol{\Omega}). \tag{6.22}$$

2. Form the inner product of Equation 6.17 with a yet-undefined function $\psi^{(1)}(\mathbf{r},\boldsymbol{\Omega})$ to obtain

$$\begin{aligned}&\left\langle \psi^{(1)}(\mathbf{r},\boldsymbol{\Omega}),\, \boldsymbol{\Omega}\cdot\nabla\delta\varphi(\mathbf{r},\boldsymbol{\Omega}) + \Sigma_t^0(\mathbf{r};\boldsymbol{\alpha})\delta\varphi(\mathbf{r},\boldsymbol{\Omega})\right\rangle \\ &= \left\langle \psi^{(1)}(\mathbf{r},\boldsymbol{\Omega}),\, \delta q(\mathbf{r};\boldsymbol{\alpha}) - \delta\Sigma_t(\mathbf{r};\boldsymbol{\alpha})\varphi^0(\mathbf{r},\boldsymbol{\Omega})\right\rangle.\end{aligned} \tag{6.23}$$

3. For a linear operator, $L^{(1)}$, define its formal adjoint operator, $A^{(1)}$, through the following relationship:

$$\left\langle \psi^{(1)}, L^{(1)}\delta\varphi\right\rangle = \left\langle \delta\varphi, A^{(1)}\psi^{(1)}\right\rangle + P^{(1)}\left(\delta\varphi, \psi^{(1)}\right), \tag{6.24}$$

where

$$L^{(1)}\delta\varphi \triangleq \boldsymbol{\Omega}\cdot\nabla\delta\varphi(\mathbf{r},\boldsymbol{\Omega}) + \Sigma_t^0(\mathbf{r};\boldsymbol{\alpha})\delta\varphi(\mathbf{r},\boldsymbol{\Omega}), \tag{6.25}$$

$$A^{(1)}\psi^{(1)} \triangleq -\mathbf{\Omega}\bullet\nabla\psi^{(1)}(\mathbf{r},\mathbf{\Omega})+\Sigma_t^0(\mathbf{r};\alpha)\psi^{(1)}(\mathbf{r},\mathbf{\Omega}), \tag{6.26}$$

and where the bilinear concomitant, $P^{(1)}(\delta\varphi,\psi^{(1)})$, on the boundary $(\partial V \times \partial\Omega)$ is obtained by applying the divergence theorem and is defined as follows:

$$\begin{aligned} P^{(1)}\left(\delta\varphi,\psi^{(1)}\right) \triangleq & \int_{\mathbf{\Omega}\cdot\mathbf{n}<0} d\Omega \int_{\partial V} |\mathbf{\Omega}\cdot\mathbf{n}|\,\delta\varphi(\mathbf{r},\mathbf{\Omega})\psi^{(1)}(\mathbf{r},\mathbf{\Omega})dA \\ & -\int_{\mathbf{\Omega}\cdot\mathbf{n}>0} d\Omega \int_{\partial V} \mathbf{\Omega}\cdot\mathbf{n}\,\delta\varphi(\mathbf{r},\mathbf{\Omega})\psi^{(1)}(\mathbf{r},\mathbf{\Omega})dA. \end{aligned} \tag{6.27}$$

Note that the superscript zero, which denotes nominal values, will be omitted henceforth in order to simplify the notation. This simplification should not cause any loss of clarity, since it will become clear from the context which quantities are to be evaluated/computed using the nominal values for the model parameters.

4. Use Equation 6.23 in conjunction with the boundary conditions given in Equation 6.18 to construct the following 1st-LASS to be satisfied by the first-level adjoint function $\psi^{(1)}(\mathbf{r},\mathbf{\Omega})$:

$$A^{(1)}\psi^{(1)} \triangleq -\mathbf{\Omega}\bullet\nabla\psi^{(1)}(\mathbf{r},\mathbf{\Omega})+\Sigma_t(\mathbf{r};\alpha)\psi^{(1)}(\mathbf{r},\mathbf{\Omega})=\Sigma_d(\mathbf{r},\mathbf{\Omega};\alpha), \tag{6.28}$$

together with the boundary condition

$$\psi^{(1)}(\mathbf{r}_s,\mathbf{\Omega})=0, \mathbf{\Omega}\bullet\mathbf{n}>0, \mathbf{r}_s\in\partial V, \tag{6.29}$$

which is selected in order to cause the bilinear concomitant $P^{(1)}(\delta\varphi,\psi^{(1)})$ in Equation 6.24 to vanish.

5. Use the 1st-LFSS defined by Equations 6.17 and 6.18 together with Equations 6.25 and 6.26 to obtain the following expression for the indirect-effect term, cf. Equation 6.16, in terms of the first-level adjoint function $\psi^{(1)}(r,\mathbf{\Omega})$:

$$\left\{\delta R(\varphi,\alpha;\psi^{(1)},\delta\alpha)\right\}_{ind} = \int dV \int_{4\pi} d\Omega\, \psi^{(1)}(\mathbf{r},\mathbf{\Omega})\left[\delta q(\mathbf{r};\alpha)-\delta\Sigma_t(\mathbf{r};\alpha)\varphi(\mathbf{r},\mathbf{\Omega})\right]. \tag{6.30}$$

The 1st-LASS is solved by using the same numerical method as used for solving the original Equations 6.1 and 6.2, except for replacing $\mathbf{\Omega}$ with $-\mathbf{\Omega}$, and recognizing that the "adjoint particles" travel backwards, that is, in the $-\mathbf{\Omega}$ direction. Furthermore, the 1st-LASS is independent of parameter variations, so it needs to be solved just once [for each particular form that the source term $\Sigma_d(\mathbf{r},\mathbf{\Omega})$ might have] to obtain the first-level adjoint function $\psi^{(1)}(\mathbf{r},\mathbf{\Omega})$. The indirect-effect term is computed efficiently once $\psi^{(1)}(\mathbf{r},\mathbf{\Omega})$ is available, by performing the integrations (quadratures) indicated in Equation 6.27.

Replacing Equations 6.30 and 6.15 in Equation 6.14 yields the following expression for the total first-order response sensitivity in terms of the first-level adjoint function $\psi^{(1)}(\mathbf{r},\mathbf{\Omega})$:

$$\delta R(\varphi,\alpha;\psi^{(1)},\delta\alpha)=\int dV\int_{4\pi} d\Omega\,\delta\Sigma_d(\mathbf{r},\mathbf{\Omega};\alpha)\varphi(\mathbf{r},\mathbf{\Omega}) + \int dV\int_{4\pi} d\Omega\,\psi^{(1)}(\mathbf{r},\mathbf{\Omega})\left[\delta q(\mathbf{r};\alpha)-\delta\Sigma_t(\mathbf{r};\alpha)\varphi(\mathbf{r},\mathbf{\Omega})\right]. \tag{6.31}$$

The partial first-order response sensitivity to a generic parameter $\boldsymbol{\alpha}_1$ is obtained from Equation 6.31 as

$$S_{m_1}^{(1)}(\varphi,\alpha;\psi^{(1)})\triangleq\frac{\partial R(\varphi,\alpha;\psi^{(1)},\delta\alpha)}{\partial\alpha_{m_1}}=\sum_{k=1}^{N_d}\frac{\partial\mu_k(\alpha)}{\partial\alpha_{m_1}}\int dV\int_{4\pi} d\Omega\, h_k(\mathbf{r},\mathbf{\Omega})\varphi(\mathbf{r},\mathbf{\Omega})$$
$$+\sum_{k=1}^{N_m}\frac{\partial Q_k(\alpha)}{\partial\alpha_{m_1}}\int dV\int_{4\pi} d\Omega\,\psi^{(1)}(r,\mathbf{\Omega})g_k(\mathbf{r})$$
$$-\sum_{k=1}^{N_m}\frac{\partial C_k(\alpha)}{\partial\alpha_{m_1}}\int dV\int_{4\pi} d\Omega\,\psi^{(1)}(r,\mathbf{\Omega})\varphi(\mathbf{r},\mathbf{\Omega})f_k(\mathbf{r}),\quad m_1=1,\ldots,N_\alpha. \tag{6.32}$$

Often the quantities $\Sigma_t(\mathbf{r};\boldsymbol{\alpha})$, $q(\mathbf{r};\boldsymbol{\alpha})$, and $\Sigma_d(\mathbf{r},\mathbf{\Omega};\boldsymbol{\alpha})$ can be represented as follows:

$$\Sigma_t(\mathbf{r};\alpha)=\sum_{j=1}^{N_m}N_j\sigma_j f_j(\mathbf{r}), \tag{6.33}$$

$$q(\mathbf{r};\alpha)=\sum_{j=1}^{N_m}N_j q_j g_j(\mathbf{r}), \tag{6.34}$$

$$\Sigma_d(\mathbf{r},\Omega;\alpha)=\sum_{k=1}^{N_d}\lambda_k h_k(\mathbf{r},\Omega), \tag{6.35}$$

where

1. N_j represents the atomic number density, σ_j denotes the microscopic cross section, and $f_j(\mathbf{r})$ denotes the spatial variation, respectively, that characterize the jth-material contained in the heterogeneous medium under consideration, while N_m denotes the total number of materials contained in this medium.
2. q_j and $g_j(\mathbf{r})$, respectively, denote the source strength and the spatial variation of the respective source contained in the jth-material within the heterogeneous medium under consideration.
3. λ_k and $h_k(\mathbf{r},\boldsymbol{\Omega})$, respectively, denote the scalar interaction coefficient of neutrons or gamma rays and the corresponding spatial and angular variation of this interaction coefficient within the detector's kth-material; the total number of distinct materials within the detector is denoted as N_d.

Thus, the variations $\delta q(\mathbf{r};\boldsymbol{\alpha})$, $\delta\Sigma_t(\mathbf{r};\boldsymbol{\alpha})$, and $\delta\Sigma_d(\mathbf{r},\boldsymbol{\Omega};\boldsymbol{\alpha})$, in the model parameters $\Sigma_t(\mathbf{r};\boldsymbol{\alpha})$, $q(\mathbf{r};\boldsymbol{\alpha})$, and $\Sigma_d(\mathbf{r},\boldsymbol{\Omega};\boldsymbol{\alpha})$, respectively, are considered to have the following expressions:

$$\delta\Sigma_t(\mathbf{r};\alpha)=\sum_{j=1}^{N_m}\left[(\delta N_j)\sigma_j+(\delta\sigma_j)N_j\right]f_j(\mathbf{r}), \tag{6.36}$$

$$\delta q(\mathbf{r};\alpha)=\sum_{j=1}^{N_m}\left[(\delta N_j)q_j+(\delta q_j)N_j\right]g_j(\mathbf{r}), \tag{6.37}$$

$$\delta\Sigma_d(\mathbf{r},\Omega;\alpha)=\sum_{k=1}^{N_D}(\delta\lambda_k)h_k(\mathbf{r},\Omega), \tag{6.38}$$

where the scalar-valued variations δN_j, $\delta\sigma_j$, δq_j, and $\delta\lambda_k$ are considered to be known. For bookkeeping purposes for this specific situation, it is convenient to consider that the model parameters N_j, σ_j, q_j, and λ_k are ordered as the components of a (column) "vector of model parameters," denoted as $\boldsymbol{\alpha}$ and defined as follows:

$$\alpha\triangleq\left[\alpha_1,\dots,\alpha_{N_\alpha}\right]^\dagger,\ \alpha_j\triangleq N_j\ for\ j=1,\dots,N_m;\ \alpha_{N_m+j}\triangleq\sigma_j\ for\ j=1,\dots,N_m;$$
$$\alpha_{2N_m+j}\triangleq q_j\ for\ j=1,\dots,N_m;\ \alpha_{3N_m+j}\triangleq\lambda_j\ for\ j=1,\dots,N_d;\ N_\alpha\triangleq 3N_m+N_d. \tag{6.39}$$

Specializing Equation 6.32 to the particular cases described by Equations 6.36 through 6.38 and identifying the terms corresponding to the various variations $\delta\boldsymbol{\alpha}$ yields the following expressions for the respective first-order partial sensitivities:

$$S_i^{(1)}(\varphi,\alpha;\psi^{(1)}) \triangleq \frac{\partial R}{\partial N_i} = -\sigma_i \int dV \int_{4\pi} d\Omega\, f_i(\mathbf{r})\psi^{(1)}(\mathbf{r},\Omega)\varphi(\mathbf{r},\Omega) + q_i \int dV \int_{4\pi} d\Omega\, \psi^{(1)}(r,\Omega) g_i(\mathbf{r}), \; for \; i=1,\ldots,N_m, \tag{6.40}$$

$$S_{i+N_m}^{(1)}(\varphi,\alpha;\psi^{(1)}) \triangleq \frac{\partial R}{\partial \sigma_i} = -N_i \int dV \int_{4\pi} d\Omega\, f_i(\mathbf{r})\psi^{(1)}(\mathbf{r},\Omega)\varphi(\mathbf{r},\Omega), \; for \; i=1,\ldots,N_m, \tag{6.41}$$

$$S_{i+2N_m}^{(1)}(\alpha;\psi^{(1)}) \triangleq \frac{\partial R}{\partial q_i} = N_i \int dV \int_{4\pi} d\Omega\, \psi^{(1)}(\mathbf{r},\Omega) g_i(\mathbf{r}), \; for \; i=1,\ldots,N_m, \tag{6.42}$$

$$S_{i+3N_m}^{(1)}(\varphi,\alpha) \triangleq \frac{\partial R}{\partial \lambda_i} = \int dV \int_{4\pi} d\Omega\, \varphi(\mathbf{r},\Omega) h_i(\mathbf{r},\Omega), \; for \; i=1,\ldots,N_d. \tag{6.43}$$

6.3 Application of the 2nd-ASAM to Compute the Second-Order Response Sensitivities to Variations in Model Parameters

The second-order response sensitivities will be denoted as $S_{m_1,m_2}^{(2)} \triangleq \dfrac{\partial^2 R}{\partial \alpha_{m_1}\partial \alpha_{m_2}}$, $m_1, m_2 = 1,\ldots,N_\alpha$, and will be obtained by applying the 2nd-ASAM developed in Chapter 3, which relies on the construction of a 2nd-LASS for each of the first-order sensitivities defined generically by Equation 6.32. Computing the G-differential of the first-order sensitivities defined in Equation 6.32 yields the following expression:

$$\delta S_{m_1}^{(1)}(\varphi,\alpha;\psi^{(1)};\delta\alpha) = \left\{\delta S_{m_1}^{(1)}(\varphi,\alpha;\psi^{(1)},\delta\alpha)\right\}_{dir} + \left\{\delta S_{m_1}^{(1)}(\varphi,\alpha;\psi^{(1)},\delta\alpha)\right\}_{ind}, \quad m_1 = 1,\ldots,N_\alpha, \tag{6.44}$$

where

$$\left\{\delta S_{m_1}^{(1)}(\varphi,\alpha;\psi^{(1)},\delta\alpha)\right\}_{dir} \triangleq \sum_{k=1}^{N_d}\sum_{m_2=1}^{N_\alpha}\frac{\partial^2\mu_j(\alpha)}{\partial\alpha_{m_1}\partial\alpha_{m_2}}\delta\alpha_{m_2}\int dV\int_{4\pi} h_k(\mathbf{r},\Omega)\varphi(\mathbf{r},\Omega)d\Omega$$

$$+\sum_{k=1}^{N_m}\sum_{m_2=1}^{N_\alpha}\frac{\partial^2 Q_j(\alpha)}{\partial\alpha_{m_1}\partial\alpha_{m_2}}\delta\alpha_{m_2}\int dV\int_{4\pi} d\Omega\,\psi^{(1)}(r,\Omega)g_k(\mathbf{r})$$

$$-\sum_{k=1}^{N_m}\sum_{m_2=1}^{N_\alpha}\frac{\partial^2 C_j(\alpha)}{\partial\alpha_{m_1}\partial\alpha_{m_2}}\delta\alpha_{m_2}\int dV\int_{4\pi} d\Omega\,\psi^{(1)}(r,\Omega)\varphi(\mathbf{r},\Omega)f_k(\mathbf{r}),\quad m_2=1,\ldots,N_\alpha, \tag{6.45}$$

and

$$\left\{\delta S_{m_1}^{(1)}(\varphi,\alpha;\psi^{(1)},\delta\alpha)\right\}_{ind} \triangleq \sum_{k=1}^{N_d}\frac{\partial\mu_k(\alpha)}{\partial\alpha_{m_1}}\int dV\int_{4\pi} h_k(\mathbf{r},\Omega)\delta\varphi(\mathbf{r},\Omega)d\Omega$$

$$+\sum_{k=1}^{N_m}\frac{\partial Q_k(\alpha)}{\partial\alpha_{m_1}}\int dV\int_{4\pi} d\Omega\,\delta\psi^{(1)}(r,\Omega)g_k(\mathbf{r})$$

$$-\sum_{k=1}^{N_m}\frac{\partial C_k(\alpha)}{\partial\alpha_{m_1}}\int dV\int_{4\pi} d\Omega\left[\delta\psi^{(1)}(r,\Omega)\varphi(\mathbf{r},\Omega)+\psi^{(1)}(r,\Omega)\delta\varphi(\mathbf{r},\Omega)\right]f_k(\mathbf{r}),$$

$$m_1=1,\ldots,N_\alpha. \tag{6.46}$$

The direct-effect term $\left\{\delta S_{m_1}^{(1)}(\varphi,\alpha;\psi^{(1)},\delta\alpha)\right\}_{dir}$ can be computed immediately. On the contrary, the indirect-effect term $\left\{\delta S_{m_1}^{(1)}(\varphi,\alpha;\psi^{(1)},\delta\alpha)\right\}_{ind}$ can be computed only after having obtained the variation $\delta\varphi(\mathbf{r},\mathbf{\Omega})$ in the forward angular flux and the variation $\delta\psi^{(1)}(r,\mathbf{\Omega})$ in the first-level adjoint function. The function $\delta\psi^{(1)}(r,\mathbf{\Omega})$ is the solution of the system of equations obtained by G-differentiating the 1st-LASS, cf. Equations 6.28 and 6.29, which yields

$$-\mathbf{\Omega}\bullet\nabla\delta\psi^{(1)}(\mathbf{r},\Omega)+\Sigma_t(\mathbf{r};\alpha)\delta\psi^{(1)}(\mathbf{r},\Omega)=\delta\Sigma_d(\mathbf{r},\Omega;\alpha)-\delta\Sigma_t(\mathbf{r};\alpha)\psi^{(1)}(\mathbf{r},\Omega), \tag{6.47}$$

$$\delta\psi^{(1)}(\mathbf{r}_s,\Omega)=0,\mathbf{\Omega}\bullet\mathbf{n}>0,\mathbf{r}_s\in\partial V. \tag{6.48}$$

It is evident from Equations 6.47 and 6.48 that the evaluation of the function $\delta\psi^{(1)}(\mathbf{r},\mathbf{\Omega})$ is just as expensive computationally as determining the variation of $\delta\varphi(\mathbf{r},\mathbf{\Omega})$ by solving the 1st-LFSS. The system comprising Equations 6.47, 6.48,

6.17, and 6.18 is called (see Chapter 3) the 2nd-LFSS. To avoid the need for solving (repeatedly) the 2nd-LFSS, the indirect-effect term $\left\{\delta S_{m_1}^{(1)}(\varphi,\alpha;\psi^{(1)},\delta\alpha)\right\}_{ind}$ will be expressed in terms of the 2nd-LASS, which will be constructed by following the general principles introduced in Chapter 3, comprising the following sequence of steps:

1. Define the inner product $\langle \mathbf{u}^{(2)}(\mathbf{r},\boldsymbol{\Omega}),\mathbf{w}^{(2)}(\mathbf{r},\boldsymbol{\Omega})\rangle$ of two vector-valued functions $\mathbf{u}^{(2)}(\mathbf{r},\boldsymbol{\Omega})\triangleq\left[u_1^{(2)}(\mathbf{r},\boldsymbol{\Omega}),u_2^{(2)}(\mathbf{r},\boldsymbol{\Omega})\right]^{\dagger}$, with $u_1^{(2)}(\mathbf{r},\boldsymbol{\Omega})\in L_2(V\times\Omega)$ $u_2^{(2)}(\mathbf{r},\boldsymbol{\Omega})\in L_2(V\times\Omega)$, and $\mathbf{w}^{(2)}(\mathbf{r},\boldsymbol{\Omega})\triangleq\left[w_1^{(2)}(\mathbf{r},\boldsymbol{\Omega}),w_2^{(2)}(\mathbf{r},\boldsymbol{\Omega})\right]^{\dagger}$, with $w_1^{(2)}(\mathbf{r},\boldsymbol{\Omega})\in L_2(V\times\Omega)$, $w_2^{(2)}(\mathbf{r},\boldsymbol{\Omega})\in L_2(V\times\Omega)$, as follows:

$$\left\langle \mathbf{u}^{(2)}(\mathbf{r},\boldsymbol{\Omega}),\mathbf{w}^{(2)}(\mathbf{r},\boldsymbol{\Omega})\right\rangle\triangleq\sum_{j=1}^{2}\int dV\int_{4\pi}d\Omega\, u_j^{(2)}(\mathbf{r},\boldsymbol{\Omega})w_j^{(2)}(\mathbf{r},\boldsymbol{\Omega}). \tag{6.49}$$

2. For the matrix-valued linear operator $\mathbf{L}^{(2)}\triangleq\begin{pmatrix} L_{11}^{(2)} & L_{12}^{(2)} \\ L_{21}^{(2)} & L_{22}^{(2)}\end{pmatrix}$, define its formal adjoint operator, $\mathbf{A}^{(2)}\triangleq\begin{pmatrix} A_{11}^{(2)} & A_{12}^{(2)} \\ A_{21}^{(2)} & A_{22}^{(2)}\end{pmatrix}$, through the following relationship

$$\left\langle \mathbf{w}^{(2)},\mathbf{L}^{(2)}\mathbf{u}^{(2)}\right\rangle=\left\langle \mathbf{u}^{(2)},\mathbf{A}^{(2)}\mathbf{w}^{(2)}\right\rangle+P^{(2)}\left(\mathbf{u}^{(2)},\mathbf{w}^{(2)}\right), \tag{6.50}$$

   where $P^{(2)}(\mathbf{u}^{(2)},\mathbf{w}^{(2)})$ denotes the corresponding bilinear concomitant on the boundary $(\partial V\times\partial\Omega)$.

3. Apply the definition provided in Equation 6.49 to form the inner product of Equations 6.47 and 6.17 with a yet-undefined function $\boldsymbol{\psi}_{m_1}^{(2)}(\mathbf{r},\boldsymbol{\Omega})\triangleq\left[\psi_{1,m_1}^{(2)}(\mathbf{r},\boldsymbol{\Omega}),\psi_{2,m_1}^{(2)}(\mathbf{r},\boldsymbol{\Omega})\right]^{\dagger}$, $\psi_{1,m_1}^{(2)}(\mathbf{r},\boldsymbol{\Omega})\in L_2(V\times\Omega)$ and $\psi_{2,m_1}^{(2)}(\mathbf{r},\boldsymbol{\Omega})\in L_2(V\times\Omega)$, to obtain the following relation:

$$\begin{aligned}
&\int dV\int_{4\pi}d\Omega\psi_{1,m_1}^{(2)}(\mathbf{r},\boldsymbol{\Omega})\left[-\boldsymbol{\Omega}\bullet\nabla\delta\psi^{(1)}(\mathbf{r},\boldsymbol{\Omega})+\Sigma_t(\mathbf{r};\alpha)\delta\psi^{(1)}(r,\boldsymbol{\Omega})\right]\\
&+\int dV\int_{4\pi}d\Omega\psi_{2,m_1}^{(2)}(\mathbf{r},\boldsymbol{\Omega})\left[\boldsymbol{\Omega}\bullet\nabla\delta\varphi(\mathbf{r},\boldsymbol{\Omega})+\Sigma_t(\mathbf{r};\alpha)\delta\varphi(\mathbf{r},\boldsymbol{\Omega})\right]\\
&=\int dV\int_{4\pi}d\Omega\psi_{1,m_1}^{(2)}(\mathbf{r},\boldsymbol{\Omega})\left[\delta\Sigma_d(\mathbf{r},\boldsymbol{\Omega};\alpha)-\delta\Sigma_t(\mathbf{r};\alpha)\psi^{(1)}(\mathbf{r},\boldsymbol{\Omega})\right]\\
&+\int dV\int_{4\pi}d\Omega\psi_{2,m_1}^{(2)}(\mathbf{r},\boldsymbol{\Omega})\left[\delta q(\mathbf{r};\alpha)-\delta\Sigma_t(\mathbf{r};\alpha)\varphi(\mathbf{r},\boldsymbol{\Omega})\right].
\end{aligned} \tag{6.51}$$

4. Apply Equation 6.50 to the left side of Equation 6.51, for the special case when $\mathbf{u}^{(2)}(\mathbf{r},\Omega) \triangleq [\delta\psi^{(1)}(\mathbf{r},\Omega), \delta\varphi(\mathbf{r},\Omega)]$, $\mathbf{w}^{(2)} \triangleq \psi^{(2)}_{m_1}(\mathbf{r},\Omega)$, $L^{(2)}_{12} = L^{(2)}_{21} \equiv 0$ $L^{(2)}_{11}\psi^{(1)} \triangleq -\Omega \bullet \nabla\psi^{(1)}(\mathbf{r},\Omega) + \Sigma_t(\mathbf{r};\alpha)\psi^{(1)}(\mathbf{r},\Omega)$, and $L^{(2)}_{22}\delta\varphi \triangleq \Omega \bullet \nabla\delta\varphi(\mathbf{r},\Omega) + \Sigma_t(\mathbf{r};\alpha)\delta\varphi(\mathbf{r},\Omega)$, to obtain:

$$\begin{aligned}
&\int dV \int_{4\pi} d\Omega \psi^{(2)}_{1,m_1}(\mathbf{r},\Omega)\left[-\Omega \bullet \nabla\delta\psi^{(1)}(\mathbf{r},\Omega) + \Sigma_t(\mathbf{r};\alpha)\delta\psi^{(1)}(r,\Omega)\right] \\
&+ \int dV \int_{4\pi} d\Omega \psi^{(2)}_{2,m_1}(\mathbf{r},\Omega)\left[\Omega \bullet \nabla\delta\varphi(\mathbf{r},\Omega) + \Sigma_t(\mathbf{r};\alpha)\delta\varphi(\mathbf{r},\Omega)\right] \\
&= \int dV \int_{4\pi} d\Omega\, \delta\psi^{(1)}(\mathbf{r},\Omega)\left[\Omega \bullet \nabla\psi^{(2)}_{1,m_1}(\mathbf{r},\Omega) + \Sigma_t(\mathbf{r};\alpha)\psi^{(2)}_{1,m_1}(\mathbf{r},\Omega)\right] \\
&+ \int dV \int_{4\pi} d\Omega\, \delta\varphi(\mathbf{r},\Omega)\left[-\Omega \bullet \nabla\psi^{(2)}_{2,m_1}(\mathbf{r},\Omega) + \Sigma_t(\mathbf{r};\alpha)\psi^{(2)}_{2,m_1}(\mathbf{r},\Omega)\right] \\
&+ P^{(2)}\left(\mathbf{u}^{(2)}, \psi^{(2)}_{m_1}\right).
\end{aligned} \tag{6.52}$$

The relation obtained in Equation 6.52 indicates that the components of the adjoint operator $\mathbf{A}^{(2)}$ are as follows: $A^{(2)}_{11}\psi^{(1)} \triangleq \Omega \bullet \nabla\psi^{(2)}_{1,m_1}(\mathbf{r},\Omega) + \Sigma_t(\mathbf{r};\alpha)\psi^{(2)}_{1,m_1}(\mathbf{r},\Omega)$, $A^{(2)}_{12} = A^{(2)}_{21} \equiv 0$, and $A^{(2)}_{22}\delta\varphi \triangleq -\Omega \bullet \nabla\psi^{(2)}_{2,m_1}(\mathbf{r},\Omega) + \Sigma_t(\mathbf{r};\alpha)\psi^{(2)}_{2,m_1}(\mathbf{r},\Omega)$.

5. Use the boundary conditions shown in Equations 6.18 and 6.48 and impose on $\psi^{(2)}_{m_1}(\mathbf{r},\Omega) \triangleq \left[\psi^{(2)}_{1,m_1}(\mathbf{r},\Omega), \psi^{(2)}_{2,m_1}(\mathbf{r},\Omega)\right]^{\dagger}$ the boundary conditions $\psi^{(2)}_{1,m_1}(\mathbf{r},\Omega) = 0, \mathbf{r}_s \in \partial V, \Omega \bullet \mathbf{n} < 0$ and $\psi^{(2)}_{1,m_1}(\mathbf{r}, \Omega) = 0$, $\mathbf{r}_s$, $\in \partial V$, $\Omega \cdot \mathbf{n} < 0$, to cause the bilinear concomitant $P^{(2)}\left(\mathbf{u}^{(2)}, \psi^{(2)}_{m_1}\right)$ in Equation 6.52 to vanish.
6. Identify the right side of Equation 6.52 with the indirect-effect term defined in Equation 6.46 to obtain the following form of the 2nd-LASS to be used for computing the functions $\psi^{(2)}_{1,m_1}(\mathbf{r},\Omega)$, and $\psi^{(2)}_{2,m_1}(\mathbf{r},\Omega)$, $m_1 = 1, \dots, N_\alpha$, which will ultimately be used to evaluate Equation 6.46:

$$\begin{aligned}
&\Omega \bullet \nabla\psi^{(2)}_{1,m_1}(\mathbf{r},\Omega) + \Sigma_t(\mathbf{r};\alpha)\psi^{(2)}_{1,m_1}(\mathbf{r},\Omega) \\
&= \sum_{k=1}^{N_m} \frac{\partial Q_k(\alpha)}{\partial \alpha_{m_1}} g_k(\mathbf{r}) - \varphi(\mathbf{r},\Omega) \sum_{k=1}^{N_m} \frac{\partial C_k(\alpha)}{\partial \alpha_{m_1}} f_k(\mathbf{r}),
\end{aligned} \tag{6.53}$$

$$\psi^{(2)}_{1,m_1}(\mathbf{r},\Omega) = 0, \mathbf{r}_s \in \partial V, \Omega \bullet \mathbf{n} < 0, \tag{6.54}$$

$$\begin{aligned}
&-\Omega \bullet \nabla\psi^{(2)}_{2,m_1}(\mathbf{r},\Omega) + \Sigma_t(\mathbf{r};\alpha)\psi^{(2)}_{2,m_1}(\mathbf{r},\Omega) \\
&= \sum_{k=1}^{N_d} \frac{\partial \mu_k(\alpha)}{\partial \alpha_{m_1}} h_k(\mathbf{r},\Omega) - \psi^{(1)}(r,\Omega) \sum_{k=1}^{N_m} \frac{\partial C_k(\alpha)}{\partial \alpha_{m_1}} f_k(\mathbf{r}),
\end{aligned} \tag{6.55}$$

$$\psi^{(2)}_{2,m_1}(\mathbf{r},\mathbf{\Omega}) = 0, \mathbf{\Omega}\bullet\mathbf{n} > 0, \mathbf{r}_s \in \partial V. \tag{6.56}$$

7. Use Equations 6.51 and 6.52 together with Equations 6.45, 6.46, and 6.19 through 6.21 in Equation 6.44 to obtain the following expression for $\delta S^{(1)}_{m_1}(\varphi,\alpha;\psi^{(1)};\psi^{(2)}_{1,m_1},\psi^{(2)}_{2,m_1};\delta\alpha)$:

$$\begin{aligned}
&\delta S^{(1)}_{m_1}(\varphi,\alpha;\psi^{(1)};\psi^{(2)}_{1,m_1},\psi^{(2)}_{2,m_1};\delta\alpha) = \left\{\delta S^{(1)}_{m_1}(\varphi,\alpha;\psi^{(1)},\delta\alpha)\right\}_{dir} \\
&-\sum_{m_2=1}^{N_\alpha}\sum_{j=1}^{N_m}\frac{\partial C_j(\alpha)}{\partial \alpha_{m_2}}\delta\alpha_{m_2}\int dV\int_{4\pi} d\Omega\left[\psi^{(2)}_{1,m_1}(\mathbf{r},\mathbf{\Omega})\psi^{(1)}(\mathbf{r},\mathbf{\Omega}) + \psi^{(2)}_{2,m_1}(\mathbf{r},\mathbf{\Omega})\varphi(\mathbf{r},\mathbf{\Omega})\right]f_j(\mathbf{r}) \\
&+\sum_{m_2=1}^{N_\alpha}\sum_{j=1}^{N_m}\frac{\partial Q_j(\alpha)}{\partial \alpha_{m_2}}\delta\alpha_{m_2}\int dV\int_{4\pi} d\Omega\,\psi^{(2)}_{2,m_1}(\mathbf{r},\mathbf{\Omega})g_j(\mathbf{r}) \\
&+\sum_{m_2=1}^{N_\alpha}\sum_{j=1}^{N_d}\frac{\partial \mu_j(\alpha)}{\partial \alpha_{m_2}}\delta\alpha_{m_2}\int dV\int_{4\pi} d\Omega\,\psi^{(2)}_{1,m_1}(\mathbf{r},\mathbf{\Omega})h_j(\mathbf{r},\mathbf{\Omega}),\ m_1 = 1,\ldots,N_\alpha.
\end{aligned} \tag{6.57}$$

The second-order mixed partial sensitivities, $S^{(2)}_{m_1,m_2} \triangleq \dfrac{\partial^2 R}{\partial\alpha_{m_1}\partial\alpha_{m_2}}$, m_1, $m_2 = 1,\ldots,N_\alpha$, of the response with respect to the model parameters are determined by identifying in Equation 6.57 the expressions multiplying the variations $\delta\alpha_{m_2}$; this identification yields

$$\begin{aligned}
S^{(2)}_{m_1,m_2} &\triangleq \frac{\partial^2 R}{\partial\alpha_{m_1}\partial\alpha_{m_2}} = \sum_{j=1}^{N_m}\frac{\partial Q_j(\alpha)}{\partial \alpha_{m_2}}\int dV\int_{4\pi} d\Omega\,\psi^{(2)}_{2,m_1}(\mathbf{r},\mathbf{\Omega})g_j(\mathbf{r}) \\
&+\sum_{k=1}^{N_m}\frac{\partial^2 Q_j(\alpha)}{\partial\alpha_{m_1}\partial \alpha_{m_2}}\int dV\int_{4\pi} d\Omega\,\psi^{(1)}(r,\mathbf{\Omega})g_k(\mathbf{r}) \\
&+\sum_{j=1}^{N_d}\frac{\partial \mu_j(\alpha)}{\partial \alpha_{m_2}}\int dV\int_{4\pi} d\Omega\,\psi^{(2)}_{1,m_1}(\mathbf{r},\mathbf{\Omega})h_j(\mathbf{r},\mathbf{\Omega}) \\
&+\sum_{k=1}^{N_d}\frac{\partial^2 \mu_j(\alpha)}{\partial\alpha_{m_1}\partial \alpha_{m_2}}\int dV\int_{4\pi} h_k(\mathbf{r},\mathbf{\Omega})\varphi(\mathbf{r},\mathbf{\Omega})d\Omega \\
&-\sum_{j=1}^{N_m}\frac{\partial C_j(\alpha)}{\partial \alpha_{m_2}}\int dV\int_{4\pi} d\Omega\left[\psi^{(2)}_{1,m_1}(\mathbf{r},\mathbf{\Omega})\psi^{(1)}(\mathbf{r},\mathbf{\Omega}) + \psi^{(2)}_{2,m_1}(\mathbf{r},\mathbf{\Omega})\varphi(\mathbf{r},\mathbf{\Omega})\right]f_j(\mathbf{r}) \\
&-\sum_{k=1}^{N_m}\frac{\partial^2 C_j(\alpha)}{\partial\alpha_{m_1}\partial \alpha_{m_2}}\int dV\int_{4\pi} d\Omega\,\psi^{(1)}(r,\mathbf{\Omega})\varphi(\mathbf{r},\mathbf{\Omega})f_k(\mathbf{r}),\quad m_1,m_2 = 1,\ldots,N_\alpha.
\end{aligned} \tag{6.58}$$

In Sections 6.3.1 through 6.3.4, below, the expression obtained in Equation 6.58 will be specialized for the particular case described by Equations 6.40 through 6.43.

6.3.1 Computation of the Second-Order Sensitivities $S_{i,j}^{(2)} \triangleq \frac{\partial S_i^{(1)}}{\partial N_i \, \partial \alpha_j}$, $i = 1,...,N_m$, $j = 1,...,N_\alpha$

The second-order sensitivities $S_{i,j}^{(2)} \triangleq \frac{\partial S_i^{(1)}}{\partial N_i \, \partial \alpha_j}$, $i=1,...,N_m$, $j=1,...,N_\alpha$, are obtained by computing the G-differential of the first-order sensitivities defined in Equation 6.40, which yields the following expression:

$$\delta S_i^{(1)}(\varphi,\alpha;\psi^{(1)};\delta\alpha) = \left\{\delta S_i^{(1)}(\varphi,\alpha;\psi^{(1)},\delta\alpha)\right\}_{dir} + \left\{\delta S_i^{(1)}(\varphi,\alpha;\psi^{(1)},\delta\alpha)\right\}_{ind}, \; i=1,...,N_m, \tag{6.59}$$

where

$$\begin{aligned}\left\{\delta S_j^{(1)}(\varphi,\alpha;\psi^{(1)},\delta\alpha)\right\}_{dir} &\triangleq -\delta\sigma_i \int dV \int_{4\pi} d\Omega \, f_i(\mathbf{r})\psi^{(1)}(\mathbf{r},\Omega)\varphi(\mathbf{r},\Omega) \\ &+ \delta q_i \int dV \int_{4\pi} d\Omega \; \psi^{(1)}(r,\Omega) g_i(\mathbf{r}), \; i=1,...,N_m,\end{aligned} \tag{6.60}$$

and

$$\begin{aligned}\left\{\delta S_i^{(1)}(\varphi,\alpha;\psi^{(1)},\delta\alpha)\right\}_{ind} &\triangleq -\sigma_i \int dV \int_{4\pi} d\Omega \, f_i(\mathbf{r}) \left[\delta\psi^{(1)}(\mathbf{r},\Omega)\varphi(\mathbf{r},\Omega) + \psi^{(1)}(\mathbf{r},\Omega)\delta\varphi(\mathbf{r},\Omega)\right] \\ &+ q_i \int dV \int_{4\pi} d\Omega \, \delta\psi^{(1)}(r,\Omega) g_i(\mathbf{r}), \;\; i=1,...,N_m.\end{aligned} \tag{6.61}$$

Comparing Equations 6.60 and 6.61 to Equations 6.45 and 6.46, respectively, and following the general procedure outlined in Section 6.3 (which led to the general result given Equation 6.57) yields the following expression for the right side of Equation 6.59:

$$\begin{aligned}&\delta S_i^{(1)}(\varphi,\alpha;\psi^{(1)};\psi_{1,i}^{(2)},\psi_{2,i}^{(2)};\delta\alpha) = \\ &-\int dV \int_{4\pi} d\Omega \left[\psi_{1,i}^{(2)}(\mathbf{r},\Omega)\psi^{(1)}(\mathbf{r},\Omega) + \psi_{2,i}^{(2)}(\mathbf{r},\Omega)\varphi(\mathbf{r},\Omega)\right]\delta\Sigma_t(\mathbf{r}) \\ &+\int dV \int_{4\pi} d\Omega \psi_{1,i}^{(2)}(\mathbf{r},\Omega)\delta\Sigma_d(\mathbf{r},\Omega) + \int dV \int_{4\pi} d\Omega \psi_{2,i}^{(2)}(\mathbf{r},\Omega)\delta q(\mathbf{r}) \\ &-\delta\sigma_i \int dV \int_{4\pi} d\Omega \, f_i(\mathbf{r})\psi^{(1)}(\mathbf{r},\Omega)\varphi(\mathbf{r},\Omega) \\ &+\delta q_i \int dV \int_{4\pi} d\Omega \, g_i(\mathbf{r})\psi^{(1)}(\mathbf{r},\Omega), \;\; i=1,...,N_m,\end{aligned} \tag{6.62}$$

where the second-level adjoint functions $\psi_{1,i}^{(2)}(\mathbf{r},\mathbf{\Omega})$ and $\psi_{2,i}^{(2)}(\mathbf{r},\mathbf{\Omega})$ are the solutions of the following 2nd-LASS:

$$\mathbf{\Omega}\bullet\nabla\psi_{1,i}^{(2)}(\mathbf{r},\mathbf{\Omega})+\Sigma_t(\mathbf{r};\alpha)\psi_{1,i}^{(2)}(\mathbf{r},\mathbf{\Omega})=-\sigma_i f_i(\mathbf{r})\varphi(\mathbf{r},\mathbf{\Omega})+q_i g_i(\mathbf{r}),\quad i=1,\ldots,N_m, \tag{6.63}$$

$$\psi_{1,i}^{(2)}(\mathbf{r},\mathbf{\Omega})=0,\mathbf{r}_s\in\partial V,\mathbf{\Omega}\bullet\mathbf{n}<0, \tag{6.64}$$

$$-\mathbf{\Omega}\bullet\nabla\psi_{2,i}^{(2)}(\mathbf{r},\mathbf{\Omega})+\Sigma_t(\mathbf{r};\alpha)\psi_{2,i}^{(2)}(\mathbf{r},\mathbf{\Omega})=-\sigma_i f_i(\mathbf{r})\psi^{(1)}(\mathbf{r},\mathbf{\Omega}),\ i=1,\ldots,N_m, \tag{6.65}$$

$$\psi_{2,i}^{(2)}(\mathbf{r},\mathbf{\Omega})=0,\mathbf{\Omega}\bullet\mathbf{n}>0,\mathbf{r}_s\in\partial V. \tag{6.66}$$

Replacing Equations 6.36 through 6.38 in Equation 6.62 and identifying the expressions multiplying the variations δN_j, $\delta\sigma_j$, δq_j, and $\delta\lambda_j$ yields the following expressions for (part of) the second-order mixed partial sensitivities of the response with respect to the model parameters:

$$\begin{aligned}S_{i,j}^{(2)}\triangleq\frac{\partial^{(2)}R}{\partial N_i\,\partial N_j}&=-\sigma_j\int dV\int_{4\pi}d\Omega\left[\psi_{1,i}^{(2)}(\mathbf{r},\mathbf{\Omega})\psi^{(1)}(\mathbf{r},\mathbf{\Omega})+\psi_{2,i}^{(2)}(\mathbf{r},\mathbf{\Omega})\varphi(\mathbf{r},\mathbf{\Omega})\right]f_j(\mathbf{r}),\\&+q_j\int dV\int_{4\pi}d\Omega\psi_{2,i}^{(2)}(\mathbf{r},\mathbf{\Omega})g_j(\mathbf{r}),\quad i,j=1,\ldots,N_m,\end{aligned} \tag{6.67}$$

$$\begin{aligned}S_{i,j+N_m}^{(2)}\triangleq\frac{\partial^{(2)}R}{\partial N_i\,\partial\sigma_j}&=-N_j\int dV\int_{4\pi}d\Omega\left[\psi_{1,i}^{(2)}(\mathbf{r},\mathbf{\Omega})\psi^{(1)}(\mathbf{r},\mathbf{\Omega})-\psi_{2,i}^{(2)}(\mathbf{r},\mathbf{\Omega})\varphi(\mathbf{r},\mathbf{\Omega})\right]f_j(\mathbf{r})\\&-\delta_{ij}\int dV\int_{4\pi}d\Omega f_i(\mathbf{r})\psi^{(1)}(\mathbf{r},\mathbf{\Omega})\varphi(\mathbf{r},\mathbf{\Omega}),\quad i,j=1,\ldots,N_m,\end{aligned} \tag{6.68}$$

$$\begin{aligned}S_{i,j+2N_m}^{(2)}\triangleq\frac{\partial^{(2)}R}{\partial N_i\,\partial q_j}&=N_j\int dV\int_{4\pi}d\Omega\psi_{2,i}^{(2)}(\mathbf{r},\mathbf{\Omega})g_j(\mathbf{r})\\&+\delta_{ij}N_j\int dV\int_{4\pi}d\Omega g_i(\mathbf{r})\psi^{(1)}(\mathbf{r},\mathbf{\Omega}),\ i,j=1,\ldots,N\end{aligned} \tag{6.69}$$

$$S_{i,j+3N_m}^{(2)}\triangleq\frac{\partial^{(2)}R}{\partial N_i\,\partial\lambda_j}=\int dV\int_{4\pi}d\Omega\psi_{1,i}^{(2)}(\mathbf{r},\mathbf{\Omega})h_j(\mathbf{r},\mathbf{\Omega}),i=1,\ldots,N_m,j=1,\ldots,N_d \tag{6.70}$$

6.3.2 Computation of the Second-Order Sensitivities $\mathbf{S}^{(2)}_{i+N_m,j} \triangleq \frac{\partial \mathbf{S}^{(1)}_i}{\partial \sigma_i \, \partial \alpha_j}$, $i = 1,...,N_m$, $j = 1,...,N_\alpha$

The second-order sensitivities $S^{(2)}_{i+N_m,j} \triangleq \frac{\partial S^{(1)}_i}{\partial \sigma_i \, \partial \alpha_j}$, $i = 1,...,N_m$, $j = 1,...,N_\alpha$, are obtained by computing the G-differential of the first-order sensitivities defined in Equation 6.41, which yields the following expression:

$$\delta S^{(1)}_{i+N_m}(\varphi,\alpha;\psi^{(1)},\delta\alpha) = \left\{\delta S^{(1)}_{i+N_m}(\varphi,\alpha;\psi^{(1)},\delta\alpha)\right\}_{dir} + \left\{\delta S^{(1)}_{i+N_m}(\varphi,\alpha;\psi^{(1)},\delta\alpha)\right\}_{ind}, \; i = 1,...,N_m, \tag{6.71}$$

where

$$\left\{\delta S^{(1)}_{i+N_m}(\varphi,\alpha;\psi^{(1)},\delta\alpha)\right\}_{dir} \triangleq -\delta N_i \int dV \int_{4\pi} d\Omega \, f_i(\mathbf{r}) \psi^{(1)}(\mathbf{r},\mathbf{\Omega}) \varphi(\mathbf{r},\mathbf{\Omega}), \; i = 1,...,N_m, \tag{6.72}$$

and

$$\left\{\delta S^{(1)}_{i+N_m}(\varphi,\alpha;\psi^{(1)},\delta\alpha)\right\}_{ind} \triangleq -N_i \int dV \int_{4\pi} d\Omega \, f_i(\mathbf{r}) \delta\psi^{(1)}(\mathbf{r},\mathbf{\Omega}) \varphi(\mathbf{r},\mathbf{\Omega}) - N_i \int dV \int_{4\pi} d\Omega \, f_i(\mathbf{r}) \psi^{(1)}(\mathbf{r},\mathbf{\Omega}) \delta\varphi(\mathbf{r},\mathbf{\Omega}), \quad i = 1,...,N_m. \tag{6.73}$$

Comparing Equations 6.72 and 6.73 to Equations 6.45 and 6.46, respectively, and following the general procedure outlined in Section 6.3 (which led to the general result given in Equation 6.57) yields the following expression for the right side of Equation 6.71:

$$\begin{aligned}\delta S^{(1)}_{i+N_m}(\varphi,\alpha;\psi^{(1)};\psi^{(2)}_{1,i+N_m},\psi^{(2)}_{2,i+N_m};\delta\alpha) &= -\delta N_i \int dV \int_{4\pi} d\Omega \, f_i(\mathbf{r}) \psi^{(1)}(\mathbf{r},\mathbf{\Omega}) \varphi(\mathbf{r},\mathbf{\Omega}) \\ &- \int dV \int_{4\pi} d\Omega \left[\psi^{(2)}_{1,i+N_m}(\mathbf{r},\mathbf{\Omega}) \psi^{(1)}(\mathbf{r},\mathbf{\Omega}) + \psi^{(2)}_{2,i+N_m}(\mathbf{r},\mathbf{\Omega}) \varphi(\mathbf{r},\mathbf{\Omega})\right] \delta\Sigma_t(\mathbf{r}) \\ &+ \int dV \int_{4\pi} d\Omega \psi^{(2)}_{1,i+N_m}(\mathbf{r},\mathbf{\Omega}) \delta\Sigma_d(\mathbf{r},\mathbf{\Omega};\alpha) \\ &+ \int dV \int_{4\pi} d\Omega \psi^{(2)}_{2,i+N_m}(\mathbf{r},\mathbf{\Omega}) \delta q(\mathbf{r};\alpha), \quad i = 1,...,N_m.\end{aligned} \tag{6.74}$$

The second-level adjoint functions $\psi^{(2)}_{1,i+N_m}$ and $\psi^{(2)}_{2,i+N_m}$ are the solutions of the following 2nd-LASS:

$$\boldsymbol{\Omega} \bullet \nabla \psi^{(2)}_{1,i+N_m}(\mathbf{r},\boldsymbol{\Omega}) + \Sigma_t(\mathbf{r};\boldsymbol{\alpha})\psi^{(2)}_{1,i+N_m}(\mathbf{r},\boldsymbol{\Omega}) = -N_i f_i(\mathbf{r})\varphi(\mathbf{r},\boldsymbol{\Omega}), \; i = 1,\ldots,N_m, \tag{6.75}$$

$$\psi^{(2)}_{1,i+N_m}(\mathbf{r},\boldsymbol{\Omega}) = 0, \; \boldsymbol{\Omega} \bullet \mathbf{n} < 0, \; \mathbf{r}_s \in \partial V, \tag{6.76}$$

$$-\boldsymbol{\Omega} \bullet \nabla \psi^{(2)}_{2,i+N_m}(\mathbf{r},\boldsymbol{\Omega}) + \Sigma_t(\mathbf{r};\boldsymbol{\alpha})\psi^{(2)}_{2,i+N_m}(\mathbf{r},\boldsymbol{\Omega}) = -N_i f_i(\mathbf{r})\psi^{(1)}(\mathbf{r},\boldsymbol{\Omega}), \; i = 1,\ldots,N_m, \tag{6.77}$$

$$\psi^{(2)}_{2,i+N_m}(\mathbf{r},\boldsymbol{\Omega}) = 0, \; \boldsymbol{\Omega} \bullet \mathbf{n} > 0, \; \mathbf{r}_s \in \partial V. \tag{6.78}$$

Replacing Equations 6.36 through 6.38 in Equation 6.74 and identifying the expressions multiplying the variations δN_j, $\delta\sigma_j$, δq_j, and $\delta\lambda_j$ yields

$$S^{(2)}_{i+N_m,j} \triangleq \frac{\partial^{(2)} R}{\partial \sigma_i \partial N_j} =$$

$$-\sigma_j \int dV \int_{4\pi} d\Omega \left[\psi^{(2)}_{1,i+N_m}(\mathbf{r},\boldsymbol{\Omega})\psi^{(1)}(\mathbf{r},\boldsymbol{\Omega}) + \psi^{(2)}_{2,i+N_m}(\mathbf{r},\boldsymbol{\Omega})\varphi(\mathbf{r},\boldsymbol{\Omega}) \right] f_j(\mathbf{r})$$

$$+q_j \int dV \int_{4\pi} d\Omega \psi^{(2)}_{2,i+N_m}(\mathbf{r},\boldsymbol{\Omega}) g_j(\mathbf{r}) \tag{6.79}$$

$$-\delta_{ij} \int dV \int_{4\pi} d\Omega \, f_i(\mathbf{r})\psi^{(1)}(\mathbf{r},\boldsymbol{\Omega})\varphi(\mathbf{r},\boldsymbol{\Omega}), \; i,j = 1,\ldots,N_m,$$

$$S^{(2)}_{i+N_m,j+N_m} \triangleq \frac{\partial^{(2)} R}{\partial \sigma_i \partial \sigma_j}$$

$$= -N_j \int dV \int_{4\pi} d\Omega \left[\psi^{(2)}_{1,i+N_m}(\mathbf{r},\boldsymbol{\Omega})\psi^{(1)}(\mathbf{r},\boldsymbol{\Omega}) + \psi^{(2)}_{2,i+N_m}(\mathbf{r},\boldsymbol{\Omega})\varphi(\mathbf{r},\boldsymbol{\Omega}) \right] f_j(\mathbf{r}), \tag{6.80}$$

$i,j = 1,\ldots,N_m,$

$$S^{(2)}_{i+N_m,j+2N_m} \triangleq \frac{\partial^{(2)} R}{\partial \sigma_i \partial q_j} = N_j \int dV \int_{4\pi} d\Omega \psi^{(2)}_{2,i+N_m}(\mathbf{r},\boldsymbol{\Omega}) g_j(\mathbf{r}), \quad i,j = 1,\ldots,N \tag{6.81}$$

$$S^{(2)}_{i+N_m,j+3N_m} \triangleq \frac{\partial^{(2)} R}{\partial \sigma_i \, \partial \lambda_j} = \int dV \int_{4\pi} d\Omega \psi^{(2)}_{1,i+N_m}(\mathbf{r},\Omega) h_j(\mathbf{r},\Omega), \tag{6.82}$$

$$i = 1,\ldots,N_m, \; j = 1,\ldots,N_D.$$

6.3.3 Computation of the Second-Order Sensitivities $S^{(2)}_{i+2N_m,j} \triangleq \dfrac{\partial S_i^{(1)}}{\partial q_i \, \partial \alpha_j}, \; i = 1,\ldots,N_m, \; j = 1,\ldots,N_\alpha$

The second-order sensitivities $S^{(2)}_{i+2N_m,j} \triangleq \dfrac{\partial S_i^{(1)}}{\partial q_i \, \partial \alpha_j}, \; i = 1,\ldots,N_m, \; j = 1,\ldots,N_\alpha$, are obtained by computing the G-differential of Equation 6.42, which yields the following expression:

$$\delta S^{(1)}_{i+2N_m}(\varphi,\alpha;\psi^{(1)},\delta\alpha) = \left\{\delta S^{(1)}_{i+2N_m}(\varphi,\alpha;\psi^{(1)},\delta\alpha)\right\}_{dir} + \left\{\delta S^{(1)}_{i+2N_m}(\varphi,\alpha;\psi^{(1)},\delta\alpha)\right\}_{ind}, \; i = 1,\ldots,N_m, \tag{6.83}$$

where

$$\left\{\delta S^{(1)}_{i+N_m}(\varphi,\alpha;\psi^{(1)},\delta\alpha)\right\}_{dir} \triangleq \delta N_i \int dV \int_{4\pi} d\Omega \; \psi^{(1)}(\mathbf{r},\Omega) g_i(\mathbf{r}), \; i = 1,\ldots,N_m, \tag{6.84}$$

and

$$\left\{\delta S^{(1)}_{i+2N_m}(\varphi,\alpha;\psi^{(1)},\delta\alpha)\right\}_{ind} = N_i \int dV \int_{4\pi} d\Omega \; \delta\psi^{(1)}(\mathbf{r},\Omega) g_i(\mathbf{r}), \; \textit{for} \; i = 1,\ldots,N_m. \tag{6.85}$$

Comparing Equations 6.84 and 6.85 to Equations 6.45 and 6.46, respectively, and following the general procedure outlined in the foregoing (which led to the general result given Equation 6.57) yields the following expression for the right side of Equation 6.85:

$$\left\{\delta S^{(1)}_{i+2N_m}(\alpha;\psi^{(1)};\psi^{(2)}_{1,i+2N_m};\delta\alpha)\right\}_{ind} = -\int dV \int_{4\pi} d\Omega \psi^{(2)}_{1,i+2N_m}(\mathbf{r},\Omega) \psi^{(1)}(\mathbf{r},\Omega) \delta\Sigma_t(\mathbf{r};\alpha) + \int dV \int_{4\pi} d\Omega \psi^{(2)}_{1,i+2N_m}(\mathbf{r},\Omega) \delta\Sigma_d(\mathbf{r},\Omega;\alpha), \tag{6.86}$$

where the second-level adjoint function $\psi^{(2)}_{1,i+2N_m}$ is the solution of the following 2nd-LASS:

$$\Omega \bullet \nabla \psi^{(2)}_{1,i+2N_m}(\mathbf{r},\Omega) + \Sigma_t(\mathbf{r};\alpha) \psi^{(2)}_{1,i+2N_m}(\mathbf{r},\Omega) = N_i \, g_i(\mathbf{r}), \; i = 1,\ldots,N_m, \tag{6.87}$$

$$\psi^{(2)}_{1,i+2N_m}(\mathbf{r},\boldsymbol{\Omega})=0,\mathbf{r}_s\in\partial V,\boldsymbol{\Omega}\bullet\mathbf{n}<0. \tag{6.88}$$

Note also that $\psi^{(2)}_{2,i+2N_m}\equiv 0$. Replacing Equations 6.36 through 6.38 in Equation 6.86 and identifying the expressions multiplying the variations δN_j, $\delta\sigma_j$, δq_j, and $\delta\lambda_j$ yields

$$\begin{aligned}S^{(2)}_{i+2N_m,j}&\triangleq\frac{\partial^{(2)}R}{\partial q_i\,\partial N_j}=-\sigma_j\int dV\int_{4\pi}d\Omega\psi^{(2)}_{1,i+2N_m}(\mathbf{r},\boldsymbol{\Omega})\psi^{(1)}(\mathbf{r},\boldsymbol{\Omega})f_j(\mathbf{r})\\&+\delta_{ij}\int dV\int_{4\pi}d\Omega\,\psi^{(1)}(\mathbf{r},\boldsymbol{\Omega})g_i(\mathbf{r}),\quad i,j=1,\ldots,N_m,\end{aligned}\tag{6.89}$$

$$\begin{aligned}S^{(2)}_{i+2N_m,j+N_m}&\triangleq\frac{\partial^{(2)}R}{\partial q_i\,\partial\sigma_j}\\&=-N_j\int dV\int_{4\pi}d\Omega\psi^{(2)}_{1,i+2N_m}(\mathbf{r},\boldsymbol{\Omega})\psi^{(1)}(\mathbf{r},\boldsymbol{\Omega})f_j(\mathbf{r}),\quad i,j=1,\ldots,N_m,\end{aligned}\tag{6.90}$$

$$S^{(2)}_{i+2N_m,j+2N_m}\triangleq\frac{\partial^{(2)}R}{\partial q_i\,\partial q_j}=0,\quad i,j=1,\ldots,N_m,\tag{6.91}$$

$$\begin{aligned}S^{(2)}_{i+2N_m,j+3N_m}&\triangleq\frac{\partial^{(2)}R}{\partial q_i\,\partial\lambda_j}\\&=\int dV\int_{4\pi}d\Omega\psi^{(2)}_{1,i+2N_m}(\mathbf{r},\boldsymbol{\Omega})h_j(\mathbf{r},\boldsymbol{\Omega}),\quad i=1,\ldots,N_m,\ j=1,\ldots,N_D.\end{aligned}\tag{6.92}$$

6.3.4 Computation of the Second-Order Sensitivities $\mathbf{S}^{(2)}_{i+3N_m,j}\triangleq\dfrac{\partial S^{(1)}_i}{\partial\lambda_i\,\partial\alpha_j}$, $i=1,\ldots,N_d$, $j=1,\ldots,N_\alpha$

Note from Equation 6.43 that the sensitivities $S^{(1)}_{i+3N_m}(\varphi,\alpha)$ do not display explicit dependence on the model parameters. The second-order sensitivities $S^{(2)}_{i+3N_m,j}\triangleq\dfrac{\partial S^{(1)}_i}{\partial\lambda_i\,\partial\alpha_j}$, $i=1,\ldots,N_d$, $j=1,\ldots,N_\alpha$, are obtained by computing the G-differential of Equation 6.43, which yields the following expression:

$$\delta S^{(1)}_{i+3N_m}(\alpha;\delta\varphi,\delta\alpha)=\int dV\int_{4\pi}d\Omega\,\delta\varphi(\mathbf{r},\boldsymbol{\Omega})h_i(\mathbf{r},\boldsymbol{\Omega}),\ for\ i=1,\ldots,N_d.\tag{6.93}$$

Comparing Equation 6.93 to Equation 6.41 indicates that, by following the procedure outlined in Sections 6.3.1 through 6.3.3, the indirect-effect term in Equation 6.93 will ultimately have the expression

$$\delta S^{(1)}_{i+3N_m}\left(\alpha;\varphi,\psi^{(2)}_{2,i+3N_m};\delta\alpha\right)=-\int dV\int_{4\pi}d\Omega\psi^{(2)}_{2,i+3N_m}(\mathbf{r},\Omega)\varphi(\mathbf{r},\Omega)\delta\Sigma_t(\mathbf{r})$$
$$+\int dV\int_{4\pi}d\Omega\psi^{(2)}_{2,i+3N_m}(\mathbf{r},\Omega)\delta q(\mathbf{r}), \tag{6.94}$$

where the second-level adjoint function $\psi^{(2)}_{2,i+3N_m}$ is the solution of the following 2nd-LASS:

$$-\Omega\bullet\nabla\psi^{(2)}_{2,i+3N_m}(\mathbf{r},\Omega)+\Sigma_t(\mathbf{r})\psi^{(2)}_{2,i+3N_m}(\mathbf{r},\Omega)=h_i(\mathbf{r},\Omega),\ i=1,\ldots,N_d, \tag{6.95}$$

$$\psi^{(2)}_{2,i+3N_m}(\mathbf{r},\Omega)=0,\mathbf{r}_s\in\partial V,\Omega\bullet\mathbf{n}>0. \tag{6.96}$$

Note that $\psi^{(2)}_{1,i+3N_m}\equiv 0$. Replacing Equations 6.36 through 6.38 in Equation 6.94 and identifying the expressions multiplying the variations δN_j, $\delta\sigma_j$, δq_j, and $\delta\lambda_j$ yields

$$S^{(2)}_{i+3N_m,j}\triangleq\frac{\partial^{(2)}R}{\partial\lambda_i\,\partial N_j}=-\sigma_j\int dV\int_{4\pi}d\Omega\psi^{(2)}_{2,i+3N_m}(\mathbf{r},\Omega)\varphi(\mathbf{r},\Omega)f_j(\mathbf{r}),$$
$$+q_j\int dV\int_{4\pi}d\Omega\psi^{(2)}_{2,i+3N_m}(\mathbf{r},\Omega)g_j(\mathbf{r}),\ i=1,\ldots,N_d,\ j=1,\ldots,N_m, \tag{6.97}$$

$$S^{(2)}_{i+3N_m,j+N_m}\triangleq\frac{\partial^{(2)}R}{\partial\lambda_i\,\partial\sigma_j}$$
$$=-N_j\int dV\int_{4\pi}d\Omega\psi^{(2)}_{2,i+3N_m}(\mathbf{r},\Omega)\varphi(\mathbf{r},\Omega)f_j(\mathbf{r}),\ i=1,\ldots,N_d,\ j=1,\ldots,N_m, \tag{6.98}$$

$$S^{(2)}_{i+3N_m,j+2N_m}\triangleq\frac{\partial^{(2)}R}{\partial\lambda_i\,\partial q_j}$$
$$=N_j\int dV\int_{4\pi}d\Omega\psi^{(2)}_{2,i+3N_m}(\mathbf{r},\Omega)g_j(\mathbf{r}),\quad i=1,\ldots,N_d,\quad j=1,\ldots,N_m, \tag{6.99}$$

$$S^{(2)}_{i+3N_m,j+3N_m}\triangleq\frac{\partial^{(2)}R}{\partial\lambda_i\,\partial\lambda_j}=0,\quad i=1,\ldots,N_d,\ j=1,\ldots,N_d. \tag{6.100}$$

6.4 Concluding Remarks

The following conclusions can be drawn based on the results that have been presented in this chapter:

1. A single 1st-LASS needs to be solved in order to compute all first-order response sensitivities, to all N_α model parameters.
2. For each model parameter, a single 2nd-LASS needs to be solved for computing the corresponding mixed second-order sensitivities. Hence, computing all of the $N_\alpha(N_\alpha+1)/2$ second-order sensitivities could require solving at most N_α 2nd-LASS.
3. The solution of each of the 2nd-LASS is a two-component vector-valued second-level adjoint function, except for the 2nd-LASS that corresponds to state variables that appear linearly in the response under consideration, in which case the vector-valued second-level adjoint function may have a null component.
4. Solving each of the 2nd-LASS involves the inversion of the same operators as need to be inverted for solving the original transport equation and/or the 1st-LASS. Only the various source terms on the right sides of the 2nd-LASS differ from each other and from the forward and/or 1st-LASS. Therefore, the same software can be used to solve both the 1st-LASS and the 2nd-LASS.
5. The computation of the second-order sensitivities involves the evaluations of integrals of the same form as those needed for computing the first-order sensitivities. Therefore, the same software can be used for computing both the first-order and second-order sensitivities.
6. Each of the mixed second-order sensitivities is computed *twice*, using two distinct second-level adjoint functions. Consequently, the 2nd-ASAM possesses an inherent *solution verification* mechanism that enables and ensures the accuracy verification of the solutions of all of the 2nd-LASS.
7. As expected, the angular flux solution of the ray-tracing problem is *separately* linear in the source strengths and detector interaction coefficients. This fact has been confirmed by the vanishing of the respective second-order unmixed sensitivities, as demonstrated by the results presented in Equations 6.70 and 6.79, respectively. In such cases, each of the respective vector-valued second-level adjoint functions will have an identically null component. This fact has been confirmed by the vanishing of the respective second-order unmixed sensitivities, as demonstrated by the results presented in Equations 6.91 and 6.100, respectively. In such cases, each of the respective vector-valued second-level adjoint functions will have an

identically null component. Similar results have been obtained for the benchmark problem modeling the linear neutron diffusion equation presented in Chapter 5.

If the monoenergetic radiation propagates through a homogeneous slab of thickness a comprising a mixture of isotopes, then Equation 6.1 takes on the following form for the uncollided angular flux, $\phi(z, \omega)$:

$$\omega \frac{d\varphi(z,\omega)}{dz} + \varphi(z,\omega)\sum_{j=1}^{N_m} C_j(\alpha) f_j = \sum_{j=1}^{N_m} Q_j(\alpha) g_j, 0 \le z \le a, \quad 0 \le \omega \le 1, \tag{6.101}$$

where ω denotes the cosine of the angle between the gamma ray's direction and the z-axis. It is evident that Equation 6.101 has the same form as Equation 2.1, with the following correspondences: $z \to t$, $a \to t_f$, $\phi(z,\omega) \to \rho(t)$, $\sum_{j=1}^{N_m} C_j(\alpha) f_j \to \sum_{i=1}^{N} w_i \beta_i$, and $\sum_{j=1}^{N_m} Q_j(\alpha) g_j \to q \sum_{i=1}^{N} w_i \beta_i$. Thus, the radiation transport equation in Equation 6.101 provides yet another context in which a relatively simple model may nevertheless comprise sufficiently many model parameters to render the 2nd-ASAM as the only viable methodology for computing exactly the response sensitivities to all of the model parameters.

7

The 2nd-ASAM for Nonlinear Systems

Chapter 3 presented the 2nd-ASAM for linear systems, in which the equations modeling the system were linear in the dependent variables. Following the material originally developed by Cacuci (2016b), this chapter presents the 2nd-ASAM for *nonlinear* systems, in which the equations modeling the system are nonlinear in the dependent variables. The results that will be obtained in this chapter will comprise, as particular cases, the results that were derived in Chapter 3.

7.1 Mathematical Modeling of a General Nonlinear System

A nonlinear physical system is modeled mathematically by means of N_u coupled *nonlinear* equations, which can be written in operator form as follows:

$$\mathbf{N}\left[\alpha(\mathbf{x}),\mathbf{u}(\mathbf{x})\right]=\mathbf{Q}\left[\alpha(\mathbf{x})\right], \quad \mathbf{x}\in\Omega_x. \tag{7.1}$$

The quantities appearing in the above equations are defined as in Chapter 3, except that $\mathbf{N}\left[\boldsymbol{\alpha}(\mathbf{x}),\mathbf{u}(\mathbf{x})\right]\triangleq\left[N_1(\boldsymbol{\alpha},\mathbf{u}),\ldots,N_{N_u}(\boldsymbol{\alpha},\mathbf{u})\right]$ denotes a N_u-component column vector whose components are *operators* (including differential, difference, integral, distribution, and infinite matrices) acting *nonlinearly* on $\mathbf{u}$ and $\boldsymbol{\alpha}$. All of the equalities in this work are considered to hold in the weak ("distributional") sense, since the right sides ("sources") of Equation 7.1 usually contain distributions ("generalized functions/functionals"), particularly Dirac distributions and derivatives and/or integrals thereof.

In view of the definitions given above, $\mathbf{N}(\boldsymbol{\alpha},\mathbf{u})$ represents the mapping $\mathbf{N}:\mathcal{D}\subset\mathcal{E}\rightarrow\mathcal{E}_Q$, where $\mathcal{D}=\mathcal{D}_u\oplus\mathcal{D}_\alpha$, $\mathcal{D}_u\subset\mathcal{E}_u$, $\mathcal{D}_\alpha\subset\mathcal{E}_\alpha$, and $\mathcal{E}=\mathcal{E}_u\oplus\mathcal{E}_\alpha$. The spaces $\mathcal{E}_u$ and $\mathcal{E}_\alpha$ are considered to be normed linear vector spaces. Note that an arbitrary element $\mathbf{e}\in\mathcal{E}$ is of the form $\mathbf{e}=(\boldsymbol{\alpha},\mathbf{u})$. If differential operators appear in Equation 7.1, then a corresponding set of boundary and/or initial conditions (which are essential to define the domain of $\mathcal{D}$) must also be given; these boundary and/or initial conditions are represented in operator form as

$$\left[\mathbf{B}(\alpha,\mathbf{u})-\mathbf{C}(\alpha)\right]_{\partial\Omega_x}=\mathbf{0}, \quad \mathbf{x}\in\partial\Omega_x, \tag{7.2}$$

where $\partial\Omega_x$ denotes the boundary of Ω_x, where the operator $\mathbf{B}(\boldsymbol{\alpha},\mathbf{u})$ acts *nonlinearly* on both the state functions, $\mathbf{u}$, and on the model parameters, $\boldsymbol{\alpha}$, and where $\mathbf{C}(\boldsymbol{\alpha})$ denotes an operator that acts nonlinearly on $\boldsymbol{\alpha}$.

The vector-valued function $\mathbf{u}(\mathbf{x})$ is considered to be the unique nontrivial solution of the physical problem described by Equations 7.1 and 7.2. As in Chapter 3, the system response (i.e., result of interest) associated with the problem modeled by Equations 7.1 and 7.2 is considered to be a scalar-valued real *nonlinear* functional of $(\boldsymbol{\alpha},\mathbf{u})$, which will be denoted as $R(\boldsymbol{\alpha},\mathbf{u})$ and which can be generally represented in operator form as

$$R(\boldsymbol{\alpha},\mathbf{u}):\mathcal{D}_R \subset \mathcal{E} \rightarrow \mathcal{F}, \tag{7.3}$$

where $\mathcal{F}$ denotes the field of real scalars.

The nominal solution of Equations 7.1 and 7.2 is denoted as $\mathbf{u}^0(\mathbf{x})$ and is obtained by solving these equations using the nominal parameter values $\boldsymbol{\alpha}^0(\mathbf{x})$. In other words, $\mathbf{u}^0(\mathbf{x})$ and $\boldsymbol{\alpha}^0(\mathbf{x})$ satisfy the following equations:

$$\mathbf{N}\left[\boldsymbol{\alpha}^0(\mathbf{x}),\mathbf{u}^0(\mathbf{x})\right]=\mathbf{Q}\left[\boldsymbol{\alpha}^0(\mathbf{x})\right], \quad \mathbf{x}\in\Omega_x, \tag{7.4}$$

$$\left[\mathbf{B}\left(\boldsymbol{\alpha}^0,\mathbf{u}^0\right)-\mathbf{C}\left(\boldsymbol{\alpha}^0\right)\right]_{\partial\Omega_x}=\mathbf{0}, \quad \mathbf{x}\in\partial\Omega_x. \tag{7.5}$$

Equations 7.4 and 7.5 represent the *base-case* or *nominal* state of the physical system. The *nominal solution*, $\mathbf{u}^0(\mathbf{x})$, of these equations is subsequently used to compute the *nominal value* $R(\mathbf{e}^0)$, $\mathbf{e}^0 \triangleq (\boldsymbol{\alpha}^0,\mathbf{u}^0)$, of the response $R(\mathbf{e})$; as before, the superscript 0 denotes nominal values.

7.2 The 1st-LASS for Computing Exactly and Efficiently the First-Order Sensitivities

The general forward and adjoint sensitivity analysis procedures for computing exactly and efficiently the *first-order sensitivities of nonlinear responses associated with general nonlinear systems* were formulated by Cacuci (1981a, 1981b), which provide the basis for the developments that will be presented in this chapter. The sensitivity of the model response considered in Equation 7.3 to variations $\mathbf{h} \triangleq (\delta\boldsymbol{\alpha}, \delta\mathbf{u})$ in the model parameters and state functions in a neighborhood around $\mathbf{e}^0 = (\boldsymbol{\alpha}^0, \mathbf{u}^0) \in \mathcal{E}$ is given by the first-order G-differential, $\delta R(\mathbf{e}^0;\mathbf{h})$, which is defined as

$$\delta R\left(\mathbf{e}^0;\mathbf{h}\right)\triangleq\left\{\frac{d}{d\varepsilon}\left[R\left(\mathbf{e}^0+\varepsilon\mathbf{h}\right)\right]\right\}_{\varepsilon=0}\triangleq\lim_{\varepsilon\to 0}\frac{R\left(\mathbf{e}^0+\varepsilon\mathbf{h}\right)-R\left(\mathbf{e}^0\right)}{\varepsilon}, \tag{7.6}$$

for $\varepsilon \in \mathcal{F}$ and all (i.e., arbitrary) vectors $\mathbf{h} \in \mathcal{E} = \mathcal{E}_\alpha \times \mathcal{E}_u$ in a neighborhood $(\mathbf{e}^0 + \varepsilon\mathbf{h})$ around $\mathbf{e}^0 = (\boldsymbol{\alpha}^0, \mathbf{u}^0) \in \mathcal{E}$. The G-differential $\delta R(\mathbf{e}^0;\mathbf{h})$ is an operator defined on the same domain as $R(\mathbf{e})$, has the same range as $R(\mathbf{e})$, and provides the total first-order sensitivity of $R(\mathbf{e})$ to variations in the model's parameters and state functions. The G-differential $\delta R(\mathbf{e}^0;\mathbf{h})$ satisfies the relation $R(\mathbf{e}^0 + \varepsilon\mathbf{h}) - R(\mathbf{e}^0) = \delta R(\mathbf{e}^0;\mathbf{h}) + \Delta(\mathbf{h})$, with $\lim_{\varepsilon \to 0}\left[\Delta(\varepsilon\mathbf{h})\right]/\varepsilon = 0$.

As was discussed in Chapter 3, the computability of the G-differential $\delta R(\mathbf{e}^0;\mathbf{h})$ is assured when $\delta R(\mathbf{e}^0;\mathbf{h})$ is linear in $\mathbf{h} \triangleq (\delta\boldsymbol{\alpha}, \delta\mathbf{u})$ in a neighborhood $(\mathbf{e}^0 + \varepsilon\mathbf{h})$ around $\mathbf{e}^0 = (\mathbf{u}^0, \boldsymbol{\alpha}^0) \in \mathcal{E}$, in which case $\delta R(\mathbf{e}^0;\mathbf{h})$ can be written in the form

$$\delta R\left(\boldsymbol{\alpha}^0, \mathbf{u}^0; \mathbf{h}\right) = \left\{\frac{\partial R(\boldsymbol{\alpha}, \mathbf{u})}{\partial \mathbf{u}}\right\}_{(\alpha^0, \mathbf{u}^0)} \delta\mathbf{u} + \left\{\frac{\partial R(\boldsymbol{\alpha}, \mathbf{u})}{\partial \boldsymbol{\alpha}}\right\}_{(\alpha^0, \mathbf{u}^0)} \delta\boldsymbol{\alpha}, \tag{7.7}$$

where the quantities $\partial R(\boldsymbol{\alpha}, \mathbf{u})/\partial\mathbf{u} \triangleq \left[\partial R(\boldsymbol{\alpha}, \mathbf{u})/\partial u_1, \ldots, \partial R(\boldsymbol{\alpha}, \mathbf{u})/\partial u_{N_u}\right]$ and, respectively, $\partial R(\boldsymbol{\alpha}, \mathbf{u})/\partial\boldsymbol{\alpha} \triangleq \left[\partial R(\boldsymbol{\alpha}, \mathbf{u})/\partial \alpha_1, \ldots, \partial R(\boldsymbol{\alpha}, \mathbf{u})/\partial \alpha_{N_\alpha}\right]$ denote the partial G-derivatives at $\mathbf{e}^0$ of $R(\mathbf{e})$ with respect to $\mathbf{u}$ and $\boldsymbol{\alpha}$. As in Chapter 3, it is convenient to refer to the quantities $[\partial R(\boldsymbol{\alpha}, \mathbf{u})/\partial\boldsymbol{\alpha}]\delta\boldsymbol{\alpha}$ and $[\partial R(\boldsymbol{\alpha}, \mathbf{u})/\partial\mathbf{u}]\delta\mathbf{u}$ as the *direct-effect term* and the *indirect-effect term*, respectively, because the direct-effect term $[\partial R(\boldsymbol{\alpha}, \mathbf{u})/\partial\boldsymbol{\alpha}]\delta\boldsymbol{\alpha}$ can be evaluated immediately, while the indirect-effect term $[\partial R(\boldsymbol{\alpha}, \mathbf{u})/\partial\mathbf{u}]\delta\mathbf{u}$ can be quantified only after determining $\delta\mathbf{u}$ as function of $\delta\boldsymbol{\alpha}$, since the system's state vector $\mathbf{u}$ and parameters $\boldsymbol{\alpha}$ are related to each other through Equations 7.1 and 7.2.

The relationship between $\delta\mathbf{u}$ and $\delta\boldsymbol{\alpha}$ is determined by taking the G-differentials at $\mathbf{e}^0$ of Equations 7.1 and 7.2, which yields

$$\mathbf{F}_{11}^{(1)}\left(\boldsymbol{\alpha}^0, \mathbf{u}^0\right)\delta\mathbf{u} = \mathbf{Q}^{(1)}\left(\boldsymbol{\alpha}^0, \mathbf{u}^0; \delta\boldsymbol{\alpha}\right), \quad \mathbf{x} \in \Omega_x, \tag{7.8}$$

$$\mathbf{B}_F^{(1)}\left(\boldsymbol{\alpha}^0, \mathbf{u}^0; \delta\boldsymbol{\alpha}, \delta\mathbf{u}\right) = \mathbf{0}, \quad \mathbf{x} \in \partial\Omega_x, \tag{7.9}$$

where the superscript (1) indicates *first-level*, the letter F (used as "operator" and "subscript") indicates *forward*, the letter B indicates *boundary and/or initial conditions*, and where the following definitions were used:

$$\mathbf{F}_{11}^{(1)}(\boldsymbol{\alpha}, \mathbf{u}) \triangleq \left\{\frac{\partial \mathbf{N}(\boldsymbol{\alpha}, \mathbf{u})}{\partial \mathbf{u}}\right\}; \quad \mathbf{Q}^{(1)}(\boldsymbol{\alpha}, \mathbf{u}; \delta\boldsymbol{\alpha}) \triangleq \frac{\partial\left[\mathbf{Q}(\boldsymbol{\alpha}) - \mathbf{N}(\boldsymbol{\alpha}, \mathbf{u})\right]}{\partial \boldsymbol{\alpha}}\delta\boldsymbol{\alpha}; \quad \mathbf{x} \in \Omega_x, \tag{7.10}$$

$$\begin{aligned}\mathbf{B}_F^{(1)}\left(\boldsymbol{\alpha}^0, \mathbf{u}^0; \delta\boldsymbol{\alpha}, \delta\mathbf{u}\right) \triangleq & \left\{\frac{\partial \mathbf{B}(\boldsymbol{\alpha}, \mathbf{u})}{\partial \mathbf{u}}\right\}_{(\alpha^0, \mathbf{u}^0)} \delta\mathbf{u} \\ & + \left\{\frac{\partial\left[\mathbf{B}(\boldsymbol{\alpha}, \mathbf{u}) - \mathbf{C}(\boldsymbol{\alpha})\right]}{\partial \boldsymbol{\alpha}}\right\}_{(\alpha^0, \mathbf{u}^0)} \delta\boldsymbol{\alpha}, \quad \mathbf{x} \in \partial\Omega_x.\end{aligned} \tag{7.11}$$

The partial G-derivatives $\partial\mathbf{N}(\boldsymbol{\alpha},\mathbf{u})/\partial\mathbf{u}$ and $\partial\mathbf{N}(\boldsymbol{\alpha},\mathbf{u})/\partial\boldsymbol{\alpha}$ appearing in Equation 7.10 are matrices of the form

$$\frac{\partial\mathbf{N}(\alpha,\mathbf{u})}{\partial\mathbf{u}} \triangleq \begin{pmatrix} \frac{\partial N_1(\alpha,\mathbf{u})}{\partial u_1} & \cdots & \frac{\partial N_1(\alpha,\mathbf{u})}{\partial u_{N_u}} \\ \vdots & \ddots & \vdots \\ \frac{\partial N_{N_u}(\alpha,\mathbf{u})}{\partial u_1} & \cdots & \frac{\partial N_{N_u}(\alpha,\mathbf{u})}{\partial u_{N_u}} \end{pmatrix},$$
$$\frac{\partial\mathbf{N}(\alpha,\mathbf{u})}{\partial\alpha} \triangleq \begin{pmatrix} \frac{\partial N_1(\alpha,\mathbf{u})}{\partial\alpha_1} & \cdots & \frac{\partial N_1(\alpha,\mathbf{u})}{\partial\alpha_{N_\alpha}} \\ \vdots & \ddots & \vdots \\ \frac{\partial N_{N_u}(\alpha,\mathbf{u})}{\partial\alpha_1} & \cdots & \frac{\partial N_{N_u}(\alpha,\mathbf{u})}{\partial\alpha_{N_\alpha}} \end{pmatrix}. \tag{7.12}$$

The partial G-derivatives that appear in Equations 7.11 are also matrices that have components similar to those defined in Equation 7.12.

The system comprising Equations 7.10 and 7.11 represents the *1st-Level Forward Sensitivity System* (1st-LASS). For a given vector of parameter variations $\delta\boldsymbol{\alpha}$ in a neighborhood around $\boldsymbol{\alpha}^0$, the 1st-LFSS needs to be solved to obtain $\delta\mathbf{u}$ as a function of $\delta\boldsymbol{\alpha}$. In turn, the relationship between $\delta\mathbf{u}$ and $\delta\boldsymbol{\alpha}$ would lead to the elimination of the appearance of $\delta\mathbf{u}$ from Equation 7.7. Therefore, Equation 7.7 can be expressed in the form

$$\delta R\left(\alpha^0,\mathbf{u}^0;\delta\alpha\right)=\sum_{i_1=1}^{N_\alpha} R_{i_1}^{(1)}\left(\alpha^0,\mathbf{u}^0\right)\delta\alpha_{i_1}, \tag{7.13}$$

where the quantities $R_{i_1}^{(1)}\left(\alpha^0,\mathbf{u}^0\right)$ are independent of the parameter variations $\delta\alpha_{i_1}$ and represent *the first-order partial sensitivities (first-order partial G-derivatives) of the response* $R(\mathbf{e})$ *with respect to each of the model parameters* α_i, *evaluated at the nominal values* $\mathbf{e}^0$. The partial sensitivities $R_{i_1}^{(1)}\left(\alpha^0,\mathbf{u}^0\right)$ of the response $R(\mathbf{e})$ with respect to each of the model parameters $\delta\alpha_{i_1}$ are evaluated at the nominal values $\mathbf{e}^0 = (\boldsymbol{\alpha}^0, \mathbf{u}^0)$ and are obtained by successively setting ($\delta\alpha_i = 1, \delta\alpha_j = 0$ *for* $j \neq i$, $i = 1,\ldots, N_\alpha$) in the expression of $\delta R(\mathbf{e}^0; \mathbf{h})$ given in Equation 7.7. Computing the (total) response sensitivity $\delta R(\mathbf{e}^0; \mathbf{h})$ by using the ($\delta\boldsymbol{\alpha}$-dependent) solution $\delta\mathbf{u}$ of the 1st-LFSS is called (Cacuci 1981a, 1981b) the *Forward Sensitivity Analysis Procedure* (FSAP). From the standpoint of computational costs and effort, the FSAP requires $O(N_\alpha)$ large-scale forward computations. Therefore, the FSAP is advantageous to employ only if, in the problem under consideration, the number N_r of responses of interest exceeds the number of system parameters and/or parameter variations of interest. This is rarely the case in practice, however, since most problems of practical interest are characterized by many parameters (i.e., $\boldsymbol{\alpha}$ has many components) and comparatively few responses. In such situations, it is not economical to employ the FSAP since it becomes prohibitively expensive to solve repeatedly the $\delta\boldsymbol{\alpha}$-dependent 1st-LFSS in order to determine $\delta\mathbf{u}$ for all possible variations $\delta\boldsymbol{\alpha}$ in the model parameters.

In most practical situations, the number of model parameters exceeds significantly the number of functional responses of interest, that is, $N_r \ll N_\alpha$ In such cases, the *Adjoint Sensitivity Analysis Procedure* (ASAP), developed by Cacuci (1981a, 1981b), is the most efficient method for computing exactly the first-order sensitivities since it requires only $O(N_r)$ large-scale computations. The implementation of the ASAP requires the introduction of adjoint operators, which can be defined in Banach spaces but are most useful in Hilbert spaces, which are endowed with inner products. The spaces $\mathcal{E}_u$ and $\mathcal{E}_Q$ are henceforth considered to be Hilbert spaces and denoted as $\mathcal{H}_u(\Omega_x)$ and $\mathcal{H}_Q(\Omega_x)$, respectively. The elements of $\mathcal{H}_u(\Omega_x)$ and $\mathcal{H}_Q(\Omega_x)$ are, as before, vector-valued functions defined on the open set $\Omega_x \subset \mathbb{R}^{J_x}$, with smooth boundary $\partial\Omega_x$. On $\mathcal{H}_u(\Omega_x)$, the inner product of two vectors $\mathbf{u}^{(a)} \in \mathcal{H}_u$ and $\mathbf{u}^{(b)} \in \mathcal{H}_u$ will be denoted as $\langle \mathbf{u}^{(a)}, \mathbf{u}^{(b)} \rangle_u$. Typically, the inner product $\langle \mathbf{u}^{(a)}, \mathbf{u}^{(b)} \rangle_u$ is defined as

$$\left\langle \mathbf{u}^{(a)}, \mathbf{u}^{(b)} \right\rangle_u \triangleq \int_{\Omega_x} \mathbf{u}^{(a)}(\mathbf{x}) \cdot \mathbf{u}^{(b)}(\mathbf{x})\, d\mathbf{x}. \tag{7.14}$$

The inner product on $\mathcal{H}_Q(\Omega_x)$ of two vectors $\mathbf{Q}^{(a)} \in \mathcal{H}_Q$ and $\mathbf{Q}^{(b)} \in \mathcal{H}_Q$ will be denoted as $\langle \mathbf{Q}^{(a)}, \mathbf{Q}^{(b)} \rangle_Q$ and has the same form as shown in Equation 7.14. These Hilbert spaces are customarily self-dual. In particular, the Riesz representation theorem ensures that the indirect-effect term $[\partial R(\boldsymbol{\alpha}, \mathbf{u})/\partial \mathbf{u}]\delta\mathbf{u}$ can be written in the following inner product form:

$$\left[\partial R(\boldsymbol{\alpha}, \mathbf{u})/\partial \mathbf{u}\right]\delta\mathbf{u} = \left\langle \partial R(\boldsymbol{\alpha}, \mathbf{u})/\partial \mathbf{u},\, \delta\mathbf{u} \right\rangle_u, \quad \delta\mathbf{u} \in \mathcal{H}_u. \tag{7.15}$$

The goal of the adjoint sensitivity analysis methodology is to compute the above indirect-effect term exactly and efficiently without needing to compute explicitly the variations $\delta\mathbf{u}$. This goal can be accomplished by constructing the *1st-Level Adjoint Sensitivity System* (1st-LASS), which commences by introducing an arbitrary vector $\boldsymbol{\psi}^{(1)}(\mathbf{x}) \equiv \left[\psi_1^{(1)}(\mathbf{x}), \ldots, \psi_{N_u}^{(1)}(\mathbf{x})\right] \in \mathcal{H}_Q$ that satisfies the following relation:

$$\left\langle \boldsymbol{\psi}^{(1)}, \mathbf{F}_{11}^{(1)}\left(\boldsymbol{\alpha}^0, \mathbf{u}^0\right)\delta\mathbf{u} \right\rangle_Q = \left\langle \mathbf{A}_{11}^{(1)}\left(\boldsymbol{\alpha}^0, \mathbf{u}^0\right)\boldsymbol{\psi}^{(1)}, \delta\mathbf{u} \right\rangle_u + \left\{ P^{(1)}\left(\boldsymbol{\alpha}^0, \mathbf{u}^0; \boldsymbol{\psi}^{(1)}; \delta\mathbf{u}\right) \right\}_{\partial\Omega_x}, \tag{7.16}$$

where $\mathbf{A}_{11}^{(1)}(\boldsymbol{\alpha}, \mathbf{u}) \triangleq \left[\mathbf{F}_{11}^{(1)}(\boldsymbol{\alpha}, \mathbf{u})\right]^*$ denotes the operator *formally adjoint* of $\mathbf{F}_{11}^{(1)}(\boldsymbol{\alpha}, \mathbf{u}) \triangleq \partial \mathbf{N}(\boldsymbol{\alpha}, \mathbf{u})/\partial \mathbf{u}$, that is, $\mathbf{A}_{11}^{(1)}(\boldsymbol{\alpha}, \mathbf{u})$ is defined as the $N_u \times N_u$ matrix obtained by transposing the formal adjoints of the components of the $N_u \times N_u$ matrix $\partial \mathbf{N}(\boldsymbol{\alpha}, \mathbf{u})/\partial \mathbf{u}$ in Equation 7.12. The symbol $[]^*$ indicates *adjoint*.

Furthermore, the quantity $\left\{P^{(1)}\left(\alpha,\mathbf{u};\psi^{(1)};\delta\mathbf{u}\right)\right\}_{\partial\Omega_x}$ in Equation 7.16 denotes the *associated bilinear form* evaluated on $\partial\Omega_x$. In certain situations, it might be computationally advantageous to include certain boundary components of $\left\{P^{(1)}\left(\alpha,\mathbf{u};\Psi^{(1)};\delta\mathbf{u}\right)\right\}_{\partial\Omega_x}$ into the components of $\mathbf{A}_{11}^{(1)}(\alpha,\mathbf{u})$.

The domain of $\mathbf{A}_{11}^{(1)}(\alpha,\mathbf{u})$ is determined next by selecting appropriate *adjoint boundary and/or initial conditions*, which will be denoted in operator form as

$$\mathbf{B}_A^{(1)}\left(\alpha^0,\mathbf{u}^0;\Psi^{(1)}\right)=\mathbf{0},\ \mathbf{x}\in\partial\Omega_x. \tag{7.17}$$

The above boundary conditions for $\mathbf{A}_{11}^{(1)}(\alpha,\mathbf{u})$ are usually inhomogeneous, that is, $\mathbf{B}_A^{(1)}\left(\alpha^0,\mathbf{u}^0;\mathbf{0}\right)\neq\mathbf{0}$, and are obtained by requiring that

1. They must be independent of *unknown* values of $\delta\mathbf{u}$ and $\delta\alpha$.
2. The substitution of the boundary and/or initial conditions represented by Equations 7.9 and 7.17 into the expression of $\left\{P^{(1)}\left(\alpha,\mathbf{u};\psi^{(1)};\delta\mathbf{u}\right)\right\}_{\partial\Omega_x}$ must cause all terms containing unknown values of $\delta\mathbf{u}$ to vanish.

Constructing the adjoint initial and/or boundary conditions for $\mathbf{A}_{11}^{(1)}(\alpha,\mathbf{u})$ as described above in items (1) and (2) and implementing them together with the forward adjoint boundary and/or initial conditions (represented by Equation 7.9) into Equation 7.16 reduces the bilinear concomitant $\left\{P^{(1)}\left(\alpha,\mathbf{u};\psi^{(1)};\delta\mathbf{u}\right)\right\}_{\partial\Omega_x}$ to a quantity that will contain boundary terms involving only known values of $\delta\alpha$, α^0, $\mathbf{u}^0$, and $\psi^{(1)}$; this quantity will be denoted by $\hat{P}^{(1)}\left(\alpha,\mathbf{u};\psi^{(1)};\delta\alpha\right)$. In general, $\hat{P}^{(1)}\left(\alpha,\mathbf{u};\psi^{(1)};\delta\alpha\right)$ does not automatically vanish as a result of the operations discussed in the foregoing. In certain cases, though, $\hat{P}^{(1)}\left(\alpha,\mathbf{u};\psi^{(1)};\delta\alpha\right)$ may vanish automatically or it may be forced to vanish by considering appropriately constructed extensions of $\mathbf{A}_{11}^{(1,form)}(\alpha,\mathbf{u})$; however, such extensions are seldom needed in practice.

Implementing the forward and adjoint boundary and/or initial conditions, cf. Equations 7.9 and 7.17, into Equation 7.16 will transform the latter into the form

$$\left\langle\mathbf{A}_{11}^{(1)}\left(\alpha^0,\mathbf{u}^0\right)\psi^{(1)},\delta\mathbf{u}\right\rangle_u=\left\langle\psi^{(1)},\mathbf{F}_{11}^{(1)}\left(\alpha^0,\mathbf{u}^0\right)\delta\mathbf{u}\right\rangle_Q-\hat{P}^{(1)}\left(\alpha^0,\mathbf{u}^0;\psi^{(1)};\delta\alpha\right). \tag{7.18}$$

The quantity $\mathbf{F}_{11}^{(1)}\left(\alpha^0,\mathbf{u}^0\right)\delta\mathbf{u}$ in the first term on the right side of Equation 7.18 is now replaced by the right side of Equation 7.9 to obtain

$$\left\langle\mathbf{A}_{11}^{(1)}\left(\alpha^0,\mathbf{u}^0\right)\psi^{(1)},\delta\mathbf{u}\right\rangle_u=\left\langle\psi^{(1)},\ \mathbf{Q}^{(1)}\left(\alpha^0,\mathbf{u}^0;\delta\alpha\right)\right\rangle_Q-\hat{P}^{(1)}\left(\alpha^0,\mathbf{u}^0;\psi^{(1)};\delta\alpha\right). \tag{7.19}$$

The definition of the function $\psi^{(1)}$ will now be completed by requiring that the left side of Equation 7.19 and the right side of Equation 7.15 represent the same functional, namely the *indirect-effect term* $[\partial R(\boldsymbol{\alpha},\mathbf{u})/\partial \mathbf{u}]\delta\mathbf{u}$, which yields the following relationship:

$$\left\langle \mathbf{A}_{11}^{(1)}\left(\alpha^0,\mathbf{u}^0\right)\psi^{(1)},\delta\mathbf{u}\right\rangle_u = \left\langle \partial R\left(\alpha^0,\mathbf{u}^0\right)/\partial\mathbf{u},\,\delta\mathbf{u}\right\rangle_u, \quad \delta\mathbf{u}\in\mathcal{H}_u, \tag{7.20}$$

which implies that the adjoint function $\psi^{(1)}$ is the *weak solution* (in the sense of distributions) of the equation

$$\mathbf{A}_{11}^{(1)}\left(\alpha^0,\mathbf{u}^0\right)\psi^{(1)}(\mathbf{x}) = \left\{\frac{\partial R(\alpha,\mathbf{u})}{\partial\mathbf{u}}\right\}_{\left(\alpha^0,\mathbf{u}^0\right)}, \quad \mathbf{x}\in\Omega_x, \tag{7.21}$$

while satisfying the adjoint boundary conditions represented by Equation 7.17. The Riesz representation theorem ensures that the relationship in Equation 7.21 holds uniquely.

The results obtained in Equations 7.15, 7.19, and 7.20 are now replaced in Equation 7.7 to obtain

$$\begin{aligned}\delta R\left(\mathbf{e}^0;\mathbf{h}\right) &= \left\{\frac{\partial R(\alpha,\mathbf{u})}{\partial\alpha}\right\}_{\left(\alpha^0,\mathbf{u}^0\right)}\delta\alpha + \left\langle\psi^{(1)},\mathbf{Q}^{(1)}\left(\alpha^0,\mathbf{u}^0;\delta\alpha\right)\right\rangle_Q - \hat{P}^{(1)}\left(\alpha^0,\mathbf{u}^0;\psi^{(1)};\delta\alpha\right)\\ &\triangleq DR\left(\alpha^0,\mathbf{u}^0;\psi^{(1)};\delta\alpha\right).\end{aligned} \tag{7.22}$$

As the right side of Equation 7.22 indicates, the desired elimination of all unknown values of $\delta\mathbf{u}$ from the expression of total first-order response sensitivity, $\delta R(\mathbf{e}^0;\mathbf{h})$, of $R(\mathbf{e})$ at $\mathbf{e}^0$ has been accomplished. Instead of depending on $\delta\mathbf{u}$, the first-order response differential $\delta R(\mathbf{e}^0;\mathbf{h})$ now depends on the adjoint function $\psi^{(1)}\in\mathcal{H}_Q$, a fact that is emphasized in the above expression by noting explicitly the identity $\delta R(\mathbf{e}^0;\mathbf{h}) \equiv DR(\boldsymbol{\alpha}^0, \mathbf{u}^0; \psi^{(1)}; \delta\boldsymbol{\alpha})$.

To emphasize that $\mathcal{H}_Q(\Omega_x)$ becomes the Hilbert space on which the adjoint function $\psi^{(1)}\in\mathcal{H}_Q$ is defined, *the Hilbert space $\mathcal{H}_Q$ will henceforth be denoted as $\mathcal{H}_\psi^{(1)}(\Omega_x)$, and the inner product $\langle\bullet,\bullet\rangle_Q$ will be denoted as $\langle\bullet,\bullet\rangle_{\psi^{(1)}}$, so that $\psi^{(1)}\in\mathcal{H}_\psi^{(1)}(\Omega_x)$.* This notation will be convenient for the subsequent developments to be presented in Section 7.3, where the second-level adjoint sensitivity systems will involve adjoint functions that will be designated as $\psi^{(2)}\in\mathcal{H}_\psi^{(2)}(\Omega_x)$, and where the inner products in $\mathcal{H}_\psi^{(2)}(\Omega_x)$ will be denoted as $\langle\bullet,\bullet\rangle_{\psi^{(2)}}$.

As indicated by Equation 7.22, *the total first-order response sensitivity $DR(\boldsymbol{\alpha}^0, \mathbf{u}^0; \psi^{(1)}; \delta\boldsymbol{\alpha})$ can be computed after solving Equations 7.21 and 7.17* only once *to obtain the adjoint function $\psi^{(1)}\in\mathcal{H}_\psi^{(1)}(\Omega_x)$. Equations 7.21 and 7.17 will be called the* 1st-LASS, *and its solution, $\psi^{(1)}\in\mathcal{H}_\psi^{(1)}(\Omega_x)$, will be called the* first-level adjoint function. It is very important to note that the 1st-LASS is independent of the variation $\delta\mathbf{u}$ in the original state functions $\mathbf{u}$ and is also independent of variations $\delta\boldsymbol{\alpha}$ in the model parameters $\boldsymbol{\alpha}$.

By successively setting ($\delta\alpha_i = 1, \delta\alpha_j = 0$ *for* $j \neq i, i = 1, \dots, N_\alpha$) in the expression of $DR(\boldsymbol{\alpha}^0, \mathbf{u}^0; \boldsymbol{\psi}^{(1)}; \delta\boldsymbol{\alpha})$ given in Equation 7.22, we obtain by means of simple integrations the individual sensitivities of the response $R(\mathbf{e})$ with respect to each of the model parameters α_i, evaluated at the nominal values $\mathbf{e}^0 = (\boldsymbol{\alpha}^0, \mathbf{u}^0)$.

In the vast majority of practical situations, the G-differential $\delta\mathbf{Q}(\boldsymbol{\alpha}^0;\delta\boldsymbol{\alpha})$ of the source term $\mathbf{Q}(\boldsymbol{\alpha}^0)$ is linear in $\delta\boldsymbol{\alpha}$, which implies that

$$\mathbf{Q}^{(1)}\left(\alpha^0, \mathbf{u}^0; \delta\alpha\right) = \sum_{i_1=1}^{N_\alpha} \frac{\partial\left[\mathbf{Q}\left(\alpha^0\right) - \mathbf{N}\left(\alpha^0, \mathbf{u}^0\right)\right]}{\partial\alpha_{i_1}} \delta\alpha_{i_1}. \tag{7.23}$$

Since $\hat{P}\left(\alpha^0, \mathbf{u}^0; \psi^{(1)}; \delta\alpha\right)$ is also linear in $\delta\boldsymbol{\alpha}$, it follows that it can be expressed as

$$\hat{P}^{(1)}\left(\alpha^0, \mathbf{u}^0; \boldsymbol{\psi}^{(1)}; \delta\alpha\right) = \sum_{i_1=1}^{N_\alpha} \frac{\partial\hat{P}^{(1)}\left(\alpha^0, \mathbf{u}^0; \boldsymbol{\psi}^{(1)}\right)}{\partial\alpha_{i_1}} \delta\alpha_{i_1}. \tag{7.24}$$

Therefore, Equation 7.22 can be expressed in the form

$$DR\left(\alpha^0, \mathbf{u}^0; \psi^{(1)}; \delta\alpha\right) = \sum_{i_1=1}^{N_\alpha} R_{i_1}^{(1)}\left(\alpha^0, \mathbf{u}^0; \psi^{(1)}\right) \delta\alpha_{i_1}, \tag{7.25}$$

where the quantities $R_{i_1}^{(1)}\left(\alpha^0; \mathbf{u}^0; \psi^{(1)}\right)$ denote the *first-order partial sensitivities (G-derivatives)* of the response $R(\mathbf{e})$ with respect to each of the model parameters α_i, evaluated at $\boldsymbol{\alpha}^0$, $\mathbf{u}^0$, $\boldsymbol{\psi}^{(1)}$, and are given by the following expression:

$$\begin{aligned} R_{i_1}^{(1)}\left(\alpha, \mathbf{u}; \boldsymbol{\psi}^{(1)}\right) &\triangleq \frac{\partial R(\alpha, \mathbf{u})}{\partial\alpha_{i_1}} - \frac{\partial\hat{P}^{(1)}\left(\alpha, \mathbf{u}; \boldsymbol{\psi}^{(1)}\right)}{\partial\alpha_{i_1}} \\ &+ \int_{\Omega_x} \boldsymbol{\psi}^{(1)}(\mathbf{x}) \cdot \left\{\frac{\partial\left[\mathbf{Q}(\alpha) - \mathbf{N}(\alpha, \mathbf{u})\right]}{\partial\alpha_{i_1}}\right\} d\mathbf{x}, \quad i_1 = 1, \dots, N_\alpha. \end{aligned} \tag{7.26}$$

For subsequent generalizations to higher levels and orders, the standardized notation for the 1st-LASS is summarized below, keeping in mind that the equalities in respective operator equations hold in the weak, distributional sense:

$$\mathbf{A}_{11}^{(1)}(\alpha, \mathbf{u})\psi^{(1)}(\mathbf{x}) = \Theta^{(1)}(\alpha, \mathbf{u}), \quad \Theta^{(1)}(\alpha, \mathbf{u}) \triangleq \frac{\partial R(\alpha, \mathbf{u})}{\partial\mathbf{u}}, \quad \mathbf{x} \in \Omega_x, \quad \psi^{(1)} \in \mathcal{H}_\psi^{(1)}(\Omega_x), \tag{7.27}$$

$$\mathbf{B}_A^{(1)}\left(\alpha,\mathbf{u};\psi^{(1)}\right)=0,\ \mathbf{x}\in\partial\Omega_x, \tag{7.28}$$

where the superscript (1) indicates first-level, the letter A (used as "operator" and "subscript") indicates *adjoint*, the letter B indicates *boundary and/or initial conditions*.

Cacuci (1981b) has also provided the *adjoint sensitivity analysis methodology for computing the first-order sensitivities of function-valued operator responses*, $\mathbf{R}(\mathbf{e})$, as opposed to the scalar-valued response considered in Equation 7.3. In such a case, $\mathbf{R}(\mathbf{e})$ can be represented by the spectral (generalized Fourier) expansion

$$\mathbf{R}\left(\mathbf{e}^0\right)=\sum_{n\in N}\left\langle\mathbf{R}\left(\mathbf{e}^0\right),\varphi_n(\mathbf{x})\right\rangle_R\varphi_n(\mathbf{x}), \tag{7.29}$$

where the set $\{\boldsymbol{\varphi}_n(\mathbf{x})\}$, $n\in N$, is an orthonormal basis for the Hilbert space $\mathcal{H}_R(\Omega_R)$; the set index N may be finite or infinite. The first-order G-derivative of Equation 7.29 yields

$$\begin{aligned}\delta\mathbf{R}\left(\alpha^0,\mathbf{u}^0;\delta\mathbf{u};\delta\alpha\right)&\triangleq\left\{\frac{d}{d\varepsilon}\left[\mathbf{R}\left(\mathbf{e}^0+\varepsilon\mathbf{h}\right)\right]\right\}_{\varepsilon=0}\\&=\left\{\frac{\partial\mathbf{R}(\alpha,\mathbf{u})}{\partial\alpha}\right\}_{\left(\alpha^0,\mathbf{u}^0\right)}\delta\alpha+\left\{\frac{\partial\mathbf{R}(\alpha,\mathbf{u})}{\partial\mathbf{u}}\right\}_{\left(\alpha^0,\mathbf{u}^0\right)}\delta\mathbf{u}\\&=\left\{\frac{\partial\mathbf{R}(\alpha,\mathbf{u})}{\partial\alpha}\right\}_{\left(\alpha^0,\mathbf{u}^0\right)}\delta\alpha+\sum_{n\in N}\left\langle\frac{\partial\mathbf{R}}{\partial\mathbf{u}}\delta\mathbf{u},\varphi_n\right\rangle_R\varphi_n.\end{aligned} \tag{7.30}$$

As noted in Equation 7.30 above, the *direct-effect term*, $\{\partial\mathrm{R}(\alpha,\mathbf{u})/\partial\alpha\}_{(\alpha^0,\mathbf{u}^0)}\delta\alpha$, can be computed directly when applying to the response $\mathbf{R}(\mathbf{e})$ the definition of the G-differential, without needing to evaluate its spectral expansion. Furthermore, each of the *generalized Fourier coefficients* $\left\langle\{\partial\mathbf{R}(\alpha,\mathbf{u})/\partial\mathbf{u}\}_{(\alpha^0,\mathbf{u}^0)}\partial\mathbf{u},\varphi_n\right\rangle_R$ will be considered, in turn, to provide a source term for the adjoint sensitivity system, just as has been done for the functional $[\partial R(\boldsymbol{\alpha},\mathbf{u})/\partial\mathbf{u}]\delta\mathbf{u}=\langle\partial R(\boldsymbol{\alpha},\mathbf{u})/\partial\mathbf{u},\delta\mathbf{u}\rangle_u$ that was defined in Equation 7.7. Repeating the mathematical derivations from Equation 7.8 to Equation 7.22 yields the following result for computing the first-order total differential $\delta\mathbf{R}\left(\alpha^0,\mathbf{u}^0;\psi_1^{(1)}(\mathbf{x}),\ldots,\psi_N^{(1)}(\mathbf{x});\delta\alpha\right)$ of the function-valued operator response $\mathbf{R}(\boldsymbol{\alpha},\mathbf{u})$:

$$\begin{aligned}&\delta\mathbf{R}\left(\alpha^0,\mathbf{u}^0;\psi_1^{(1)}(\mathbf{x}),\ldots,\psi_N^{(1)}(\mathbf{x});\delta\alpha\right)=\left\{\frac{\partial\mathbf{R}(\alpha,\mathbf{u})}{\partial\alpha}\right\}_{\left(\alpha^0,\mathbf{u}^0\right)}\delta\alpha+\\&\sum_{n\in N}\left\{\left\langle\psi_n^{(1)}(\mathbf{x}),\frac{\partial\left[\mathbf{Q}(\alpha)-\mathbf{N}(\alpha,\mathbf{u})\right]}{\partial\alpha}\delta\alpha\right\rangle_Q-\hat{P}^{(1)}\left(\delta\alpha,\psi_n^{(1)}(\mathbf{x});\alpha^0\right)\right\}\varphi_n,\end{aligned} \tag{7.31}$$

where each adjoint function $\psi_n^{(1)}(\mathbf{x}),\ n \in N$, is the solution of the following 1st-LASS:

$$\mathbf{A}_{11}^{(1)}\left(\alpha^0,\mathbf{u}^0\right)\psi_n^{(1)}(\mathbf{x}) = \left\{\frac{\partial \mathbf{R}(\alpha,\mathbf{u})}{\partial \mathbf{u}}\right\}_{\left(\alpha^0,\mathbf{u}^0\right)} \varphi_n(\mathbf{x}), \quad \mathbf{x} \in \Omega_x, \quad n \in N, \tag{7.32}$$

$$\mathbf{B}_A^{(1)}\left[\alpha^0,\mathbf{u}^0;\psi_n^{(1)}(\mathbf{x})\right] = \mathbf{0}, \ \mathbf{x} \in \partial\Omega_x. \tag{7.33}$$

As Equations 7.32 and 7.33 indicate, the adjoint system must be solved anew, with a different source on the right side of Equation 7.32, for each $n \in N$. From the foregoing considerations, it is evident that *the orthonormal basis* $\{\boldsymbol{\varphi}_n(\mathbf{x})\}$, $n \in N$, *must be chosen such as to minimize an* a priori *user-selected error criterion, to ensure that the spectral expansion in Equation 7.29 represents the known nominal value of* $\mathbf{R}(\boldsymbol{\alpha},\mathbf{u})$ *within the selected error criterion with a minimal number, N, of terms in the expansion*. The selection of a specific error criterion and of a specific basis $\{\boldsymbol{\varphi}_n\}$, $n \in N$, will differ from problem to problem, but the procedures and considerations for performing this selection (e.g., using classical Fourier expansion, orthogonal and/or chaos polynomials, wavelets, collocation, and pseudo-spectral methods) are well known.

7.3 The 2nd-LASS for Computing Exactly and Efficiently the Second-Order Sensitivities of Scalar-Valued Responses for Nonlinear Systems

The *second-order total sensitivity* of the model response $R(\mathbf{e})$ to variations $\mathbf{h} \triangleq (\delta\boldsymbol{\alpha},\delta\mathbf{u})$ in the model parameters and state functions around $\mathbf{e}^0 = (\boldsymbol{\alpha}^0,\mathbf{u}^0)$ is given by the *second-order G-differential* of $R(\mathbf{e})$ at $\mathbf{e}^0 = (\boldsymbol{\alpha}^0,\mathbf{u}^0)$ in the directions $\mathbf{h} \triangleq (\delta\boldsymbol{\alpha},\delta\mathbf{u})$. This *second-order G-differential* of $R(\mathbf{e})$ is assumed to exist and will be denoted as $\delta^{(2)}R(\mathbf{e}^0;\mathbf{h})$. By definition, $\delta^{(2)}R(\mathbf{e}^0;\mathbf{h})$ is defined inductively as the total first-variation (or first-order differential) of the first-order total response variation (or first-order differential) as follows:

$$\begin{aligned}\delta^{(2)}R\left(\mathbf{e}^0;\mathbf{h}_1,\mathbf{h}_2\right) &\triangleq \left\{\frac{d}{d\varepsilon_1}\left[\delta R\left(\mathbf{e}^0+\varepsilon_1\mathbf{h}_1;\mathbf{h}_2\right)\right]\right\}_{\varepsilon_1=0} \\ &= \left\{\frac{\partial^2}{\partial\varepsilon_1\,\partial\varepsilon_2}\left[R\left(\mathbf{e}^0+\varepsilon_1\mathbf{h}_1+\varepsilon_2\mathbf{h}_2\right)\right]\right\}_{\varepsilon_1=\varepsilon_2=0},\end{aligned} \tag{7.34}$$

for $\varepsilon_1,\varepsilon_2 \in \mathbf{F}$, and all (i.e., arbitrary) vectors of variations $\mathbf{h}_i \in \mathscr{E}$; $i = 1, 2$; around $\mathbf{e}^0 = (\boldsymbol{\alpha}^0,\mathbf{u}^0) \in \mathscr{E}$. In order for the second-order response sensitivities (i.e., second-order partial G-derivatives) to model parameters to exist, the first-order sensitivities must satisfy the conditions generally expressed in Equations 3.7 and 3.8, in Chapter 3. Under these conditions, which will henceforth be assumed to be satisfied, the total second-order response differential $\delta^{(2)}R(\mathbf{e}^0;\mathbf{h})$ is bilinear in the parameter variations $\delta\alpha_i$, so that Equation 7.34 will ultimately be expressed in the following form:

$$\delta^{(2)}R\left(\mathbf{e}^0;\mathbf{h}_1,\mathbf{h}_2\right)=\sum_{i_2,i_1=1}^{N_\alpha} R^{(2)}_{i_2 i_1}\delta\alpha_{i_2}\delta\alpha_{i_1}. \tag{7.35}$$

In Equation 7.35, the quantities $R^{(2)}_{i_2 i_1}$ do *not* depend on the parameter variations and represent the *mixed second-order partial sensitivities (G-derivatives)* of the model response $R(\mathbf{e})$ at $\mathbf{e}^0$. The exact and efficient computation of the second-order partial sensitivities (G-derivatives) $R^{(2)}_{i_2 i_1}$ will be the objective of the developments to be presented in this section.

As Equation 7.26 indicates, the first-order partial sensitivities are functionals of the form $R^{(1)}_{i_1}\left(\alpha,\mathbf{u};\psi^{(1)}\right)\in\mathcal{H}_\alpha\oplus\mathcal{H}_u(\Omega_x)\oplus\mathcal{H}^{(1)}_\psi(\Omega_x)\to\mathcal{F}$. The product Hilbert space $\mathcal{H}^{(1)}_\psi(\Omega_x)\oplus\mathcal{H}_u(\Omega_x)$ will play a prominent role in the derivations to follow. For notational convenience, therefore, it will be denoted as $\mathcal{H}^{(2)}_\psi(\Omega_x)\triangleq\mathcal{H}^{(1)}_\psi(\Omega_x)\oplus\mathcal{H}_u(\Omega_x)$. An arbitrary vector in $\mathcal{H}^{(2)}_\psi(\Omega_x)$ will be denoted as $\varphi^{(2)}\triangleq\left(\mathbf{u};\psi^{(1)}\right)\in\mathcal{H}_u(\Omega_x)\oplus\mathcal{H}^{(1)}_\psi(\Omega_x)$, with the respective nominal value denoted as $\varphi^{(2,0)}\triangleq\left(\mathbf{u}^0;\psi^{(1,0)}\right)\in\mathcal{H}_u(\Omega_x)\oplus\mathcal{H}^{(1)}_\psi(\Omega_x)$. A vector of arbitrary variations around $\varphi^{(2,0)}\triangleq\left(\mathbf{u}^0;\psi^{(1,0)}\right)$ will be denoted as $\delta\varphi^{(2)}\triangleq\left(\delta\mathbf{u};\delta\psi^{(1)}\right)$. The arguments of $\boldsymbol{\varphi}^{(2)}$, $\boldsymbol{\varphi}^{(2,0)}$, and $\delta\boldsymbol{\varphi}^{(2)}$ may be written out explicitly whenever needed for clarity.

For each $i_1=1,\ldots,N_\alpha$, the first-order G-differential $\delta R^{(1)}_{i_1}\left(\alpha^0,\varphi^{(2,0)};\delta\alpha,\delta\varphi^{(2)}\right)$ of $R^{(1)}_{i_1}\left(\alpha;\varphi^{(2)}\right)$ can be computed in the neighborhood of the point $(\boldsymbol{\alpha}^0,\boldsymbol{\varphi}^{(2,0)})$, in the usual manner, namely:

$$\begin{aligned}\delta R^{(1)}_{i_1}\left(\alpha^0,\varphi^{(2,0)};\delta\alpha,\delta\varphi^{(2)}\right)&\triangleq\left\{\frac{d}{d\varepsilon}R^{(1)}_{i_1}\left(\alpha^0+\varepsilon\delta\alpha,\mathbf{u}^0+\varepsilon\delta\mathbf{u};\psi^{(1,0)}+\varepsilon\delta\psi^{(1)}\right)\right\}_{\varepsilon=0}\\&=\left\{\frac{\partial R^{(1)}_{i_1}}{\partial\alpha}\delta\alpha+\frac{\partial R^{(1)}_{i_1}}{\partial\mathbf{u}}\delta\mathbf{u}+\frac{\partial R^{(1)}_{i_1}}{\partial\psi^{(1)}}\delta\psi^{(1)}\right\}_{\left(\alpha^0,\varphi^{(2,0)}\right)}.\end{aligned} \tag{7.36}$$

The above relation can be expressed as

$$\delta R^{(1)}_{i_1}\left(\alpha^0,\varphi^{(2,0)};\delta\alpha,\delta\varphi^{(2)}\right)\triangleq\left(\delta R^{(1)}_{i_1}\right)_{direct}+\left(\delta R^{(1)}_{i_1}\right)_{indirect}, \tag{7.37}$$

where $\left(\delta R_{i_1}^{(1)}\right)_{direct}$ denotes the *direct-effect term*, while $\left(\delta R_{i_1}^{(1)}\right)_{indirect}$ denotes the *indirect-effect term*. The *direct-effect term* $\left(\delta R_{i_1}^{(1)}\right)_{direct}$ is defined as follows:

$$\left(\delta R_{i_1}^{(1)}\right)_{direct} \triangleq \left(\frac{\partial R_{i_1}^{(1)}}{\partial \boldsymbol{\alpha}}\right)_{\left(\alpha^0,\varphi^{(2,0)}\right)} \delta\alpha = \sum_{i_2=1}^{N_\alpha} \frac{\partial R_{i_1}^{(1)}\left(\boldsymbol{\alpha}^0,\boldsymbol{\varphi}^{(2,0)}\right)}{\partial \alpha_{i_2}} \delta\alpha_{i_2},\ i_1 = 1,\ldots,N_\alpha, \quad (7.38)$$

where

$$\frac{\partial R_{i_1}^{(1)}\left(\boldsymbol{\alpha}^0,\boldsymbol{\varphi}^{(2,0)}\right)}{\partial \alpha_{i_2}} \triangleq \frac{\partial}{\partial \alpha_{i_2}} \left\{ \frac{\partial R(\boldsymbol{\alpha},\mathbf{u})}{\partial \alpha_{i_1}} + \left\langle \boldsymbol{\psi}^{(1)}, \frac{\partial\left[\mathbf{Q}(\boldsymbol{\alpha})-\mathbf{N}(\boldsymbol{\alpha},\mathbf{u})\right]}{\partial \alpha_{i_1}} \right\rangle_{\psi^{(1)}} - \frac{\partial \hat{P}^{(1)}\left(\boldsymbol{\alpha},\mathbf{u};\boldsymbol{\psi}^{(1)}\right)}{\partial \alpha_{i_1}} \right\}_{\left(\alpha^0,\varphi^{(2,0)}\right)}, \quad i_1, i_2 = 1,\ldots,N_\alpha. \quad (7.39)$$

Since the *direct-effect term* $\left(\delta R_{i_1}^{(1)}\right)_{direct}$ is independent of variations $\delta\boldsymbol{\varphi}^{(2)} \triangleq \left(\delta\mathbf{u};\delta\boldsymbol{\psi}^{(1)}\right)$, it *can be computed immediately* at this stage, without needing any additional large-scale computations.

On the contrary, the *indirect-effect term* $\left(\delta R_{i_1}^{(1)}\right)_{indirect}$ is defined from Equations 7.36 and 7.37 as follows:

$$\left(\delta R_{i_1}^{(1)}\right)_{indirect} \triangleq \left\{ \frac{\partial R_{i_1}^{(1)}}{\partial \boldsymbol{\psi}^{(1)}} \delta\boldsymbol{\psi}^{(1)} + \frac{\partial R_{i_1}^{(1)}}{\partial \mathbf{u}} \delta\mathbf{u} \right\}_{\left(\alpha^0,\varphi^{(2,0)}\right)} = \left\{ \left(\delta\boldsymbol{\psi}^{(1)},\delta\mathbf{u}\right) \begin{pmatrix} \boldsymbol{\Theta}_{1,i_1}^{(2)}\left(\boldsymbol{\alpha},\boldsymbol{\varphi}^{(2)}\right) \\ \boldsymbol{\Theta}_{2,i_1}^{(2)}\left(\boldsymbol{\alpha},\boldsymbol{\varphi}^{(2)}\right) \end{pmatrix} \right\}_{\left(\alpha^0,\varphi^{(2,0)}\right)}, \quad (7.40)$$

where

$$\boldsymbol{\Theta}_{1,i_1}^{(2)}\left(\boldsymbol{\alpha},\boldsymbol{\varphi}^{(2)}\right) \triangleq \frac{\partial R_{i_1}^{(1)}\left(\boldsymbol{\alpha},\mathbf{u};\boldsymbol{\psi}^{(1)}\right)}{\partial \boldsymbol{\psi}^{(1)}} = \frac{\partial}{\partial \alpha_{i_1}} \left[\mathbf{Q}(\boldsymbol{\alpha}) - \mathbf{N}(\boldsymbol{\alpha},\mathbf{u}) - \frac{\partial \hat{P}^{(1)}\left(\boldsymbol{\alpha},\mathbf{u};\boldsymbol{\psi}^{(1)}\right)}{\partial \boldsymbol{\psi}^{(1)}} \right], \quad (7.41)$$

and

$$\boldsymbol{\Theta}_{2,i_1}^{(2)}\left(\boldsymbol{\alpha},\boldsymbol{\varphi}^{(2)}\right) \triangleq \frac{\partial R_{i_1}^{(1)}\left(\boldsymbol{\alpha},\mathbf{u};\boldsymbol{\psi}^{(1)}\right)}{\partial \mathbf{u}} = \frac{\partial}{\partial \alpha_{i_1}} \left[\boldsymbol{\Theta}^{(1)}(\boldsymbol{\alpha},\mathbf{u}) - \boldsymbol{\Lambda}\left(\boldsymbol{\alpha},\mathbf{u};\boldsymbol{\psi}^{(1)}\right) - \frac{\partial \hat{P}^{(1)}\left(\boldsymbol{\alpha},\mathbf{u};\boldsymbol{\psi}^{(1)}\right)}{\partial \mathbf{u}} \right], \quad (7.42)$$

with

$$\Lambda\left(\alpha,\mathbf{u};\boldsymbol{\psi}^{(1)}\right)\triangleq\left(\Lambda_1,\ldots,\Lambda_{N_u}\right);\,\Lambda_k\left(\alpha,\mathbf{u};\boldsymbol{\psi}^{(1)}\right)\triangleq\sum_{j=1}^{N_u}\psi_j^{(1)}(\mathbf{x})\frac{\partial N_{jk}(\alpha,\mathbf{u})}{\partial\mathbf{u}};\;k=1,\ldots,N_u. \tag{7.43}$$

The *indirect-effect term*, $\left(\delta R_{i_1}^{(1)}\right)_{indirect}$, *cannot be computed at this stage yet, since the vectors of variations* $\delta\mathbf{u}$ and $\delta\boldsymbol{\psi}^{(1)}$ *are not completely known*. The relation between *the vectors of variations* $\delta\mathbf{u}$ and $\delta\boldsymbol{\psi}^{(1)}$ needs to be determined next by noting that the vector of variations $\delta\boldsymbol{\psi}^{(1)}$ will be the solution of the system of operator equations that will result after G-differentiating the 1st-LASS given by Equations 7.27 and 7.28 in the neighborhood ($\boldsymbol{\alpha}^0 + \varepsilon\delta\boldsymbol{\alpha}$, $\mathbf{u}^0 + \varepsilon\delta\mathbf{u}$; $\boldsymbol{\psi}^{(1,0)} + \varepsilon\delta\boldsymbol{\psi}^{(1)}$) of the point ($\boldsymbol{\alpha}^0$, $\boldsymbol{\varphi}^{(2,0)}$). Performing this G-differentiation yields

$$\left[\frac{d}{d\varepsilon}\left\{\mathbf{A}_{11}^{(1)}\left(\alpha^0+\varepsilon\delta\alpha,\mathbf{u}^0+\varepsilon\delta\mathbf{u}\right)\left(\psi^{(1,0)}+\varepsilon\delta\psi^{(1)}\right)-\Theta^{(1)}\left(\alpha^0+\varepsilon\delta\alpha,\mathbf{u}^0+\varepsilon\delta\mathbf{u}\right)\right\}\right]_{\varepsilon=0}$$
$$=\mathbf{0},\;\mathbf{x}\in\Omega_x, \tag{7.44}$$

$$\left[\frac{d}{d\varepsilon}\left\{\mathbf{B}_A^{(1)}\left(\alpha^0+\varepsilon\delta\alpha,\mathbf{u}^0+\varepsilon\delta\mathbf{u};\psi^{(1,0)}+\varepsilon\delta\psi^{(1)}\right)\right\}\right]_{\varepsilon=0}=\mathbf{0},\;\mathbf{x}\in\partial\Omega_x. \tag{7.45}$$

Carrying out the differentiations and evaluations indicated in Equations 7.44 and 7.45 yields the following system of operator equations:

$$\mathbf{F}_{11}^{(2)}\left(\alpha^0;\mathbf{u}^0\right)\delta\psi^{(1)}+\mathbf{F}_{12}^{(2)}\left(\alpha^0;\mathbf{u}^0\right)\delta\mathbf{u}=\mathbf{Q}^{(2)}\left(\alpha^0,\varphi^{(2,0)};\delta\alpha\right), \tag{7.46}$$

$$\delta\mathbf{B}_A^{(1)}\left(\alpha^0,\varphi^{(2,0)};\delta\alpha,\delta\psi^{(1)}\right)=\mathbf{0},\;\mathbf{x}\in\partial\Omega_x, \tag{7.47}$$

where the following definitions were used:

$$\mathbf{F}_{11}^{(2)}(\alpha,\mathbf{u})\triangleq\mathbf{A}_{11}^{(1)}(\alpha,\mathbf{u}),\quad\mathbf{F}_{12}^{(2)}(\alpha;\mathbf{u})\triangleq\frac{\partial\left[\mathbf{A}_{11}^{(1)}(\alpha,\mathbf{u})\psi^{(1)}\right]}{\partial\mathbf{u}}-\frac{\partial\Theta^{(1)}(\alpha,\mathbf{u})}{\partial\mathbf{u}},$$
$$\mathbf{Q}^{(2)}\left(\alpha,\varphi^{(2)};\delta\alpha\right)\triangleq\frac{\partial\left[\Theta^{(1)}(\alpha,\mathbf{u})-\mathbf{A}_{11}^{(1)}(\alpha,\mathbf{u})\psi^{(1)}\right]}{\partial\alpha}\delta\alpha. \tag{7.48}$$

As Equations 7.46 and 7.47 indicate, the vector of variations $\delta\boldsymbol{\psi}^{(1)}$ will also depend on the parameter variations $\delta\boldsymbol{\alpha}$, just as the vector of variations $\delta\mathbf{u}$ did. To construct a well-posed system of equations for obtaining both $\delta\boldsymbol{\psi}^{(1)}$ and $\delta\mathbf{u}$

in terms of $\delta\boldsymbol{\alpha}$, the 1st-LFSS given by Equations 7.8 and 7.9 is combined with Equations 7.46 and 7.47 to obtain the following block-matrix system:

$$\begin{pmatrix} \mathbf{F}_{11}^{(2)}\left(\alpha^0,\mathbf{u}^0\right) & \mathbf{F}_{12}^{(2)}\left(\alpha^0,\mathbf{u}^0\right) \\ \mathbf{0} & \mathbf{F}_{11}^{(1)}\left(\alpha^0,\mathbf{u}^0\right) \end{pmatrix}\begin{pmatrix} \delta\psi^{(1)} \\ \delta\mathbf{u} \end{pmatrix} = \begin{pmatrix} \mathbf{Q}^{(2)}\left(\alpha^0,\varphi^{(2,0)};\delta\alpha\right) \\ \mathbf{Q}^{(1)}\left(\alpha^0,\mathbf{u}^0;\delta\alpha\right) \end{pmatrix}, \quad \mathbf{x}\in\Omega_x, \tag{7.49}$$

together with the following boundary and/or initial conditions:

$$\mathbf{B}_F^{(2)}\left(\alpha^0,\varphi^{(2,0)};\delta\alpha,\delta\varphi^{(2)}\right) \triangleq \begin{pmatrix} \delta\mathbf{B}_A^{(1)}\left(\alpha^0,\varphi^{(2,0)};\delta\alpha,\delta\varphi^{(2)}\right) \\ \mathbf{B}_F^{(1)}\left(\alpha^0,\mathbf{u}^0;\delta\alpha,\delta\mathbf{u}\right) \end{pmatrix} = \mathbf{0}, \ \mathbf{x}\in\partial\Omega_x. \tag{7.50}$$

Equations 7.49 and 7.50 constitute the *2nd-Level Forward Sensitivity System* (2nd-LFSS). They can be solved, in principle, to obtain the vectors $\delta\mathbf{u}$ and $\delta\boldsymbol{\psi}^{(1)}$ as functions of the vector of parameter variations $\delta\boldsymbol{\alpha}$ and could subsequently be replaced in Equation 7.40 to determine the indirect-effect term $\left(\delta R_{i_1}^{(1)}\right)_{indirect}$. This procedure would be impractical to use for large-scale systems, since it would require $O\left(N_\alpha^2\right)$ large-scale computations.

To circumvent the need for solving the 2nd-LFSS, the alternative *2nd-Level Adjoint Sensitivity System* (2nd-LASS) will be developed in the following by using the same guiding principles as those underlying the development of the 1st-LASS. The goal of the 2nd-LASS is to compute the indirect-effect term $\left(\delta R_{i_1}^{(1)}\right)_{indirect}$ in a manner that circumvents the need for computing explicitly the vectors of variations $\delta\boldsymbol{\psi}^{(1)}$ and $\delta\mathbf{u}$ by solving the 2nd-LFSS.

To construct the 2nd-LASS, recall that an element in the Hilbert space $\mathcal{H}_\psi^{(2)}(\Omega_x) \triangleq \mathcal{H}_\psi^{(1)}(\Omega_x) \oplus \mathcal{H}_u(\Omega_x)$ is a two-block vector of the form $\psi_{i_1}^{(2)} \triangleq \left(\psi_{1,i_1}^{(2)},\psi_{2,i_1}^{(2)}\right) \in \mathcal{H}_\psi^{(2)}(\Omega_x)$, with $\psi_{1,i_1}^{(2)} \in \mathcal{H}_\psi^{(1)}(\Omega_x)$, $\psi_{2,i_1}^{(2)} \in \mathcal{H}_u(\Omega_x)$. Since Hilbert spaces are self-dual, the spaces $\mathcal{H}_\psi^{(1)}(\Omega_x)$ and $\mathcal{H}_u(\Omega_x)$ actually coincide, but the distinct notation will be kept in this section for clarity. The inner product $\langle\bullet,\bullet\rangle_{\psi^{(2)}}$ between any two vectors $\mathbf{a}_{i_1}^{(2)} \triangleq \left(\mathbf{a}_{1,i_1}^{(2)},\mathbf{a}_{2,i_1}^{(2)}\right) \in \mathcal{H}_\psi^{(2)}(\Omega_x)$ and $\mathbf{b}_{i_1}^{(2)} \triangleq \left(\mathbf{b}_{1,i_1}^{(2)},\mathbf{b}_{2,i_1}^{(2)}\right) \in \mathcal{H}_\psi^{(2)}(\Omega_x)$ is defined as follows:

$$\left\langle \mathbf{a}_{i_1}^{(2)},\mathbf{b}_{i_1}^{(2)} \right\rangle_{\psi^{(2)}} \triangleq \int_{\Omega_x} \left(\mathbf{a}_{1,i_1}^{(2)}\cdot\mathbf{b}_{1,i_1}^{(2)} + \mathbf{a}_{2,i_1}^{(2)}\cdot\mathbf{b}_{2,i_1}^{(2)}\right) dx = \sum_{j=1}^{2}\int_{\Omega_x}\left(\mathbf{a}_{j,i_1}^{(2)}\cdot\mathbf{b}_{j,i_1}^{(2)}\right) dx. \tag{7.51}$$

Using the above definition, we construct the inner product of Equation 7.49 with a yet-undefined vector $\psi_{i_1}^{(2)} \triangleq \left(\psi_{1,i_1}^{(2)},\psi_{2,i_1}^{(2)}\right) \in \mathcal{H}_\psi^{(2)}(\Omega_x)$ to obtain

$$\left\langle \left(\psi_{1,i_1}^{(2)}, \psi_{2,i_1}^{(2)}\right), \begin{pmatrix} \mathbf{F}_{11}^{(2)}\left(\alpha^0, \mathbf{u}^0\right) & \mathbf{F}_{12}^{(2)}\left(\alpha^0; \mathbf{u}^0\right) \\ 0 & \mathbf{F}_{11}^{(1)}\left(\alpha^0; \mathbf{u}^0\right) \end{pmatrix} \begin{pmatrix} \delta\psi^{(1)} \\ \delta\mathbf{u} \end{pmatrix} \right\rangle_{\psi^{(2)}}$$

$$= \left\langle \left(\psi_{1,i_1}^{(2)}, \psi_{2,i_1}^{(2)}\right), \begin{pmatrix} \mathbf{Q}^{(2)}\left(\alpha^0, \varphi^{(2,0)}; \delta\alpha\right) \\ \mathbf{Q}^{(1)}\left(\alpha^0, \mathbf{u}^0; \delta\alpha\right) \end{pmatrix} \right\rangle_{\psi^2} \tag{7.52}$$

$$= \int_{\Omega_x} \left[\psi_{1,i_1}^{(2)} \cdot \mathbf{Q}^{(2)}\left(\alpha^0, \varphi^{(2,0)}; \delta\alpha\right) + \psi_{2,i_1}^{(2)} \cdot \mathbf{Q}^{(1)}\left(\alpha^0, \mathbf{u}^0; \delta\alpha\right) \right] dx, \quad \mathbf{x} \in \Omega_x.$$

The inner product on the left side of Equation 7.52 is now further transformed by using the definition of the adjoint operator to obtain

$$\left\langle \left(\psi_{1,i_1}^{(2)}, \psi_{2,i_1}^{(2)}\right), \begin{pmatrix} \mathbf{F}_{11}^{(2)}\left(\alpha^0, \mathbf{u}^0\right) & \mathbf{F}_{12}^{(2)}\left(\alpha^0, \mathbf{u}^0\right) \\ 0 & \mathbf{F}_{11}^{(1)}\left(\alpha^0, \mathbf{u}^0\right) \end{pmatrix} \begin{pmatrix} \delta\psi^{(1)} \\ \delta\mathbf{u} \end{pmatrix} \right\rangle_{\psi^{(2)}}$$

$$= \left\langle \left(\delta\psi^{(1)}, \delta\mathbf{u}\right), \begin{pmatrix} \left[\mathbf{F}_{11}^{(2)}\left(\alpha^0, \mathbf{u}^0\right)\right]^* & 0 \\ \left[\mathbf{F}_{12}^{(2)}\left(\alpha^0, \mathbf{u}^0\right)\right]^* & \left[\mathbf{F}_{11}^{(1)}\left(\alpha^0, \mathbf{u}^0\right)\right]^* \end{pmatrix} \begin{pmatrix} \psi_{1,i_1}^{(2)} \\ \psi_{2,i_1}^{(2)} \end{pmatrix} \right\rangle_{\psi^{(2)}} \tag{7.53}$$

$$+ \left\{ P^{(2)}\left[\alpha^0, \varphi^{(2,0)}; \psi_{1,i_1}^{(2,0)}, \psi_{2,i_1}^{(2,0)}; \delta\mathbf{u}; \delta\psi^{(1)}\right]\right\}_{\partial\Omega_x}.$$

In the above equation, the symbol $[]^*$ indicates *adjoint*, as before, while the quantity $\left\{ P^{(2)}\left[\alpha^0, \varphi^{(2,0)}; \psi_{1,i_1}^{(2,0)}, \psi_{2,i_1}^{(2,0)}; \delta\mathbf{u}; \delta\psi^{(1)}\right]\right\}_{\partial\Omega_x}$ denotes the corresponding bilinear concomitant on $\mathbf{x} \in \partial\Omega_x$. As before, the superscript zero denotes nominal values. The domains of the adjoint operators appearing on the right side of Equation 7.53 can now be specified by requiring that the function $\psi_{i_1}^{(2)} \triangleq \left(\psi_{1,i_1}^{(2)}(\mathbf{x}), \psi_{2,i_1}^{(2)}(\mathbf{x})\right) \in \mathcal{H}_{\psi}^{(2)}(\Omega_x)$ satisfies adjoint boundary conditions denoted as

$$\mathbf{B}_A^{(2)}\left(\alpha^0, \varphi^{(2,0)}; \psi_{i_1}^{(2,0)}\right) = \mathbf{0}, \quad \mathbf{x} \in \partial\Omega_x. \tag{7.54}$$

The above adjoint boundary and/or initial conditions for $\psi_{i_1}^{(2)} \triangleq \left(\psi_{1,i_1}^{(2)}, \psi_{2,i_1}^{(2)}\right) \in \mathcal{H}_{\psi}^{(2)}(\Omega_x)$ are determined following the same principles as described in Section 7.2, namely that

1. They must be independent of the *unknown* values of $\delta\alpha$; $\delta\mathbf{u}$; $\delta\boldsymbol{\psi}^{(1)}$.

2. The substitution of the boundary and/or initial conditions represented by Equations 7.50 and 7.54 into the expression of $\left\{P^{(2)}\left[\alpha^0,\varphi^{(2,0)};\psi^{(2,0)};\delta\mathbf{u};\delta\psi^{(1)}\right]\right\}_{\partial\Omega_x}$ must cause all terms containing unknown values of $\delta\mathbf{u}$ and $\delta\mathbf{\Psi}^{(1)}$ to vanish, which will reduce the bilinear concomitant to an expression that will contain only known values of $\delta\boldsymbol{\alpha}$; this expression will be denoted as $\hat{P}^{(2)}\left(\alpha^0,\varphi^{(2,0)};\psi^{(2,0)};\delta\alpha\right)$. In general, $\hat{P}^{(2)}\left(\alpha^0,\varphi^{(2,0)};\psi^{(2,0)};\delta\alpha\right)$ does not automatically vanish as a result of these operations.

Implementing the second-level forward and adjoint boundary and initial conditions, namely Equations 7.50 and 7.54, into Equation 7.53 will transform the latter into the form

$$\left\langle \left(\delta\psi^{(1)},\delta\mathbf{u}\right), \begin{pmatrix} \left[\mathbf{F}_{11}^{(2)}\right]^* & 0 \\ \left[\mathbf{F}_{12}^{(2)}\right]^* & \left[\mathbf{F}_{11}^{(1)}\right]^* \end{pmatrix} \begin{pmatrix} \psi_{1,i_1}^{(2)} \\ \psi_{2,i_1}^{(2)} \end{pmatrix} \right\rangle_{\psi^{(2)}}$$

$$= \left\langle \left(\psi_{1,i_1}^{(2)},\psi_{2,i_1}^{(2)}\right), \begin{pmatrix} \mathbf{F}_{11}^{(2)} & \mathbf{F}_{12}^{(2)} \\ \mathbf{0} & \mathbf{F}_{11}^{(1)} \end{pmatrix} \begin{pmatrix} \delta\psi^{(1)} \\ \delta\mathbf{u} \end{pmatrix} \right\rangle_{\psi^{(2)}} - \hat{P}^{(2)}\left(\alpha^0,\varphi^{(2,0)};\psi_{i_1}^{(2,0)};\delta\alpha\right). \tag{7.55}$$

The expression containing the variations $\delta\mathbf{u}$ and $\delta\boldsymbol{\psi}^{(1)}$ in the first term on the right side of Equation 7.55 is now replaced by using the right side of Equation 7.49, which transforms Equation 7.55 into the form

$$\left\langle \left(\delta\psi^{(1)},\delta\mathbf{u}\right), \begin{pmatrix} \left[\mathbf{F}_{11}^{(2)}\right]^* & 0 \\ \left[\mathbf{F}_{12}^{(2)}\right]^* & \left[\mathbf{F}_{11}^{(1)}\right]^* \end{pmatrix} \begin{pmatrix} \psi_{1,i_1}^{(2)} \\ \psi_{2,i_1}^{(2)} \end{pmatrix} \right\rangle_{\psi^{(2)}}$$

$$= \left\langle \left(\psi_{1,i_1}^{(2)},\psi_{2,i_1}^{(2)}\right), \begin{pmatrix} \mathbf{Q}^{(2)}\left(\alpha^0,\varphi^{(2,0)};\delta\alpha\right) \\ \mathbf{Q}^{(1)}\left(\alpha^0,\mathbf{u}^0;\delta\alpha\right) \end{pmatrix} \right\rangle_{\psi^{(2)}} \tag{7.56}$$

$$-\hat{P}^{(2)}\left(\alpha^0,\varphi^{(2,0)};\psi_{i_1}^{(2,0)};\delta\alpha\right).$$

The definition of the second-level adjoint function $\psi_{i_1}^{(2)}$ will now be completed by requiring that the left side of Equation 7.56 and the right side of

Equation 7.40 represent the same functional, namely the *indirect-effect term*, $\left(\delta R_{i_1}^{(1)}\right)_{indirect}$, which yields the following relationship:

$$\left\langle \left(\delta\boldsymbol{\psi}^{(1)},\delta\mathbf{u}\right), \begin{pmatrix} \left[\mathbf{F}_{11}^{(2)}\left(\boldsymbol{\alpha}^0,\mathbf{u}^0\right)\right]^* & 0 \\ \left[\mathbf{F}_{12}^{(2)}\left(\boldsymbol{\alpha}^0,\mathbf{u}^0\right)\right]^* & \left[\mathbf{F}_{11}^{(1)}\left(\boldsymbol{\alpha}^0,\mathbf{u}^0\right)\right]^* \end{pmatrix} \begin{pmatrix} \boldsymbol{\psi}_{1,i_1}^{(2)} \\ \boldsymbol{\psi}_{2,i_1}^{(2)} \end{pmatrix} \right\rangle_{\psi^{(2)}}$$
$$= \left\langle \left(\delta\boldsymbol{\psi}^{(1)},\delta\mathbf{u}\right), \begin{pmatrix} \boldsymbol{\Theta}_{1,i_1}^{(2)}\left(\boldsymbol{\alpha}^0,\boldsymbol{\varphi}^{(2,0)}\right) \\ \boldsymbol{\Theta}_{2,i_1}^{(2)}\left(\boldsymbol{\alpha}^0,\boldsymbol{\varphi}^{(2,0)}\right) \end{pmatrix} \right\rangle_{\psi^{(2)}}. \tag{7.57}$$

Equation 7.57 implies that the second-level adjoint function $\boldsymbol{\psi}_{i_1}^{(2)}(\mathbf{x}) \triangleq \left(\boldsymbol{\psi}_{1,i_1}^{(2)}(\mathbf{x}),\boldsymbol{\psi}_{2,i_1}^{(2)}(\mathbf{x})\right) \in \mathcal{H}_{\psi}^{(2)}(\Omega_x)$ is the *weak solution* (in the sense of distributions) of the following operator equation:

$$\begin{pmatrix} \mathbf{A}_{11}^{(2)}\left(\boldsymbol{\alpha}^0,\mathbf{u}^0\right) & \mathbf{0} \\ \mathbf{A}_{21}^{(2)}\left(\boldsymbol{\alpha}^0,\mathbf{u}^0\right) & \mathbf{A}_{22}^{(2)}\left(\boldsymbol{\alpha}^0,\mathbf{u}^0\right) \end{pmatrix} \begin{pmatrix} \boldsymbol{\psi}_{1,i_1}^{(2)}(\mathbf{x}) \\ \boldsymbol{\psi}_{2,i_1}^{(2)}(\mathbf{x}) \end{pmatrix}$$
$$= \begin{pmatrix} \boldsymbol{\Theta}_{1,i_1}^{(2)}\left(\boldsymbol{\alpha}^0,\boldsymbol{\varphi}^{(2,0)}\right) \\ \boldsymbol{\Theta}_{2,i_1}^{(2)}\left(\boldsymbol{\alpha}^0,\boldsymbol{\varphi}^{(2,0)}\right) \end{pmatrix}, \quad \mathbf{x}\in\Omega_x,\ i_1=1,\dots,N_\alpha, \tag{7.58}$$

where the second-level adjoint operators $\mathbf{A}_{ij}^{(2)}$, $i,j=1,2$, are defined as follows:

$$\mathbf{A}_{11}^{(2)}(\boldsymbol{\alpha},\mathbf{u}) \triangleq \left[\mathbf{F}_{11}^{(2)}(\boldsymbol{\alpha},\mathbf{u})\right]^* = \left[\mathbf{A}_{11}^{(1)}(\boldsymbol{\alpha},\mathbf{u})\right]^* = \mathbf{F}_{11}^{(1)}(\boldsymbol{\alpha},\mathbf{u}),$$
$$\mathbf{A}_{21}^{(2)}(\boldsymbol{\alpha},\mathbf{u}) \triangleq \left[\mathbf{F}_{12}^{(2)}(\boldsymbol{\alpha},\mathbf{u})\right]^*, \quad \mathbf{A}_{22}^{(2)}(\boldsymbol{\alpha},\mathbf{u}) \triangleq \left[\mathbf{F}_{11}^{(1)}(\boldsymbol{\alpha},\mathbf{u})\right]^* = \mathbf{A}_{11}^{(1)}(\boldsymbol{\alpha},\mathbf{u}). \tag{7.59}$$

The Riesz representation theorem ensures that the relationship in Equation 7.58 holds uniquely. Of course, the second-level adjoint function $\boldsymbol{\psi}_{i_1}^{(2)}(\mathbf{x}) \triangleq \left(\boldsymbol{\psi}_{1,i_1}^{(2)}(\mathbf{x}),\boldsymbol{\psi}_{2,i_1}^{(2)}(\mathbf{x})\right) \in \mathcal{H}_{\psi}^{(2)}(\Omega_x)$ satisfies the second-level adjoint boundary conditions represented by Equation 7.54.

The results obtained in Equations 7.38, 7.40, 7.56, and 7.57 are now replaced in Equation 7.37 to obtain

$$
\begin{aligned}
&\delta R_{i_1}^{(1)}\left(\alpha^0,\varphi^{(2,0)};\delta\alpha,\delta\mathbf{u},\delta\psi^{(1)}\right)\\
&=\sum_{i_2=1}^{N_\alpha}\frac{\partial R_{i_1}^{(1)}\left(\alpha^0,\varphi^{(2,0)}\right)}{\partial\alpha_{i_2}}\delta\alpha_{i_2}\\
&+\left\langle\left(\psi_{1,i_1}^{(2)},\psi_{2,i_1}^{(2)}\right),\begin{pmatrix}\mathbf{Q}^{(2)}\left(\alpha^0,\varphi^{(2,0)};\delta\alpha\right)\\ \mathbf{Q}^{(1)}\left(\alpha^0,\mathbf{u}^0;\delta\alpha\right)\end{pmatrix}\right\rangle_{\psi^{(2)}}\\
&-\hat{P}^{(2)}\left(\alpha^0,\varphi^{(2,0)};\psi_{i_1}^{(2,0)};\delta\alpha\right),\quad i_1=1,\ldots,N_\alpha.
\end{aligned}
\tag{7.60}
$$

Noting that the quantities $\mathbf{Q}^{(2)}(\boldsymbol{\alpha}^0, \boldsymbol{\varphi}^{(2,0)}; \delta\boldsymbol{\alpha})$ and $\hat{P}_2\left(\alpha^0,\varphi^{(2,0)};\psi_{i_1}^{(2,0)};\delta\alpha\right)$ are both linear in $\delta\boldsymbol{\alpha}$, it follows that they can be expressed in the following forms:

$$
\mathbf{Q}^{(2)}\left(\alpha^0,\varphi^{(2,0)};\delta\alpha\right)=\sum_{i_2=1}^{N_\alpha}\left\{\frac{\partial\left[\Theta^{(1)}(\alpha,\mathbf{u})-\mathbf{A}_{11}^{(1)}(\alpha,\mathbf{u})\psi^{(1)}\right]}{\partial\alpha_{i_2}}\delta\alpha_{i_2}\right\}_{\left(\alpha^0,\varphi^{(2,0)}\right)},
\tag{7.61}
$$

and

$$
\hat{P}^{(2)}\left(\alpha^0,\varphi^{(2,0)};\psi^{(2,0)};\delta\alpha\right)=\sum_{i_2=1}^{N_\alpha}\frac{\partial\hat{P}^{(2)}\left(\alpha^0,\varphi^{(2,0)};\psi_{1,i_1}^{(2,0)},\psi_{2,i_1}^{(2,0)}\right)}{\partial\alpha_{i_2}}\delta\alpha_{i_2}.
\tag{7.62}
$$

Using the relations expressed by Equations 7.61, 7.62, and 7.23 in Equation 7.60, it follows that Equation 7.60 can be expressed in the following form:

$$
\begin{aligned}
&\delta R_{i_1}^{(1)}\left(\alpha^0,\varphi^{(2,0)};\delta\alpha,\delta\mathbf{u},\delta\psi^{(1)}\right)\\
&=\sum_{i_2=1}^{N_\alpha}R_{i_1i_2}^{(2)}\left(\alpha^0,\varphi^{(2,0)};\psi_{1,i_1}^{(2,0)},\psi_{i,i_1}^{(2,0)}\right)\delta\alpha_{i_2},\quad i_1=1,\ldots,N_\alpha,
\end{aligned}
\tag{7.63}
$$

where

$$R_{i_1 i_2}^{(2)}\left(\alpha,\varphi^{(2)};\psi_{1,i_1}^{(2)},\psi_{i,i_1}^{(2)}\right)\triangleq\frac{\partial}{\partial\alpha_{i_2}}\left[R_{i_1}^{(1)}\left(\alpha,\varphi^{(2)}\right)-\hat{P}^{(2)}\left(\alpha,\varphi^{(2)};\psi_{1,i_1}^{(2)},\psi_{2,i_1}^{(2)}\right)\right]$$
$$+\int_{\Omega_x}\left\{\psi_{1,i_1}^{(2)}(\mathbf{x})\cdot\left[\frac{\partial\left[\Theta^{(1)}(\alpha,\mathbf{u})-\mathbf{A}_{11}^{(1)}(\alpha,\mathbf{u})\psi^{(1)}\right]}{\partial\alpha_{i_2}}\right]\right. \tag{7.64}$$
$$\left.+\psi_{2,i_1}^{(2)}(\mathbf{x})\cdot\left[\frac{\partial\left[\mathbf{Q}(\alpha)-\mathbf{N}(\alpha,\mathbf{u})\right]}{\partial\alpha_{i_2}}\right]\right\}dx,$$
$$\text{for}\ \ i_1,i_2=1,\ldots,N_\alpha.$$

As Equation 7.63 indicates, *each of the G-differentials* $\delta R_{i_1}^{(1)}$ ($\boldsymbol{\alpha}^0$, $\boldsymbol{\varphi}^{(2,0)}$; $\delta\boldsymbol{\alpha}$, $\delta\mathbf{u}$, $\delta\boldsymbol{\psi}^{(1)}$) *contains one complete row (running on the index* $i_2 = 1,\ldots, N_\alpha$) *of the mixed second-order partial sensitivities,* $R_{i_1 i_2}^{(2)}\left(\alpha,\varphi^{(2)};\psi_{1,i_1}^{(2)},\psi_{i,i_1}^{(2)}\right)$, which can be computed exactly and efficiently after solving *only once* the 2nd-LASS, namely Equations 7.58 and 7.54, to obtain the *second-level adjoint function* $\psi_{i_1}^{(2)}(\mathbf{x})\triangleq\left(\psi_{1,i_1}^{(2)}(\mathbf{x}),\psi_{2,i_1}^{(2)}(\mathbf{x})\right)\in\mathcal{H}_{\Psi}^{(2)}(\Omega_x)$, for each index $i_1 = 1,\ldots, N_\alpha$.

It is also important to note that the construction and solution of the 2nd-LASS requires very little effort beyond that already invested in solving the original forward system, namely Equations 7.1 and 7.2, to compute the state variable $\mathbf{u}(\mathbf{x})$ and for solving the 1st-LASS, cf. Equations 7.27 and 7.28, to compute the first-level adjoint function $\boldsymbol{\psi}^{(1)}(\mathbf{x})$. This fact becomes readily apparent by using in Equation 7.58 the relations given in Equation 7.59, to recast the 2nd-LASS, cf. Equation 7.58, into the following form:

$$\mathbf{F}_{11}^{(1)}(\alpha,\mathbf{u})\psi_{1,i_1}^{(2)}(\mathbf{x})=\Theta_{1,i_1}^{(2)}\left(\alpha,\varphi^{(2)}\right),\ \mathbf{x}\in\Omega_x,\ \ i_1=1,\ldots,N_\alpha, \tag{7.65}$$

$$\mathbf{A}_{11}^{(1)}(\alpha,\mathbf{u})\psi_{2,i_1}^{(2)}(\mathbf{x})=\Theta_{2,i_1}^{(2)}\left(\alpha,\varphi^{(2)}\right)-\mathbf{A}_{21}^{(2)}(\alpha,\mathbf{u})\psi_{1,i_1}^{(2)},\ \mathbf{x}\in\Omega_x,\ i_1=1,\ldots,N_\alpha. \tag{7.66}$$

Since Equation 7.65 is linear in $\psi_{1,i_1}^{(2)}(\mathbf{x})$, solving it requires less computational effort than solving the original Equation 7.1 for determining $\mathbf{u}(\mathbf{x})$, because, in contradistinction to Equation 7.65, Equation 7.1 is nonlinear in $\mathbf{u}(\mathbf{x})$. Furthermore, solving Equation 7.66 to obtain $\psi_{2,i_1}^{(2)}(\mathbf{x})$ is equivalent to solving Equation 7.27, with the same left side as for the 1st-LASS but with a different source term on the right side of Equation 7.66 by comparison to the right side of Equation 7.27. Thus, the same computer program and solvers can be used for solving the 2nd-LASS as were already used for solving the original system of Equations 7.1 and 7.2 and for solving the 1st-LASS.

Only the sources for the respective operator equations and boundary conditions for the 2nd-LASS will differ from those used to solve the 1st-LASS and the original system of equations. For this reason, the 2nd-LASS (which comprises Equations 7.65, 7.66, and 7.54) was designated as "second-level" rather than "second-order" adjoint sensitivity system, since the 2nd-LASS does not involve any explicit second-order G-derivatives of the operators underlying the original system but involves the inversion of the same operators that needed to be inverted for solving the 1st-LASS and the original system of equations.

Note that the 2nd-LASS is independent of parameter variations $\delta\boldsymbol{\alpha}$, as well as of the variations $\delta\mathbf{u}$ and $\delta\boldsymbol{\psi}^{(1)}$. Thus, *the exact computation of all of the partial second-order sensitivities,* $\left\{\partial^2 R(\boldsymbol{\alpha},\mathbf{u})/\partial\alpha_j\,\partial\alpha_i\right\}_{(\alpha^0,\mathbf{u}^0)}$, $i,j = 1,\ldots,N_\alpha$, *requires at most* N_α *large-scale (adjoint) computations using the 2nd-LASS,* rather than $O\left(N_\alpha^2\right)$ large-scale computations as would be required by forward methods.

Another important characteristic of using the 2nd-LASS is the flexibility it provides for prioritizing the computation of the second-order sensitivities: only the second-order partial sensitivities of the response $R(\mathbf{e})$ that are deemed important will need to be computed; the unimportant ones can be deliberately neglected while knowing the error incurred by neglecting them. The computation of the second-order sensitivities should be based on the priority ranking of the relative magnitudes of the first-order sensitivities: the second-order mixed G-derivatives (sensitivities) corresponding to the largest relative first-order response sensitivity should be computed first; the second largest relative first-order response sensitivity should be considered next, and so on. Computing second-order partial sensitivities that correspond to vanishing first-order sensitivities may also be of interest, since vanishing first-order sensitivities may indicate critical points of the response in the phase-space of model parameters.

It is also important to note that *by solving the 2nd-LASS* N_α*-times*, the "off-diagonal" second-order mixed sensitivities $R_{i_1 i_2}^{(2)}\left(\boldsymbol{\alpha}^0,\boldsymbol{\varphi}^{(2,0)};\boldsymbol{\psi}_{1,i_1}^{(2,0)},\boldsymbol{\psi}_{2,i_1}^{(2,0)}\right)$ *will be computed twice, in two different ways, thereby providing an intrinsic (numerical) verification that the second-level adjoint functions* $\boldsymbol{\psi}_{1,i_1}^{(2)}(\mathbf{x})$ *and* $\boldsymbol{\psi}_{2,i_1}^{(2)}(\mathbf{x})$ *are computed accurately*. Finally, in view of Equations 7.34 and 7.64, the expression of the second-order G-differential, $\delta^{(2)}R(\boldsymbol{\alpha}^0, \mathbf{u}^0; \delta\boldsymbol{\alpha}_1, \delta\boldsymbol{\alpha}_2)$, of the response $R(\mathbf{e})$ at $\mathbf{e}^0 = (\boldsymbol{\alpha}^0, \mathbf{u}^0)$ in a neighborhood $(\mathbf{e}^0 + \varepsilon\mathbf{h})$ around $\mathbf{e}^0$ is given by

$$\delta^{(2)}R\left(\boldsymbol{\alpha}^0,\mathbf{u}^0;\delta\alpha_1,\delta\alpha_2\right)=\sum_{i_1=1}^{N_\alpha}\sum_{i_2=1}^{N_\alpha}R_{i_1 i_2}^{(2)}\left(\boldsymbol{\alpha}^0,\boldsymbol{\varphi}^{(2,0)};\boldsymbol{\psi}_{1,i_1}^{(2,0)},\boldsymbol{\psi}_{2,i_1}^{(2,0)}\right)\delta\alpha_{i_1}\delta\alpha_{i_2}. \qquad (7.67)$$

As Equation 7.67 indicates, the second-order partial sensitivities are functionals of the form $R_{i_1 i_2}^{(2)}\left(\boldsymbol{\alpha},\boldsymbol{\varphi}^{(2)};\boldsymbol{\psi}_{i_1}^{(2)}\right)\in\mathcal{H}_\alpha\oplus\mathcal{H}_\psi^{(2)}(\Omega_x)\oplus\mathcal{H}_\psi^{(2)}(\Omega_x)=\mathcal{H}_\alpha\oplus 2\otimes\mathcal{H}_\psi^{(2)}(\Omega_x)\rightarrow\mathcal{F}$.

7.4 Concluding Remarks

This chapter has presented the 2nd-ASAM for computing exactly and efficiently the second-order functional derivatives of system responses (i.e., "system performance parameters") to the system's model parameters. As has been shown in this chapter, for *each functional-type response* of interest produced by a physical system comprising N_α parameters and N_r responses, the 2nd-ASAM *requires one large-scale computation using the 1st-LASS for obtaining all of the first-order sensitivities* $\left\{\partial R(\mathbf{u},\alpha)/\partial\alpha_i\right\}_{\left(\mathbf{u}^0,\alpha^0\right)}$, $i=1,\ldots,N_\alpha$, followed by at most N_α *large-scale computations using the 2nd-LASS for obtaining exactly all of the second-order sensitivities* $\left\{\partial^2 R(\mathbf{u},\alpha)/\partial\alpha_j\,\partial\alpha_i\right\}_{\left(\mathbf{u}^0,\alpha^0\right)}$, $i,j=1,\ldots,N_\alpha$. In contradistinction, forward methods would require $\left(N_\alpha^2/2+3N_\alpha/2\right)$ large-scale computations for obtaining all of the first- and second-order sensitivities, for all N_r system responses.

It has also been shown that constructing and solving the 2nd-LASS requires very little additional effort beyond the construction of the 1st-LASS needed for computing the first-order sensitivities. Furthermore, due to the symmetry properties of the second-order sensitivities $\left\{\partial^2 R(\mathbf{u},\alpha)/\partial\alpha_j\,\partial\alpha_i\right\}_{\left(\mathbf{u}^0,\alpha^0\right)}$, $j=1,\ldots,N_\alpha$, the 2nd-ASAM encompasses an automatic, inherent, "solution verification" of the correctness and accuracy of the second-level adjoint functions $\psi_i^{(2)}\equiv\left(\psi_{i1}^{(2)},\psi_{i2}^{(2)}\right)$ needed for the efficient and exact computation of the second-order sensitivities.

Since real Hilbert spaces provide the natural mathematical setting for computational purposes, the 2nd-ASAM presented in this chapter was set in *real* (as opposed to complex) Hilbert spaces. However, this setting does *not* restrict, in any way, the generality of the concepts presented here, since the 2nd-ASAM can be readily set in complex Hilbert spaces by simply changing some terminology, without affecting its substance.

8

Application of the 2nd-ASAM to a Nonlinear Heat Conduction Problem

This chapter presents an application of the *2nd-ASAM for nonlinear systems* developed in Chapter 7 to a benchmark problem that models the steady-state heat conduction within a cylindrical test section for performing heat transfer experiments, which contains electrically heated rods and is filled with liquid lead–bismuth eutectic (LBE). The material presented in this chapter follows the material originally developed by Cacuci (2016c).

8.1 Mathematical Modeling of Heated Cylindrical Test Section

A cylindrical test section for performing heat transfer experiments contains electrically heated rods and is filled with liquid LBE. The length of the cylindrical test section is $\ell = 1.7$ m and its radius is $a = 15$ cm. The thermal conductivity, denoted as $k(T)$, of the LBE is considered to depend linearly on the temperature, having the functional form

$$k(T) = k_0(1 + cT), \tag{8.1}$$

where the nominal values (which will be denoted by using a superscript zero) of the quantities k_0 and c are $k_0^0 = 4.3663\,(\mathrm{Wm^{-1}K^{-1}})$ and $c^0 = 2.8844 \times 10^{-3}(\mathrm{K^{-1}})$. For simplicity, the electrically heated rods are considered to provide a volumetrically distributed heat source of nominal strength $Q^0 = 1.11 \times 10^4(\mathrm{Wm^{-3}})$. The test section is insulated on its lateral surface. The temperature at the bottom of the test section is kept at a constant nominal temperature $T_a^0 = 400\,\mathrm{K}$. At the top of the section, at $z = \ell/2$, heat is removed by a heat exchanger operating at a constant heat flux, $q(\mathrm{Wm^{-2}})$, having a nominal value $q^0 = 7.44 \times 10^3(\mathrm{Wm^{-2}})$.

To model (mathematically) the heat conduction process inside the test section described above, it is convenient to take the center of the coordinate systems as the center of the cylinder, so that the test section extends in the axial (vertical) direction from $-\ell/2 \le z \le \ell/2$. Since the test section is insulated on its radial surface and since the length of the cylindrical test section is much greater than its radius, the temperature variation in the radial direction can be neglected in comparison to the temperature variations in the axial direction, for the purposes of this illustrative problem. Hence, the axial temperature distribution, $T(z)$, in the LBE can be modeled by the following nonlinear heat conduction model:

$$\frac{d}{dz}\left[k(T)\frac{dT(z)}{dz}\right]+Q=0,\ -\frac{\ell}{2}<z<\frac{\ell}{2}, \tag{8.2}$$

$$\left[k(T)\frac{dT}{dz}\right]_{z=\ell/2}=-q,\ at\ z=\frac{\ell}{2}, \tag{8.3}$$

$$T(z)=T_a,\ at\ z=-\frac{\ell}{2}. \tag{8.4}$$

Notably, both Equations 8.2 and 8.3 are nonlinear in $T(z)$, thereby rendering the above heat conduction benchmark problem ideally suited for illustrating the application of the 2nd-ASAM to the general case of *nonlinear differential equations subject to nonlinear boundary conditions.*

The solution of the above system of nonlinear differential equations can be solved by using Kirchhoff's transformation to obtain the solution

$$T(z)=\left[-1+\sqrt{(1+cT_a)^2+2c\tau(z)}\right]\Big/c, \tag{8.5}$$

where the function $\tau(z)$ is defined as

$$\tau(z)\equiv\frac{Q}{2k_0}\left(\frac{3\ell^2}{4}+z\ell-z^2\right)-\frac{q}{k_0}\left(z+\frac{\ell}{2}\right). \tag{8.6}$$

Note that $T(z)$ attains its maximum value, denoted as $T(z_{max})$, at the location

$$z_{max}=\frac{Q\ell-2q}{2Q}, \tag{8.7}$$

where it has the expression

$$T(z_{max})=\left[-1+\sqrt{(1+cT_a)^2+2c\tau(z_{max})}\right]\Big/c, \tag{8.8}$$

with $\tau(z_{max})$ given by

$$\tau(z_{max})=\frac{(Q\ell-q)^2}{2k_0Q}. \tag{8.9}$$

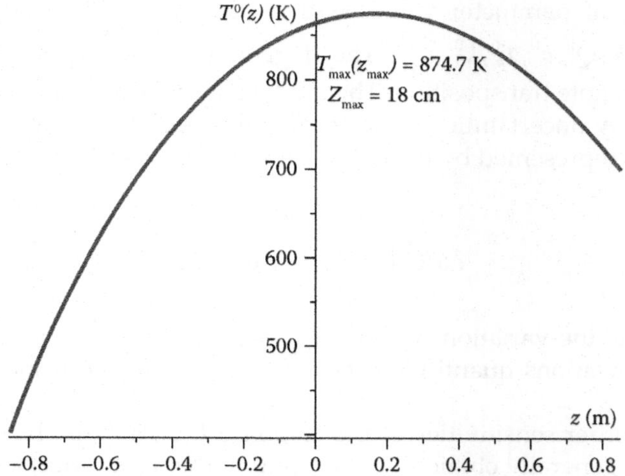

FIGURE 8.1
Variation of the nominal temperature, $T(z)$.

For the nominal parameter values provided in the foregoing, the axial variation of the nominal temperature, $T^0(z)$, is depicted in Figure 8.1. In particular, $T^0(z)$ takes on the values $T^0(-\ell/2) = 400\,\text{K}$ at the bottom of the test section and $T^0(\ell/2) = 700\,\text{K}$ at the top of the test section.

It is of interest to install thermocouples within the test section in order to measure the temperatures at various axial locations. A typical thermocouple response would indicate the temperature, $T(z_r)$, at some response location, denoted as $z_r, -\ell/2 \le z_r \le \ell/2$. Such a response is mathematically represented in the form

$$R(T;\alpha) \triangleq T(z_r) = \int_{-\ell/2}^{\ell/2} T(z)\delta(z - z_r)\,dz, \tag{8.10}$$

where $\delta(z - z_r)$ is the customary Dirac delta-functional. A particularly important location for installing a measuring thermocouple is the location $z_{max} = 18$ cm, where the maximum temperature $T(z_{max}) = 874.4\text{K}$ occurs.

8.2 Application of the 2nd-ASAM for Computing the First-Order Sensitivities

The model ("input") parameters in this problem are considered to be the (five) components of the (column) vector $\boldsymbol{\alpha}$ defined as follows:

$$\alpha \triangleq (\alpha_1, \alpha_2, \alpha_3, \alpha_4, \alpha_5)^\dagger \triangleq (Q, q, T_a, k_0, c)^\dagger. \tag{8.11}$$

The nominal parameter values will be denoted using a superscript 0, namely $\boldsymbol{\alpha}^0 \triangleq \left(Q^0, q^0, T_a^0, k_0^0, c^0\right)^\dagger$. The dagger will be used throughout this chapter to denote transposition. The model parameters, $\boldsymbol{\alpha}$, are considered to be afflicted by uncertainties, so they can vary from their nominal values $\boldsymbol{\alpha}^0$ by amounts represented by the component of the "vector of variations," $\boldsymbol{\delta\alpha}$, defined as

$$\delta\alpha \triangleq \left(\delta Q, \delta q, \delta T_a, \delta k_0, \delta c\right)^\dagger. \tag{8.12}$$

In practice, the variations δQ, δq, δT_a, δk_0, δc are usually taken to be the standard deviations quantifying the uncertainties in the respective model parameters.

The first-order sensitivities of the response $R(\mathbf{e})$, $\mathbf{e} \triangleq (\boldsymbol{\alpha}, T)$ to the variations $\boldsymbol{\delta\alpha}$ are generally obtained by computing the (first-order) G-differential $\delta R(\mathbf{e}^0;\mathbf{h})$ of $R(\mathbf{e})$ at $\mathbf{e}^0 \triangleq (\boldsymbol{\alpha}^0, T^0)$, which is defined as

$$\delta R\left(\mathbf{e}^0;\mathbf{h}\right) \triangleq \frac{d}{d\varepsilon}\left\{R\left(\mathbf{e}^0 + \varepsilon\mathbf{h}\right)\right\}_{\varepsilon=0}, \quad \text{with } \mathbf{h} \triangleq \left(\delta\alpha, \delta T\right)^\dagger. \tag{8.13}$$

Applying the above definition to the response defined by Equation 8.10 yields the first-order differential, $\delta R(\boldsymbol{\alpha}^0, T^0; \boldsymbol{\delta\alpha})$, of the response $R(\boldsymbol{\alpha}, T)$:

$$\delta R(\alpha^0, T^0; \delta\alpha) \triangleq \frac{d}{d\varepsilon}\left[\int_{-\ell/2}^{\ell/2}\left[T^0(z) + \varepsilon\delta T(z)\right]\delta(z - z_r)\,dz\right]_{\varepsilon=0} = \int_{-\ell/2}^{\ell/2}\delta T(z)\delta(z - z_r)\,dz. \tag{8.14}$$

Next, taking the G-differential of Equations 8.2 through 8.4 yields

$$\begin{aligned}\frac{d^2}{dz^2}\left[k\left(T^0\right)\delta T(z)\right] &= -(\delta Q) + (\delta k_0)\frac{Q^0}{k_0^0} - (\delta c)k_0^0\frac{d}{dz}\left[T^0(z)\frac{dT^0(z)}{dz}\right] \\ &\triangleq Q^{(1)}(\alpha^0, T^0; \delta\alpha), \quad z \in \left(-\ell/2, \ell/2\right),\end{aligned} \tag{8.15}$$

$$\begin{aligned}\left\{\frac{d}{dz}\left[k\left(T^0\right)\delta T(z)\right]\right\}_{z=\ell/2} &= -(\delta q) + (\delta k_0)\frac{q^0}{k_0^0} - (\delta c)k_0^0\left[T^0(z)\frac{dT^0(z)}{dz}\right]_{z=\ell/2} \\ &\triangleq q_1(\alpha^0, T^0; \delta\alpha), \, at \; z = \ell/2,\end{aligned} \tag{8.16}$$

$$\delta T(z)\Big|_{z=-\ell/2} = \left(\delta T_a\right), at \;\; z = -\ell/2. \tag{8.17}$$

Next, define the Hilbert space $\mathcal{H}(\Omega) \triangleq \mathcal{L}_2(\Omega)$ to consist of all square-integrable functions $f(z)$ defined on the domain $z \in \Omega \triangleq [-\ell/2, \ell/2]$ and endowed with the inner product [for two functions, $f_1(z)$ and $f_1(z)$] as follows:

$$\langle f_1(z), f_2(z) \rangle \triangleq \int_{-\ell/2}^{\ell/2} f_1(z) f_2(z) dz. \tag{8.18}$$

Forming now the inner product of Equation 8.15 with a yet-undefined function $\Psi(z) \in \mathcal{L}_2(\Omega)$ and integrating the resulting equations twice by parts to transfer the differential operations from $\delta T(z)$ to $\Psi(z)$ yields

$$\int_{-\ell/2}^{\ell/2} \Psi(z) \frac{d^2}{dz^2}\left[k(T^0)\delta T(z)\right]dz = \int_{-\ell/2}^{\ell/2} \Psi(z) Q^{(1)}\left(\alpha^0, T^0; \delta\alpha\right)dz$$

$$= \left\{\Psi(z)\frac{d}{dz}\left[k(T^0)\delta T(z)\right] - \delta T(z) k(T^0)\frac{d\Psi}{dz}\right\}_{z=-\ell/2}^{z=\ell/2} + \int_{-\ell/2}^{\ell/2} \delta T(z)\left[k(T^0)\frac{d^2\Psi}{dz^2}\right]dz. \tag{8.19}$$

Applying the general principles outlined in Chapter 7 to Equations 8.14 through 8.19 yields the following 1st-LASS:

$$k(T^0)\frac{d^2\Psi(z)}{dz^2} = \delta(z - z_r), \quad z \in (-\ell/2, \ell/2), \tag{8.20}$$

$$\left.\frac{d\Psi}{dz}\right|_{z=\ell/2} = 0, \; at \; z = \ell/2, \tag{8.21}$$

$$\Psi(z) = 0, \; at \; z = -\ell/2. \tag{8.22}$$

The 1st-LASS above can be readily solved to obtain the *first-level adjoint function*

$$\Psi(z) = \frac{1}{k\left[T^0(z_r)\right]}\left[(z - z_r)H(z - z_r) - z - \ell/2\right], \tag{8.23}$$

where $H(z)$ is the customary Heaviside unit step functional, defined as $H(z) = 0$, *if* $z < 0$, and $H(z) = 1$, *if* $z \geq 0$. Note that the *first-level adjoint function* $\Psi(z)$ can also be interpreted to be the Green's function

$$G(z, z_r) = \frac{1}{k\left[T(z_r)\right]}\left[(z - z_r)H(z - z_r) - z - \ell/2\right], \tag{8.24}$$

for the 1st-LASS, that is, Equations 8.20 through 8.22, since the point $z_r \in [-\ell/2, \ell/2]$ is arbitrary.

The 1st-LASS depends on the response, which provides the source term $\delta(z - z_r)$ in Equation 8.20. Note that this source term does not belong to the Hilbert space $\mathcal{L}_2(\Omega)$ but belongs to the Sobolev space $\mathcal{H}_0^1(\Omega) \subset \mathcal{L}_2(\Omega)$, that is, $\Sigma_d^0 \delta(x-b) \in \mathcal{H}_0^1 \subset \mathcal{L}_2(\Omega)$, as usually encountered when computing Green's functions. By using the well-known Lax–Milgram lemma, it can be shown that the bilinear form on the right side of the last equality in Equation 8.19 is coercive, so that the 1st-LASS can be solved uniquely, as has been done to obtain the expression for the adjoint function $\Psi(z) \in \mathcal{H}(\Omega) = \mathcal{H}_0^1(\Omega) \subset \mathcal{L}_2(\Omega)$ shown in Equation 8.23. The mathematical technicalities requiring the use of Sobolev spaces stemming from the consideration of distributions, as encountered above for the 1st-LASS, will also arise in the construction of the 2nd-LASS, as will be seen in the remainder of this chapter. Thus, even though the foregoing mathematical technicalities will not be repeated in the sequel, all of the solutions to such 2nd-LASS should be interpreted in the "weak sense," in the appropriate Sobolev space. No confusion should arise, however, since the respective solutions for the 2nd-LASS will be unique and will be obtained explicitly, just as was done in Equation 8.23 for the first-level adjoint function $\Psi(z)$.

Using the results of Equations 8.15 through 8.22 in Equation 8.14 transforms the latter into the form

$$\begin{aligned} DR(\boldsymbol{\alpha}^0, T^0; \delta\boldsymbol{\alpha}) = & \int_{-\ell/2}^{\ell/2} \Psi(z) Q^{(1)}(\boldsymbol{\alpha}^0, T^0; \delta\boldsymbol{\alpha}) dz \\ & -(\delta T_a)\left[k\left(T^0\right)\frac{d\Psi}{dz}\right]_{z=-\ell/2} - \Psi\left(z=\ell/2\right) q_1(\boldsymbol{\alpha}^0, T^0; \delta\boldsymbol{\alpha}). \end{aligned} \tag{8.25}$$

Using Equations 8.15 and 8.16, respectively, to replace the expressions of $Q_1(\boldsymbol{\alpha}^0, T^0; \delta\boldsymbol{\alpha})$ and $q_1(\boldsymbol{\alpha}^0, T^0; \delta\boldsymbol{\alpha})$ in Equation 8.25 yields

$$\begin{aligned} DR(\boldsymbol{\alpha}^0, T^0; \delta\boldsymbol{\alpha}) \triangleq \sum_{i_1=1}^{5} R_{i_1}^{(1)} \delta\alpha_{i_1} = & \frac{\partial T(z_r)}{\partial Q}(\delta Q) + \frac{\partial T(z_r)}{\partial q}(\delta q) \\ & + \frac{\partial T(z_r)}{\partial T_a}(\delta T_a) + \frac{\partial T(z_r)}{\partial k_0}(\delta k_0) + \frac{\partial T(z_r)}{\partial c}(\delta c), \end{aligned} \tag{8.26}$$

where the first-order sensitivities of $T(z_r)$ to the respective model parameters have the following expressions:

$$R_1^{(1)}(\boldsymbol{\alpha}, T; \Psi) \triangleq \frac{\partial T(z_r)}{\partial Q} = -\int_{-\ell/2}^{\ell/2} \Psi(z) dz, \tag{8.27}$$

$$R_2^{(1)}(\alpha,T;\Psi) \triangleq \frac{\partial T(z_r)}{\partial q} = \Psi(\ell/2), \tag{8.28}$$

$$R_3^{(1)}(\alpha,T;\Psi) \triangleq \frac{\partial T(z_r)}{\partial T_a} = -\left[k\left(T^0\right)\frac{d\Psi}{dz}\right]_{z=-\ell/2}, \tag{8.29}$$

$$R_4^{(1)}(\alpha,T;\Psi) \triangleq \frac{\partial T(z_r)}{\partial k_0} = \frac{1}{k_0^0}\left[Q^0\int_{-\ell/2}^{\ell/2}\Psi(z)dz - q^0\Psi(\ell/2)\right], \tag{8.30}$$

$$R_5^{(1)}(\alpha,T;\Psi) \triangleq \frac{\partial T(z_r)}{\partial c} = -\int_{-\ell/2}^{\ell/2}\Psi(z)k_0^0\frac{d}{dz}\left[T^0(z)\frac{dT^0(z)}{dz}\right]dz$$
$$+\Psi(\ell/2)k_0^0\left[T^0(z)\frac{dT^0}{dz}\right]_{z=\ell/2} = k_0^0\frac{T_a^2-\left[T^0(z_r)\right]^2}{2k\left[T^0(z_r)\right]}. \tag{8.31}$$

The above expressions indicate that all of the first-order sensitivities can be computed exactly and efficiently, using simple integrations, once the adjoint function has been obtained by solving the 1st-LASS. Thus, replacing the expression of $\Psi(z)$ given into Equation 8.23 in Equations 8.27 through 8.31 and carrying out the respective integrations over z yields the following evaluated expressions for the first-order sensitivities of the response $T(z_r)$ to the model parameters:

$$\frac{\partial T(z_r)}{\partial Q} = \frac{1}{2k\left[T^0(z_r)\right]}\left(\frac{3\ell^2}{4}+z_r\ell-z_r^2\right), \tag{8.32}$$

$$\frac{\partial T(z_r)}{\partial q} = -\frac{z_r+\ell/2}{k\left[T^0(z_r)\right]}, \tag{8.33}$$

$$\frac{\partial T(z_r)}{\partial T_a} = \frac{1+c^0T_a^0}{1+c^0T^0(z_r)}, \tag{8.34}$$

$$\frac{\partial T(z_r)}{\partial k_0} = \frac{1}{k_0^0k\left[T^0(z_r)\right]}\left[-\frac{Q^0}{2}\left(\frac{3\ell^2}{4}+z_r\ell-z_r^2\right)+q^0\left(z_r+\frac{\ell}{2}\right)\right] = -\frac{\tau(z_r)}{k\left[T^0(z_r)\right]}, \tag{8.35}$$

$$\frac{\partial T(z_r)}{\partial c} = k_0^0\frac{T_a^2-\left[T^0(z_r)\right]^2}{2k\left[T^0(z_r)\right]} = \frac{T^0(z_r)-\tau(z_r)-T_a^0}{c^0\left[1+c^0T^0(z_r)\right]}. \tag{8.36}$$

One of the main uses of sensitivities is for ranking the relative importance of parameter variations in influencing variations in responses. Relative sensitivities are used for this purpose, since they are dimensionless numbers. The relative sensitivity of a response $R(\mathbf{e})$ to the ith-parameter, α_i, is defined as $S_i^{rel} \triangleq \left(\partial R/\partial \alpha_i\right)_{\mathbf{e}^0}\left[\alpha_i^0/R\left(\mathbf{e}^0\right)\right]$. The relative sensitivities of $T(z_r)$ are depicted in Figures 8.2 through 8.6 as functions of the arbitrary location $z_r \in [-\ell/2, \ell/2]$.

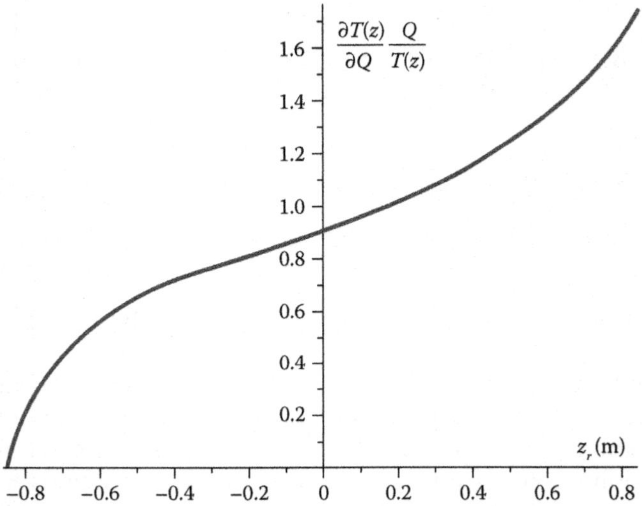

FIGURE 8.2
Relative sensitivity of $T(z_r)$ to Q.

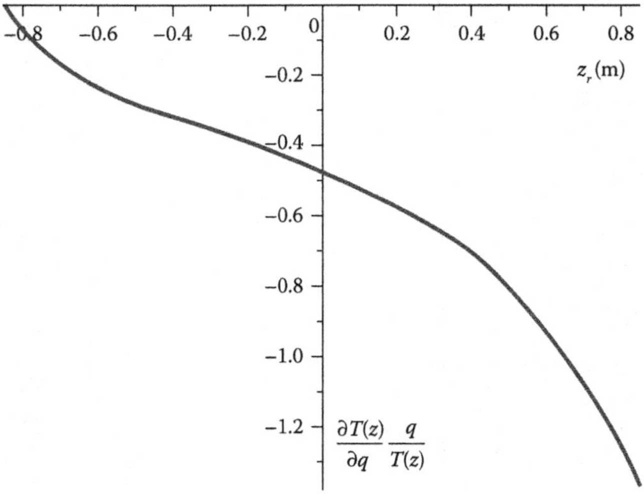

FIGURE 8.3
Relative sensitivity of $T(z_r)$ to q.

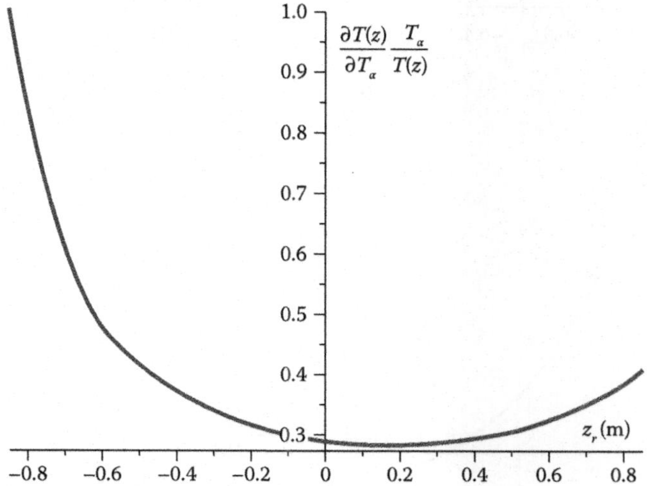

FIGURE 8.4
Relative sensitivity of $T(z_r)$ to T_a.

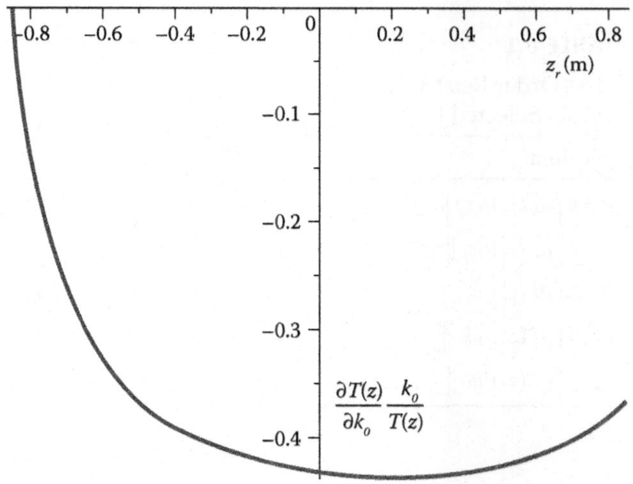

FIGURE 8.5
Relative sensitivity of $T(z_r)$ to k_0.

As Figures 8.2 through 8.6 indicate, the relative importance of the various model parameters depends on the position z_r. Ranking them by the largest attainable maximum absolute values of their relative sensitivities, the importance of the model parameters is as follows: the heat source Q, the boundary heat flux q, the ambient temperature T_a, the linear heat conductivity coefficient k_0, and the nonlinear heat conductivity coefficient c. For convenience, the numerical results of the relative sensitivities of $T(z_r)$ at several important locations z_r are presented in Table 8.1.

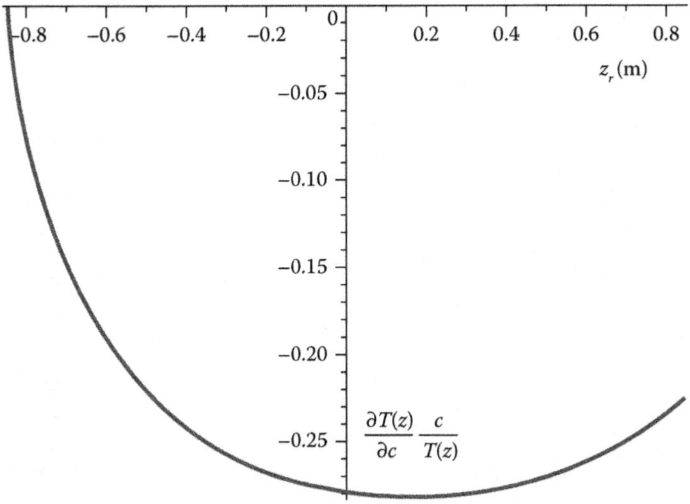

FIGURE 8.6
Relative sensitivity of $T(z_r)$ to c.

TABLE 8.1
First-Order Relative Sensitivities of the Temperature $T(z_r)$ at Selected Locations z_r

Location	$z_r = -l/2$	$z_r = z_{max}$	$z_r = \ell/2$
$R_1^{rel} \triangleq [\partial T(z_r)/\partial Q][Q/T(z_r)]$	0.0	1.0	1.72
$R_2^{rel} \triangleq [\partial T(z_r)/\partial q][q/T(z_r)]$	0.0	−0.53	−1.4
$R_3^{rel} \triangleq [\partial T(z_r)/\partial T_a][T_a/T(z_r)]$	1.0	0.3	0.42
$R_4^{rel} \triangleq [\partial T(z_r)/\partial k_0][k_0/T(z_r)]$	0.0	−0.42	−0.36
$R_5^{rel} \triangleq [\partial T(z_r)/\partial c][c/T(z_r)]$	0.0	−0.28	−0.23

8.3 Application of the 2nd-ASAM to Compute the Second-Order Sensitivities

As discussed in the general theory presented in Chapter 7, the fundamental consideration and starting point for computing the second-order response sensitivities, $R_{ij}^{(2)}(\alpha^0) \triangleq \partial^2 R/\partial\alpha_i \, \partial\alpha_j = \partial^2 R/\partial\alpha_j \, \partial\alpha_i$, is to consider the first-order sensitivities to be responses of the form $R_i^{(1)}(\alpha, T; \Psi)$, $i = 1,2,3,4,5$. This fact was explicitly indicated in the respective definitions, provided by Equations 8.27 through 8.31. Based on this fundamental consideration, the 2nd-ASAM proceeds by computing the first-order G-differential, $\delta R_i^{(1)}(\alpha, T; \Psi)$, of each of the functionals $R_i^{(1)}(\alpha,T;\Psi)$, $i=1,2,3,4,5$, at the point ($\boldsymbol{\alpha}^0$, T^0; Ψ^0) of nominal values, using the definition of the G-differential, namely:

$$\begin{aligned}\delta R_{i_1}^{(1)}\left(\alpha^0,T^0,\Psi^0;\delta T,\delta\Psi,\delta\alpha\right) &\triangleq \left\{\frac{d}{d\varepsilon}\left[R_{i_1}^{(1)}\left(\alpha^0+\varepsilon\delta\alpha,T^0+\varepsilon\delta T,\Psi^0+\varepsilon\delta\Psi\right)\right]\right\}_{\varepsilon=0} \\ &\triangleq \sum_{i_2=1}^{5} R_{i_1 i_2}^{(2)}\left(\alpha^0,T^0\right)\delta\alpha_{i_2},\ i_1=1,\ldots,5,\end{aligned} \tag{8.37}$$

for an arbitrary scalar $\varepsilon \in \mathcal{F}$ and vectors (δT, $\delta\Psi$, $\delta\alpha$) $\in \mathcal{H}(\Omega)\oplus\mathcal{H}(\Omega)\oplus\mathcal{H}_\alpha$. For the model under consideration, there will be $N_\alpha(N_\alpha + 1)/2 = 15$ distinct second-order derivatives, due to the symmetry property $\partial^2 R/\partial\alpha_i\partial\alpha_j = \partial^2 R/\partial\alpha_j\partial\alpha_i$. The sensitivities $R_{ij}^{(2)}\left(\alpha^0\right)$ will be computed next, in the order presented in Table 8.1.

8.3.1 Computation of the Second-Order Sensitivities $R_{1i}^{(2)} \triangleq \partial^2 T(z_r)/(\partial Q\,\partial\alpha_i)$

Applying the definition shown in Equation 8.37 to Equation 8.27 yields the G-differential, $\delta R_1^{(1)}\left(\alpha^0,T^0,\Psi^0;\delta T,\delta\Psi,\delta\alpha\right)$, of the first-order sensitivity $R_1^{(1)}(\alpha,T;\Psi)$, in the form

$$\delta R_1^{(1)}\left(\alpha^0,T^0,\Psi^0;\delta T,\delta\Psi,\delta\alpha\right)\triangleq\frac{d}{d\varepsilon}\left\{-\int_{-\ell/2}^{\ell/2}\left[\Psi^0(z)+\varepsilon\delta\Psi(z)\right]\right\}_{\varepsilon=0}=-\int_{-\ell/2}^{\ell/2}\delta\Psi(z)\,dz. \tag{8.38}$$

The function $\delta\Psi(z)$ in Equation 8.38 is the solution of the system of equations obtained by G-differentiating Equations 8.20 through 8.22, namely

$$\begin{aligned}&k\left(T^0\right)\frac{d^2\left[\delta\Psi(z)\right]}{dz^2}+k_0^0c^0\left[\delta T(z)\right]\frac{d^2\Psi^0(z)}{dz^2} \\ &=-\left\{(\delta k_0)\left[1+c^0T^0(z)\right]+(\delta c)k_0^0T^0(z)\right\}\frac{d^2\Psi^0(z)}{dz^2},\ z\in(-\ell/2,\ \ell/2),\end{aligned} \tag{8.39}$$

$$\left.\frac{d\left[\delta\Psi(z)\right]}{dz}\right|_{z=\ell/2}=0,\ at\ z=\ell/2, \tag{8.40}$$

$$\delta\Psi(z)=0,\ at\ \ z=-\ell/2. \tag{8.41}$$

The quantity $d^2\Psi^0(z)/dz^2$ in Equation 8.39 can be replaced by using Equation 8.20 to obtain the following form of Equation 8.39:

$$\begin{aligned}&k\left(T^0\right)\frac{d^2\left[\delta\Psi(z)\right]}{dz^2}+\delta T(z)\frac{c^0\delta(z-z_r)}{1+c^0T^0(z)} \\ &=-\left[\frac{(\delta k_0)}{k_0^0}+(\delta c)\frac{T^0(z)}{1+c^0T^0(z)}\right]\delta(z-z_r)\triangleq Q^{(2)}\left(\alpha^0,T^0;\delta\alpha\right),\ \ z\in(-\ell/2,\ \ell/2).\end{aligned} \tag{8.42}$$

As Equation 8.38 indicates, the entire contribution to $\delta R_1^{(1)}(\boldsymbol{\alpha}^0, T^0, \Psi^0; \delta T, \delta\Psi, \delta\boldsymbol{\alpha})$ stems from the indirect-effect term.

The results in Equations 8.42 and 8.15 are written in the following block-matrix form:

$$\begin{pmatrix} k(T^0)\dfrac{d^2[\;]}{dz^2} & \dfrac{c^0\delta(z-z_r)}{1+c^0T^0(z)} \\ 0 & \dfrac{d^2}{dz^2}\{k(T^0)[\;]\} \end{pmatrix} \begin{pmatrix} \delta\Psi(z) \\ \delta T(z) \end{pmatrix} = \begin{pmatrix} Q^{(2)}(\boldsymbol{\alpha}^0, T^0; \delta\boldsymbol{\alpha}) \\ Q^{(1)}(\boldsymbol{\alpha}^0, T^0; \delta\boldsymbol{\alpha}) \end{pmatrix}, \; z \in (-\ell/2, \ell/2). \tag{8.43}$$

Following the procedure outlined in Chapter 7, introduce the vector $\Psi_1^{(2)} \triangleq (\Psi_{1,1}^{(2)}, \Psi_{2,1}^{(2)}) \in \mathcal{H}_\Psi^{(2)}(\Omega) \triangleq \mathcal{H}(\Omega) \oplus \mathcal{H}(\Omega)$ and define the inner product between two functions $\Phi_1^{(2)} \triangleq [\Phi_{1,1}^{(2)}(z), \Phi_{2,1}^{(2)}(z)] \in \mathcal{H}_\Psi^{(2)}(\Omega)$ and $\Psi_1^{(2)} \triangleq [\Psi_{1,1}^{(2)}(z), \Psi_{2,1}^{(2)}(z)] \in \mathcal{H}_\Psi^{(2)}(\Omega)$ as follows:

$$\left\langle \Phi_1^{(2)}(z), \Psi_1^{(2)}(z) \right\rangle \triangleq \int_{-\ell/2}^{\ell/2} \left[\Phi_{1,1}^{(2)}(z)\Psi_{1,1}^{(2)}(z) + \Phi_{2,1}^{(2)}(z)\Psi_{2,1}^{(2)}(z) \right] dz. \tag{8.44}$$

Forming now the inner product of $\Psi_1^{(2)} \triangleq [\Psi_{1,1}^{(2)}(z), \Psi_{2,1}^{(2)}(z)]$ with Equation 8.43 yields the following sequence of equalities:

$$\begin{aligned} &\left\langle [\Psi_{1,1}^{(2)}(z), \Psi_{2,1}^{(2)}(z)] \begin{pmatrix} k(T^0)\dfrac{d^2[\;]}{dz^2} & \dfrac{c^0\delta(z-z_r)}{1+c^0T^0(z)} \\ 0 & \dfrac{d^2}{dz^2}\{k(T^0)[\;]\} \end{pmatrix} \begin{pmatrix} \delta\Psi(z) \\ \delta T(z) \end{pmatrix} \right\rangle \\ &= \left\langle [\Psi_{1,1}^{(2)}(z), \Psi_{2,1}^{(2)}(z)] \begin{pmatrix} Q^{(2)}(\boldsymbol{\alpha}^0, T^0; \delta\boldsymbol{\alpha}) \\ Q^{(1)}(\boldsymbol{\alpha}^0, T^0; \delta\boldsymbol{\alpha}) \end{pmatrix} \right\rangle \\ &= \left\langle [\delta\Psi(z), \delta T(z)] \begin{pmatrix} \dfrac{d^2}{dz^2}\{k(T^0)[\;]\} & 0 \\ \dfrac{c^0\delta(z-z_r)}{1+c^0T^0(z)} & k(T^0)\dfrac{d^2[\;]}{dz^2} \end{pmatrix} \begin{pmatrix} \Psi_{1,1}^{(2)}(z) \\ \Psi_{2,1}^{(2)}(z) \end{pmatrix} \right\rangle \\ &+ \left\{ P^{(2)}\left[\boldsymbol{\alpha}^0, T^0; \Psi_{1,1}^{(2)}, \Psi_{2,1}^{(2)}; \delta T(z), \delta\Psi(z)\right] \right\}_{\partial\Omega}, \end{aligned} \tag{8.45}$$

where the bilinear concomitant $\left\{P^{(2)}\left[\alpha^0,T^0;\Psi_{1,1}^{(2)},\Psi_{2,1}^{(2)};\delta T(z),\delta\Psi(z)\right]\right\}_{\partial\Omega}$ has the form

$$\left\{P^{(2)}\left[\alpha^0,T^0;\Psi_{1,1}^{(2)},\Psi_{2,1}^{(2)};\delta T(z),\delta\Psi(z)\right]\right\}_{\partial\Omega}\triangleq\left\{\Psi_{1,1}^{(2)}(z)k(T^0)\frac{d[\delta\Psi(z)]}{dz}\right.$$

$$\left.-\delta\Psi(z)\frac{d\left[k(T^0)\Psi_{1,1}^{(2)}(z)\right]}{dz}+\Psi_{2,1}^{(2)}(z)\frac{d\left[k(T^0)\delta T(z)\right]}{dz}-k(T^0)\delta T(z)\frac{d\Psi_{2,1}^{(2)}(z)}{dz}\right\}_{z=-\ell/2}^{z=\ell/2}. \tag{8.46}$$

The definition of the second-level adjoint function $\Psi_1^{(2)}\triangleq\left(\Psi_{1,1}^{(2)},\Psi_{2,1}^{(2)}\right)$ is now completed by requiring the term on the right side of the last equality in Equation 8.45 to represent the same functional as the right side of Equation 8.38, which yields the following 2nd-LASS:

$$\begin{pmatrix}\frac{d^2}{dz^2}\left\{k(T^0)[\ ]\right\} & 0\\ \frac{c^0\delta(z-z_r)}{1+c^0T^0(z)} & k(T^0)\frac{d^2[\ ]}{dz^2}\end{pmatrix}\begin{pmatrix}\Psi_{1,1}^{(2)}(z)\\ \Psi_{2,1}^{(2)}(z)\end{pmatrix}=\begin{pmatrix}-1\\ 0\end{pmatrix}. \tag{8.47}$$

As generally discussed in Chapter 7, the boundary conditions for the above 2nd-LASS are obtained by using the boundary conditions given in Equations 8.16, 8.17, 8.40, and 8.41 in Equation 8.46 to eliminate all unknown values of $\delta T(z)$ and $\delta\Psi(z)$, respectively, in the bilinear concomitant $\left\{P^{(2)}\left[\alpha^0,T^0;\Psi_{1,1}^{(2)},\Psi_{2,1}^{(2)};\delta T(z),\delta\Psi(z)\right]\right\}_{\partial\Omega}$. These considerations lead to the following boundary conditions for the 2nd-LASS:

$$\left.\frac{d\left[k(T^0)\Psi_{1,1}^{(2)}\right]}{dz}\right|_{z=\ell/2}=0,\ at\ z=\ell/2, \tag{8.48}$$

$$\Psi_{1,1}^{(2)}(z)=0,\ at\ z=-\ell/2, \tag{8.49}$$

$$\left.\frac{d\Psi_{2,1}^{(2)}(z)}{dz}\right|_{z=\ell/2}=0,\ at\ z=\ell/2, \tag{8.50}$$

$$\Psi_{2,1}^{(2)}(z)=0,\ at\ z=-\ell/2. \tag{8.51}$$

Inserting the above boundary conditions together with those given in Equations 8.16, 8.17, 8.40, and 8.41 into Equation 8.46 reduces the bilinear concomitant $\left\{P^{(2)}\left[\alpha^0,T^0;\Psi_{1,1}^{(2)},\Psi_{2,1}^{(2)};\delta T(z),\delta\Psi(z)\right]\right\}_{\partial\Omega}$ to the quantity $\hat{P}^{(2)}\left[\alpha^0,T^0;\Psi_{1,1}^{(2)},\Psi_{2,1}^{(2)};\delta\alpha\right]$, which has the following form:

$$\hat{P}^{(2)}\left[\alpha^0,T^0;\Psi_{1,1}^{(2)},\Psi_{2,1}^{(2)};\delta\alpha\right]\triangleq q_1\left(\alpha,T^0;\delta\alpha\right)\left[\Psi_{2,1}^{(2)}(z)\right]_{z=\ell/2}+\left(\delta T_a\right)\left[k\left(T^0\right)\frac{d\Psi_{2,1}^{(2)}(z)}{dz}\right]_{z=-\ell/2}. \tag{8.52}$$

Finally, using Equations 8.52, 8.47, and 8.38 in Equation 8.45 leads to the following expression for the second-order differential expression $\delta R_1^{(1)}\left(\alpha^0,T^0,\Psi^0;\delta T,\delta\Psi,\delta\alpha\right)$:

$$\delta R_1^{(1)}\left(\alpha^0,T^0,\Psi^0;\delta T,\delta\Psi,\delta\alpha\right)=\int_{-\ell/2}^{\ell/2}\left[\Psi_{1,1}^{(2)}(z)Q^{(2)}(\alpha^0,T^0;\delta\alpha)+\Psi_{2,1}^{(2)}(z)Q^{(1)}(\alpha^0,T^0;\delta\alpha)\right]dz$$
$$-q_1\left(\alpha,T^0;\delta\alpha\right)\left[\Psi_{2,1}^{(2)}(z)\right]_{z=\ell/2}-\left(\delta T_a\right)\left[k\left(T^0\right)\frac{d\Psi_{2,1}^{(2)}(z)}{dz}\right]_{z=-\ell/2}. \tag{8.53}$$

Solving Equations 8.47, through 8.49 yields the following expression for the second-level adjoint function $\Psi_{1,1}^{(2)}(z)$:

$$\Psi_{1,1}^{(2)}(z)=\frac{1}{2k\left[T^0(z)\right]}\left(\frac{3\ell^2}{4}+z\ell-z^2\right). \tag{8.54}$$

Solving Equations 8.47 8.50, and 8.51 yields the following expression for the second-level adjoint function $\Psi_{2,1}^{(2)}(z)$:

$$\Psi_{2,1}^{(2)}(z)=C\left(\alpha^0,T^0;z_r\right)\frac{1}{2}\left(\frac{3\ell^2}{4}+z_r\ell-z_r^2\right)\left[z+\frac{\ell}{2}-(z-z_r)H(z-z_r)\right], \tag{8.55}$$

with

$$C\left(\alpha^0,T^0;z_r\right)\triangleq\frac{k_0^0c^0}{k^3\left[T^0(z_r)\right]}. \tag{8.56}$$

Replacing $Q^{(1)}(\boldsymbol{\alpha}^0,T^0;\delta\boldsymbol{\alpha})$ and $Q^{(2)}(\boldsymbol{\alpha}^0,T^0;\delta\boldsymbol{\alpha})$ with the corresponding expressions from Equations 8.15 and 8.42, respectively, yields the following expression for the differential $\delta R_1^{(1)}\left(\boldsymbol{\alpha}^0,T^0,\Psi^0;\delta T,\delta\Psi,\delta\boldsymbol{\alpha}\right)$:

$$\delta R_1^{(1)}\left(\boldsymbol{\alpha}^0,T^0,\Psi^0;\delta T,\delta\Psi,\delta\boldsymbol{\alpha}\right)=\sum_{i_2=1}^{5}R_{1i_2}^{(2)}\left(\boldsymbol{\alpha}^0;\Psi_{1,1}^{(2)},\Psi_{2,1}^{(2)}\right)\delta\alpha_{i_2}$$

$$\triangleq\frac{\partial^2 T(z_r)}{\partial Q^2}(\delta Q)+\frac{\partial^2 T(z_r)}{\partial Q\,\partial q}(\delta q)+\frac{\partial^2 T(z_r)}{\partial Q\,\partial T_a}(\delta T_a)+\frac{\partial^2 T(z_r)}{\partial Q\,\partial k_0}(\delta k_0)+\frac{\partial^2 T(z_r)}{\partial Q\,\partial c}(\delta c), \tag{8.57}$$

where

$$R_{11}^{(2)}\left(\boldsymbol{\alpha}^0;\Psi_{1,1}^{(2)},\Psi_{2,1}^{(2)}\right)\triangleq\frac{\partial^2 T(z_r)}{\partial Q^2}=-\int_{-\ell/2}^{\ell/2}\Psi_{2,1}^{(2)}(z)\,dz, \tag{8.58}$$

$$R_{12}^{(2)}\left(\boldsymbol{\alpha}^0;\Psi_{1,1}^{(2)},\Psi_{2,1}^{(2)}\right)\triangleq\frac{\partial^2 T(z_r)}{\partial Q\,\partial q}=\left[\Psi_{2,1}^{(2)}(z)\right]_{z=\ell/2}, \tag{8.59}$$

$$R_{13}^{(2)}\left(\boldsymbol{\alpha}^0;\Psi_{1,1}^{(2)},\Psi_{2,1}^{(2)}\right)\triangleq\frac{\partial^2 T(z_r)}{\partial Q\,\partial T_a}=-\left[k\left(T^0\right)\frac{d\Psi_{2,1}^{(2)}(z)}{dz}\right]_{z=-\ell/2}, \tag{8.60}$$

$$R_{14}^{(2)}\left(\boldsymbol{\alpha}^0;\Psi_{1,1}^{(2)},\Psi_{2,1}^{(2)}\right)\triangleq\frac{\partial^2 T(z_r)}{\partial Q\,\partial k_0}$$
$$=\frac{1}{k_0^0}\left[Q^0\int_{-\ell/2}^{\ell/2}\Psi_{2,1}^{(2)}(z)\,dz-\int_{-\ell/2}^{\ell/2}\Psi_{1,1}^{(2)}(z)\,\delta(z-z_r)\,dz-q^0\left[\Psi_{2,1}^{(2)}(z)\right]_{z=\ell/2}\right], \tag{8.61}$$

$$R_{15}^{(2)}\left(\boldsymbol{\alpha}^0;\Psi_{1,1}^{(2)},\Psi_{2,1}^{(2)}\right)\triangleq\frac{\partial^2 T(z_r)}{\partial Q\,\partial c}=-k_0^0\int_{-\ell/2}^{\ell/2}\Psi_{2,1}^{(2)}(z)\frac{d}{dz}\left[T^0(z)\frac{dT^0(z)}{dz}\right]dz$$
$$-\int_{-\ell/2}^{\ell/2}\Psi_{1,1}^{(2)}(z)\frac{T^0(z)}{1+c^0T^0(z)}\delta(z-z_r)\,dz+k_0^0\left[\Psi_{2,1}^{(2)}(z)T^0(z)\frac{dT^0}{dz}\right]_{z=\ell/2} \tag{8.62}$$
$$=-k_0^0\left\{\frac{T_a^2}{2}\left[\frac{d\Psi_{2,1}^{(2)}}{dz}\right]_{z=-\ell/2}+\frac{1}{2}\int_{-\ell/2}^{\ell/2}\left[T^0(z)\right]^2\frac{d^2\Psi_{2,1}^{(2)}}{dz^2}dz+\frac{\Psi_{1,1}^{(2)}(z_r)T^0(z_r)}{k\left[T^0(z_r)\right]}\right\}.$$

Replacing Equations 8.54 and 8.55 in the above expressions and carrying out the integrations over z yields the following expressions for the above second-order sensitivities:

$$R_{11}^{(2)}\left(\boldsymbol{\alpha}^0,T^0;z_r\right)\triangleq\frac{\partial^2 T(z_r)}{\partial Q^2}=-\frac{1}{4}C\left(\boldsymbol{\alpha}^0,T^0;z_r\right)\left(\frac{3\ell^2}{4}+z_r\ell-z_r^2\right)^2, \quad (8.63)$$

$$R_{12}^{(2)}\left(\boldsymbol{\alpha}^0,T^0;z_r\right)\triangleq\frac{\partial^2 T(z_r)}{\partial Q\,\partial q}=\frac{1}{2}C\left(\boldsymbol{\alpha}^0,T^0;z_r\right)\left(\frac{3\ell^2}{4}+z_r\ell-z_r^2\right)\left(z_r+\frac{\ell}{2}\right), \quad (8.64)$$

$$R_{13}^{(2)}\left(\boldsymbol{\alpha}^0,T^0;z_r\right)\triangleq\frac{\partial^2 T(z_r)}{\partial Q\,\partial T_a}=-C\left(\boldsymbol{\alpha}^0,T^0;z_r\right)\frac{k(T_a^0)}{2}\left(\frac{3\ell^2}{4}+z_r\ell-z_r^2\right), \quad (8.65)$$

$$R_{14}^{(2)}\left(\boldsymbol{\alpha}^0,T^0;z_r\right)\triangleq\frac{\partial^2 T(z_r)}{\partial Q\,\partial k_0}$$
$$=C\left(\boldsymbol{\alpha}^0,T^0;z_r\right)\frac{1}{2}\left\{\tau^0(z_r)-\frac{1}{c^0}\left[1+c^0T^0(z_r)\right]^2\right\}\left(\frac{3\ell^2}{4}+z_r\ell-z_r^2\right), \quad (8.66)$$

$$R_{15}^{(2)}\left(\boldsymbol{\alpha}^0,T^0;z_r\right)\triangleq\frac{\partial^2 T(z_r)}{\partial Q\,\partial c}$$
$$=C\left(\boldsymbol{\alpha}^0,T^0;z_r\right)\frac{k_0^0}{2c^0}\left\{T_a^0+\tau^0(z_r)-2T^0(z_r)-c^0\left[T^0(z_r)\right]^2\right\}\left(\frac{3\ell^2}{4}+z_r\ell-z_r^2\right). \quad (8.67)$$

The second-order relative sensitivities $\left(R_{1i}^{(2)}\right)_{rel}\triangleq\left[\partial^2 T(z_r)/\partial Q\,\partial\alpha_i\right]\left[Q\alpha_i/T(z_r)\right]$ are depicted in Figures 8.7 through 8.11 as functions of the arbitrary location $z_r\in[-\ell/2,\ell/2]$.

As Figures 8.7 through 8.11 indicate, the relative importance (magnitude) of the second-order relative sensitivities $\left(R_{1i}^{(2)}\right)_{rel}\triangleq\left[\partial^2 T(z_r)/\partial Q\,\partial\alpha_i\right]\left[Q\alpha_i/T(z_r)\right]$ depends on the position z_r. Ranking them by the largest attainable maximum absolute values of these relative sensitivities, the importance of the model parameters is as follows: the heat source Q, the boundary heat flux q, the linear heat conductivity coefficient k_0, the ambient temperature T_a, and the nonlinear heat conductivity coefficient c. By comparison to the importance ranking of the first-order relative sensitivities, the model parameters k_0 and T_a have "switched places," in that the relative sensitivity $[\partial^2 T(z_r)/\partial Q\partial k_0][Qk_0/T(z_r)]$ has become about three times as large as the relative sensitivity $[\partial^2 T(z_r)/\partial Q\partial T_a][QT_a/T(z_r)]$. Furthermore, the relative sensitivity $[\partial^2 T(z_r)/\partial Q\partial c][Qc/T(z_r)]$ has become third in the importance ranking of the

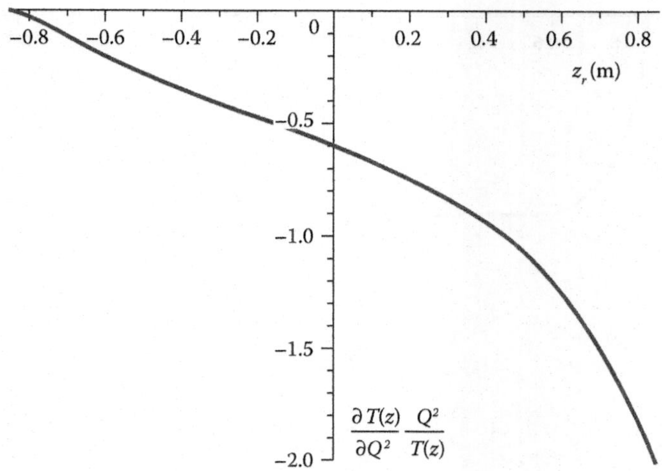

FIGURE 8.7

Relative sensitivity $\frac{\partial^2 T(z_r)}{\partial Q^2}\frac{Q^2}{T(z_r)}$.

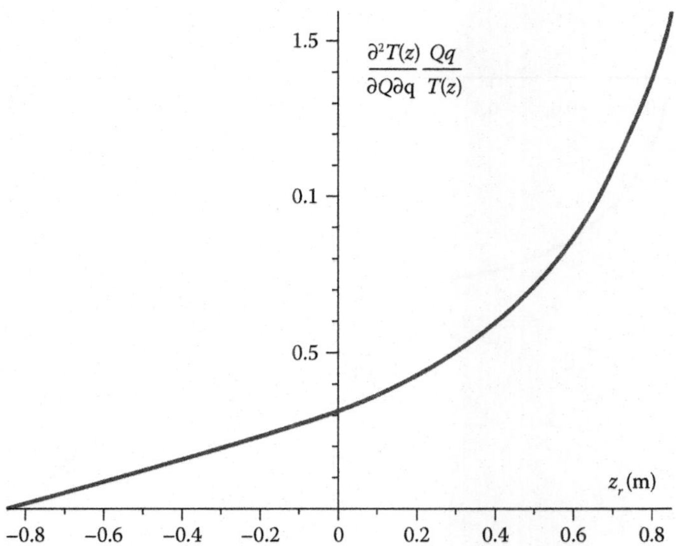

FIGURE 8.8

Relative sensitivity $\frac{\partial^2 T(z_r)}{\partial Q \partial q}\frac{Qq}{T(z_r)}$.

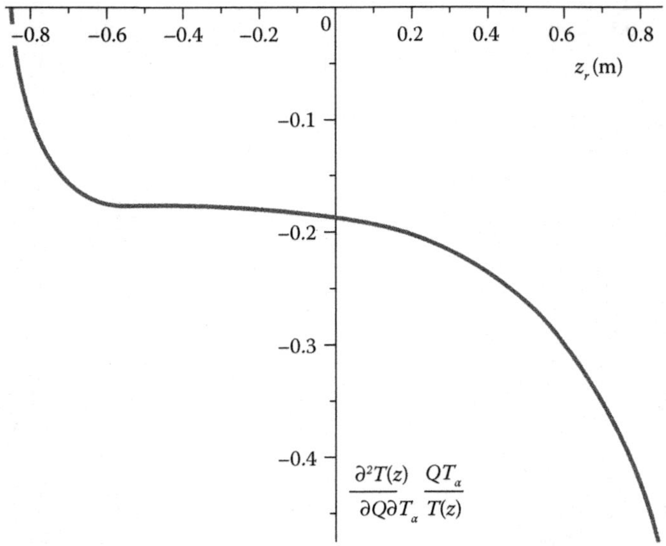

FIGURE 8.9

Relative sensitivity $\frac{\partial^2 T(z_r)}{\partial Q \partial T_a} \frac{Q T_a}{T(z_r)}$.

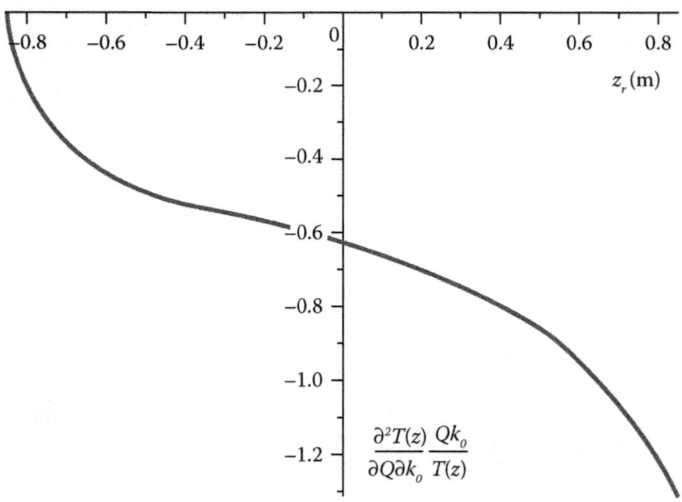

FIGURE 8.10

Relative sensitivity $\frac{\partial^2 T(z_r)}{\partial Q \partial k_0} \frac{Q k_0}{T(z_r)}$.

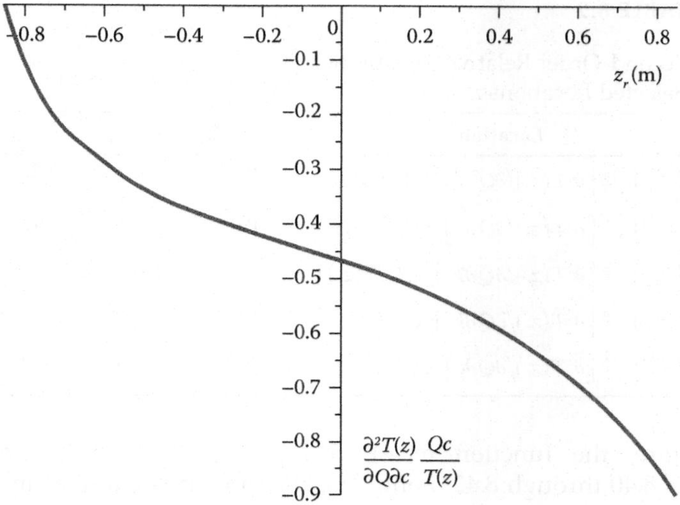

FIGURE 8.11
Relative sensitivity $\frac{\partial^2 T(z_r)}{\partial Q \partial c}\frac{Qc}{T(z_r)}$.

partial sensitivities $\left(R_{1i}^{(2)}\right)_{rel} \triangleq \left[\partial^2 T(z_r)/\partial Q \partial \alpha_i\right]\left[Q\alpha_i/T(z_r)\right]$. The specific values of the second-order relative sensitivities $\left(R_{1i}^{(2)}\right)_{rel}$ at the locations $z_r = -\ell/2$, $z_r = z_{max}$, and $z_r = \ell/2$ are presented in Table 8.2.

As Table 8.2 indicates, the ranking of the absolute values of the relative sensitivities $\left(R_{1i}^{(2)}\right)_{rel}$ is as follows: the heat source Q, the boundary heat flux q, the linear heat conductivity coefficient k_0, the nonlinear heat conductivity coefficient c, and the ambient temperature T_a.

8.3.2 Computation of the Second-Order Sensitivities $\left(R_{2i}^{(2)}\right)(\alpha^0) \triangleq \partial^2 T(z_r)/(\partial q \partial \alpha_i)$

Applying the definition shown in Equation 8.37 to Equation 8.28 yields the G-differential $\delta R_2^{(1)}\left(\alpha^0, T^0, \Psi^0; \delta T, \delta\Psi, \delta\alpha\right)$ of the first-order sensitivity $R_2^{(1)}\left(\alpha^0, T^0; \Psi^0\right)$, as follows:

$$\begin{aligned} \delta R_2^{(1)}\left(\alpha^0, T^0, \Psi^0; \delta T, \delta\Psi, \delta\alpha\right) &\triangleq \sum_{i_2=1}^{5} R_{2i_2}^{(2)}\left(\alpha^0, T^0\right)\delta\alpha_{i_2} \\ &= \int_{-\ell/2}^{\ell/2} \delta\Psi(z)\,\delta\left(z - \frac{\ell}{2}\right) dz. \end{aligned} \tag{8.68}$$

TABLE 8.2

Second-Order Relative Sensitivities $\left(R_{1i}^{(2)}\right)_{rel} \triangleq \dfrac{\partial^2 T(z_r)}{\partial Q \partial \alpha_i}\dfrac{Q\alpha_i}{T(z_r)}$ at Selected Locations z_r

Location	$z_r = -\ell/2$	$z_r = z_{max}$	$z_r = \ell/2$
$\left(R_{11}^{(2)}\right)_{rel} \triangleq \left[\partial^2 T(z_r)/\partial Q^2\right]\left[Q^2/T(z_r)\right]$	0.0	–0.7	–2.0
$\left(R_{12}^{(2)}\right)_{rel} \triangleq \left[\partial^2 T(z_r)/\partial Q \partial q\right]\left[Qq/T(z_r)\right]$	0.0	0.4	1.6
$\left(R_{13}^{(2)}\right)_{rel} \triangleq \left[\partial^2 T(z_r)/\partial Q \partial T_a\right]\left[QT_a/T(z_r)\right]$	0.0	–0.2	–0.5
$\left(R_{14}^{(2)}\right)_{rel} \triangleq \left[\partial^2 T(z_r)/\partial Q \partial k_0\right]\left[Qk_0/T(z_r)\right]$	0.0	–0.7	–1.3
$\left(R_{15}^{(2)}\right)_{rel} \triangleq \left[\partial^2 T(z_r)/\partial Q \partial c\right]\left[Qc/T(z_r)\right]$	0.0	–0.5	–0.9

As before, the function $\delta\Psi(z)$ in Equation 8.68 is the solution of Equations 8.40 through 8.42. Following the same procedure as in the previous subsection leads to the following expression for the partial-differential $\delta R_2^{(1)}\left(\alpha^0, T^0, \Psi^0; \delta T, \delta\Psi, \delta\alpha\right)$:

$$\delta R_2^{(1)}\left(\alpha^0, T^0, \Psi^0; \delta T, \delta\Psi, \delta\alpha\right)$$

$$= \int_{-\ell/2}^{\ell/2} \left[\Psi_{1,2}^{(2)}(z) Q^{(2)}(\alpha^0, T^0; \delta\alpha) + \Psi_{2,2}^{(2)}(z) Q^{(1)}(\alpha^0, T^0; \delta\alpha)\right] dz \tag{8.69}$$

$$- q_1\left(\alpha, T^0; \delta\alpha\right)\left[\Psi_{2,2}^{(2)}(z)\right]_{z=\ell/2} - (\delta T_a)\left[k\left(T^0\right)\frac{d\Psi_{2,2}^{(2)}(z)}{dz}\right]_{z=-\ell/2},$$

where the second-level adjoint function $\Psi_2^{(2)} \triangleq \left(\Psi_{1,2}^{(2)}, \Psi_{2,2}^{(2)}\right)$ is the solution of the following 2nd-LASS:

$$\begin{pmatrix} \dfrac{d^2}{dz^2}\left\{k\left(T^0\right)[\;]\right\} & 0 \\ \dfrac{c^0 \delta(z - z_r)}{1 + c^0 T^0(z)} & k\left(T^0\right)\dfrac{d^2[\;]}{dz^2} \end{pmatrix} \begin{pmatrix} \Psi_{1,2}^{(2)}(z) \\ \Psi_{2,2}^{(2)}(z) \end{pmatrix} = \begin{pmatrix} \delta\left(z - \dfrac{\ell}{2}\right) \\ 0 \end{pmatrix}, \tag{8.70}$$

$$\left.\frac{d\left[k\left(T^0\right)\Psi_{1,2}^{(2)}\right]}{dz}\right|_{z=\ell/2} = 0, \; at \; z = \ell/2, \tag{8.71}$$

$$\Psi_{1,2}^{(2)}(z) = 0, \; at \;\; z = -\ell/2. \tag{8.72}$$

$$\left.\frac{d\Psi_{2,2}^{(2)}}{dz}\right|_{z=\ell/2} = 0, \; at \; z = \ell/2, \tag{8.73}$$

$$\Psi_{2,2}^{(2)}(z)=0, \; at \;\; z=-\ell/2. \tag{8.74}$$

Solving Equations 8.70 through 8.74 yields the following expressions for the components of the second-level adjoint function $\mathbf{\Psi}_2^{(2)} \triangleq \left(\Psi_{1,2}^{(2)}, \Psi_{2,2}^{(2)}\right)$:

$$\Psi_{1,2}^{(2)}(z)=\frac{-1}{k\left[T^0(z)\right]}\left[z+\frac{\ell}{2}-\left(z-\frac{\ell}{2}\right)H\left(z-\frac{\ell}{2}\right)\right], \tag{8.75}$$

and

$$\Psi_{2,2}^{(2)}(z)=-C\left(\alpha^0,T^0;z_r\right)\left(z_r+\frac{\ell}{2}\right)\left[z+\frac{\ell}{2}-(z-z_r)H(z-z_r)\right]. \tag{8.76}$$

The second-order sensitivities $R_{2i}^{(2)}(\alpha)\triangleq \partial^2 T(z_r)/(\partial q\, \partial \alpha_i)$ will formally have the same expressions as those shown Equations 8.58 through 8.62, except that the function $\mathbf{\Psi}_1^{(2)} \triangleq \left(\Psi_{1,1}^{(2)}, \Psi_{2,1}^{(2)}\right)$ will be replaced by the function $\mathbf{\Psi}_2^{(2)} \triangleq \left(\Psi_{1,2}^{(2)}, \Psi_{2,2}^{(2)}\right)$, to obtain

$$R_{21}^{(2)}\left(\alpha^0,T^0;z_r\right)\triangleq\frac{\partial^2 T(z_r)}{\partial q\,\partial Q}=-\int_{-\ell/2}^{\ell/2}\Psi_{2,2}^{(2)}(z)dz, \tag{8.77}$$

$$R_{22}^{(2)}\left(\alpha^0,T^0;z_r\right)\triangleq\frac{\partial^2 T(z_r)}{\partial q^2}=\left[\Psi_{2,2}^{(2)}(z)\right]_{z=\ell/2}, \tag{8.78}$$

$$R_{23}^{(2)}\left(\alpha^0,T^0;z_r\right)\triangleq\frac{\partial^2 T(z_r)}{\partial q\,\partial T_a}=-\left[k\left(T^0\right)\frac{d\Psi_{2,2}^{(2)}(z)}{dz}\right]_{z=-\ell/2}, \tag{8.79}$$

$$\begin{aligned} R_{24}^{(2)}\left(\alpha^0,T^0;z_r\right)&\triangleq\frac{\partial^2 T(z_r)}{\partial q\,\partial k_0} \\ &=\frac{1}{k_0^0}\left[Q^0\int_{-\ell/2}^{\ell/2}\Psi_{2,2}^{(2)}(z)dz-\int_{-\ell/2}^{\ell/2}\Psi_{1,2}^{(2)}(z)\delta(z-z_r)dz-q^0\left[\Psi_{2,2}^{(2)}(z)\right]_{z=\ell/2}\right], \end{aligned} \tag{8.80}$$

$$\begin{aligned} R_{25}^{(2)}\left(\alpha^0,T^0;z_r\right)&\triangleq\frac{\partial^2 T(z_r)}{\partial q\,\partial c} \\ &=-k_0^0\left\{\frac{T_a^2}{2}\left[\frac{d\Psi_{2,2}^{(2)}}{dz}\right]_{z=-\ell/2}+\frac{1}{2}\int_{-\ell/2}^{\ell/2}\left[T^0(z)\right]^2\frac{d^2\Psi_{2,2}^{(2)}}{dz^2}dz+\frac{\Psi_{1,2}^{(2)}(z_r)T^0(z_r)}{k\left[T^0(z_r)\right]}\right\}. \end{aligned} \tag{8.81}$$

Replacing Equations 8.75 and 8.76 in the above expressions and carrying out the integrations over z yields the following explicit expressions for the above second-order sensitivities:

$$\frac{\partial^2 T(z_r)}{\partial q \partial Q} = -\int_{-\ell/2}^{\ell/2} \Psi_{2,2}^{(2)}(z)dz = \frac{\partial^2 T(z_r)}{\partial Q \partial q} = \left[\Psi_{2,1}^{(2)}(z)\right]_{z=\ell/2}$$
$$= \frac{1}{2} C\left(T^0, z_r; \boldsymbol{\alpha}^0\right)\left(\frac{3\ell^2}{4} + z_r \ell - z_r^2\right)\left(z_r + \frac{\ell}{2}\right), \tag{8.82}$$

$$\frac{\partial^2 T(z_r)}{\partial q^2} = -C\left(T^0, z_r; \boldsymbol{\alpha}^0\right)\left(z_r + \frac{\ell}{2}\right)^2, \tag{8.83}$$

$$\frac{\partial^2 T(z_r)}{\partial q \partial T_a} = C\left(T^0, z_r; \boldsymbol{\alpha}^0\right) k\left(T_a^0\right)\left(z_r + \frac{\ell}{2}\right), \tag{8.84}$$

$$\frac{\partial^2 T(z_r)}{\partial q \partial k_0} = C\left(T^0, z_r; \boldsymbol{\alpha}^0\right)\left\{\frac{1}{c^0}\left[1 + c^0 T^0(z_r)\right]^2 - \tau^0(z_r)\right\}\left(z_r + \frac{\ell}{2}\right), \tag{8.85}$$

$$\frac{\partial^2 T(z_r)}{\partial q \partial c} = C\left(T^0, z_r; \boldsymbol{\alpha}^0\right)\frac{k_0^0}{c^0}\left\{c^0\left[T^0(z_r)\right]^2 + 2T^0(z_r) - \tau^0(z_r) - T_a^0\right\}\left(z_r + \frac{\ell}{2}\right). \tag{8.86}$$

The symmetry of the second-order sensitivity $\partial^2 T(z_r)/\partial q \partial Q$ implies the equality between Equations 8.82 and 8.59, which in turn provides a stringent independent verification of the accuracy of computing the second-level adjoint functions $\boldsymbol{\Psi}_1^{(2)} \triangleq \left(\Psi_{1,1}^{(2)}, \Psi_{2,1}^{(2)}\right)$ and $\boldsymbol{\Psi}_2^{(2)} \triangleq \left(\Psi_{1,2}^{(2)}, \Psi_{2,2}^{(2)}\right)$.

The second-order relative sensitivities $\left(R_{2i}^{(2)}\right)_{rel} \triangleq \left[\partial^2 T(z_r)/\partial q \partial \alpha_i\right] \left[q\alpha_i/T(z_r)\right]$ are depicted in Figures 8.12 through 8.15 as functions of the arbitrary location $z_r \in [-\ell/2, \ell/2]$.

Figures 8.12 through 8.15 indicate that the relative importance (magnitude) of each of the second-order relative sensitivities $\left(R_{2i}^{(2)}\right)_{rel} \triangleq \left[\partial^2 T(z_r)/\partial q \partial \alpha_i\right]\left[q\alpha_i/T(z_r)\right]$ depends on the position z_r. Ranking the parameters by the largest attainable maximum absolute values of these relative second-order sensitivities, the importance of the model parameters is as follows: the heat source Q, the boundary heat flux q, the linear heat conductivity coefficient k_0, the nonlinear heat conductivity coefficient c, and the ambient temperature T_a. The specific values of the second-order relative sensitivities $\left(R_{2i}^{(2)}\right)_{rel} \triangleq \left[\partial^2 T(z_r)/\partial q \partial \alpha_i\right]\left[q\alpha_i/T(z_r)\right]$ at the locations $z_r = -\ell/2$, $z_r = z_{max}$, and $z_r = \ell/2$ are presented in Table 8.3.

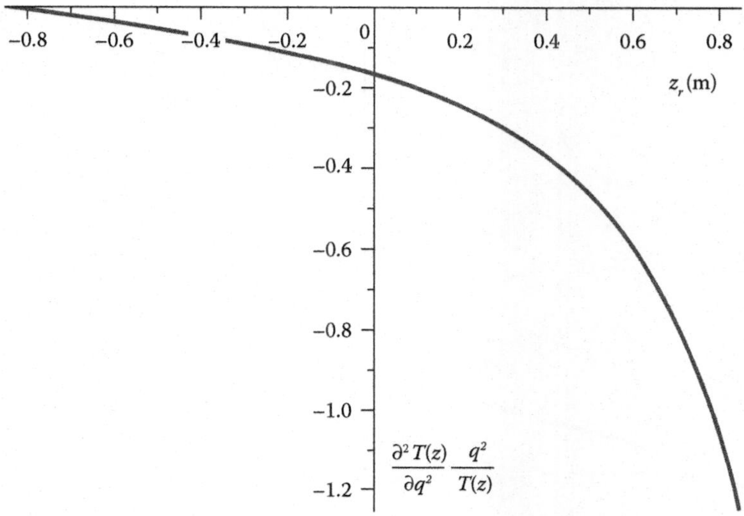

FIGURE 8.12
Relative sensitivity $\frac{\partial^2 T(z_r)}{\partial q^2}\frac{q^2}{T(z_r)}$.

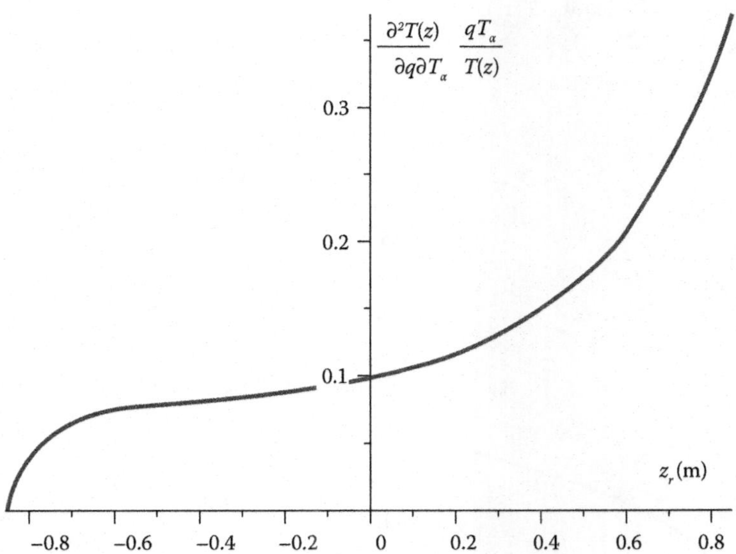

FIGURE 8.13
Relative sensitivity $\frac{\partial^2 T(z_r)}{\partial q\,\partial T_a}\frac{qT_a}{T(z_r)}$.

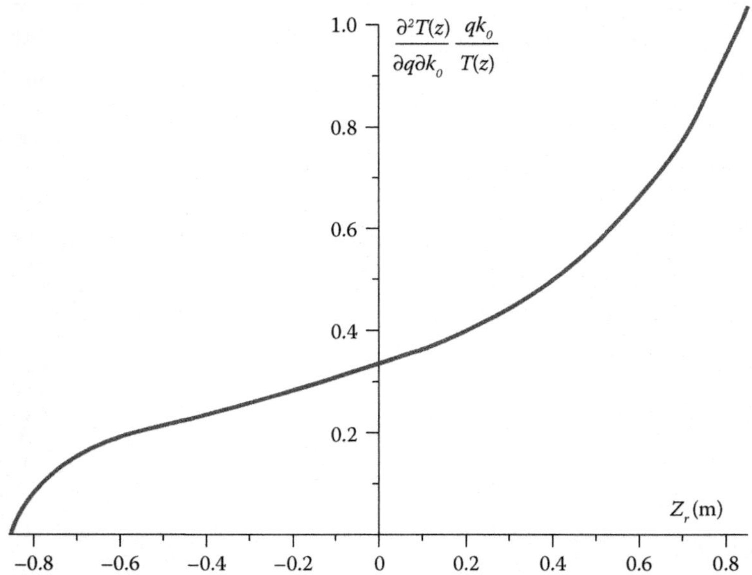

FIGURE 8.14

Relative sensitivity $\frac{\partial^2 T(z_r)}{\partial q\, \partial k_0} \frac{qk_0}{T(z_r)}$.

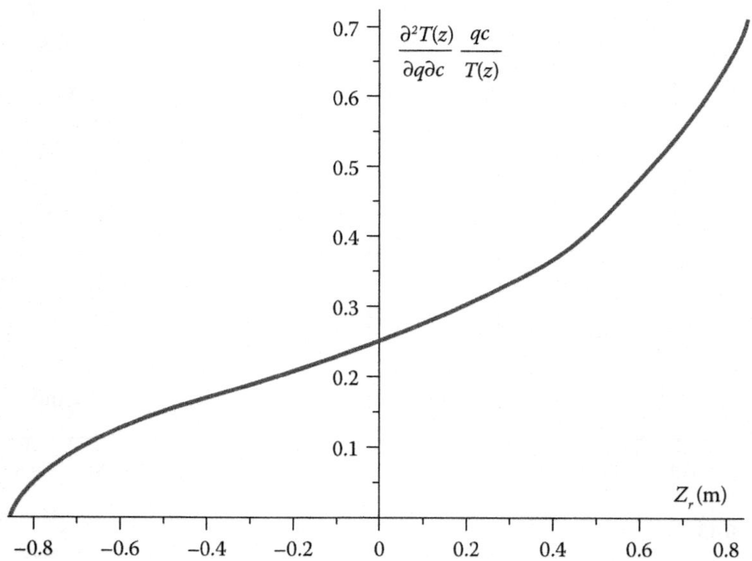

FIGURE 8.15

Relative sensitivity $\frac{\partial^2 T(z_r)}{\partial q\, \partial c} \frac{qc}{T(z_r)}$.

TABLE 8.3

Second-Order Relative Sensitivities $\left(R_{2i}^{(2)}\right)_{rel} \triangleq \frac{\partial^2 T(z_r)}{\partial q \partial \alpha_i}\frac{q\alpha_i}{T(z_r)}$ at Selected Locations z_r

Location	$z_r = -\ell/2$	$z_r = z_{max}$	$z_r = \ell/2$
$\left(R_{21}^{(2)}\right)_{rel} \triangleq \left[\partial^2 T(z_r)/\partial Q \partial q\right]\left[Qq/T(z_r)\right]$	0.0	0.4	1.6
$\left(R_{22}^{(2)}\right)_{rel} \triangleq \left[\partial^2 T(z_r)/\partial q^2\right]\left[q^2/T(z_r)\right]$	0.0	−0.2	−1.3
$\left(R_{23}^{(2)}\right)_{rel} \triangleq \left[\partial^2 T(z_r)/\partial q \partial T_a\right]\left[qT_a/T(z_r)\right]$	0.0	0.1	0.4
$\left(R_{24}^{(2)}\right)_{rel} \triangleq \left[\partial^2 T(z_r)/\partial q \partial k_0\right]\left[qk_0/T(z_r)\right]$	0.0	0.4	1.1
$\left(R_{25}^{(2)}\right)_{rel} \triangleq \left[\partial^2 T(z_r)/\partial q \partial c\right]\left[qc/T(z_r)\right]$	0.0	0.3	0.7

8.3.3 Computation of the Second-Order Sensitivities $R_{3i}^{(2)}\left(\alpha^0\right) \triangleq \partial^2 T(z_r)/(\partial T_a \partial \alpha_i)$

Applying the definition shown in Equation 8.37 to Equation 8.29 yields the G-differential $\delta R_3^{(1)}\left(\alpha^0, T^0, \Psi^0; \delta T, \delta\Psi, \delta\alpha\right)$ of the first-order sensitivity $R_3^{(1)}\left(\alpha^0, T^0; \Psi^0\right)$ as follows:

$$\delta R_3^{(1)}\left(\alpha^0, T^0, \Psi^0; \delta T, \delta\Psi, \delta\alpha\right) \triangleq \sum_{i_2=1}^{5} R_{3i_2}^{(2)}\left(\alpha^0, T^0\right)\delta\alpha_{i_2}$$

$$= \frac{d}{d\varepsilon}\left\{-\int_{-\ell/2}^{\ell/2}\left(k_0^0 + \varepsilon\delta k_0\right)\left[1 + \left(c^0 + \varepsilon\delta c\right)\left(T^0 + \varepsilon\delta T\right)\right]\frac{d\left(\Psi^0 + \varepsilon\delta\Psi\right)}{dz}\delta\left(z + \frac{\ell}{2}\right)dz\right\}_{\varepsilon=0} \quad (8.87)$$

$$= \left[\delta R_3^{(1)}\left(\alpha^0, T^0, \Psi^0; \delta T, \delta\Psi, \delta\alpha\right)\right]_{direct} + \left[\delta R_3^{(1)}\left(\alpha^0, T^0, \Psi^0; \delta T, \delta\Psi, \delta\alpha\right)\right]_{indirect},$$

where the direct-effect term is defined as

$$\left[\delta R_3^{(1)}\left(\alpha^0, T^0, \Psi^0; \delta T, \delta\Psi, \delta\alpha\right)\right]_{direct} \triangleq$$

$$= -\int_{-\ell/2}^{\ell/2}\left\{(\delta k_0)\left[1 + c^0 T^0(z)\right] + (\delta c)k_0^0 T^0(z) + k_0^0 c^0\left[\delta T(z)\right]\right\}\frac{d\Psi^0(z)}{dz}\delta\left(z + \frac{\ell}{2}\right)dz, \quad (8.88)$$

while the indirect-effect term is defined as

$$\left[\delta R_3^{(1)}\left(\alpha^0, T^0, \Psi^0; \delta T, \delta\Psi, \delta\alpha\right)\right]_{indirect} \triangleq -\int_{-\ell/2}^{\ell/2} k\left[T^0(z)\right]\frac{d\left[\delta\Psi(z)\right]}{dz}\delta\left(z + \frac{\ell}{2}\right)dz. \quad (8.89)$$

The direct-effect term defined in Equation 8.88 can be evaluated immediately by noting from Equation 8.23 that

$$\left[\frac{d\Psi(z)}{dz}\right]_{z=-\ell/2} = -\frac{1}{k\left[T^0(z_r)\right]}, \tag{8.90}$$

and by using the above result in Equation 8.88 to obtain

$$\begin{aligned}&\left[\delta R_3^{(1)}\left(\alpha^0,T^0,\Psi^0;\delta T,\delta\Psi,\delta\alpha\right)\right]_{direct}\\&=\frac{1}{k\left[T^0(z_r)\right]}\left[(\delta k_0)\left(1+c^0T_a^0\right)+(\delta c)k_0^0T_a^0+(\delta T_a)c^0k_0^0\right].\end{aligned} \tag{8.91}$$

On the contrary, the indirect-effect term defined in Equation 8.89 needs to be evaluated by constructing the corresponding 2nd-LASS for a second-level adjoint function, denoted as $\Psi_3^{(2)} \triangleq \left(\Psi_{1,3}^{(2)}, \Psi_{2,3}^{(2)}\right)$, by following the general principles of the 2nd-ASAM presented in Chapter 7. Applying these principles leads to the following expression for the indirect-effect term defined in Equation 8.89:

$$\begin{aligned}&\delta R_3^{(1)}\left(\alpha^0,T^0,\Psi^0;\delta T,\delta\Psi,\delta\alpha\right)\\&=\int_{-\ell/2}^{\ell/2}\left[\Psi_{1,3}^{(2)}(z)Q^{(2)}(\alpha^0,T^0;\delta\alpha)+\Psi_{2,3}^{(2)}(z)Q^{(1)}(\alpha^0,T^0;\delta\alpha)\right]dz\\&-q_1\left(\alpha,T^0;\delta\alpha\right)\left[\Psi_{2,3}^{(2)}(z)\right]_{z=\ell/2}-(\delta T_a)\left[k\left(T^0\right)\frac{d\Psi_{2,3}^{(2)}(z)}{dz}\right]_{z=-\ell/2},\end{aligned} \tag{8.92}$$

where the second-level adjoint function $\Psi_3^{(2)} \triangleq \left(\Psi_{1,3}^{(2)}, \Psi_{2,3}^{(2)}\right)$ is the solution of the following 2nd-LASS:

$$\begin{pmatrix}\dfrac{d^2}{dz^2}\left\{k\left(T^0\right)[\ ]\right\} & 0\\ \dfrac{c^0\delta(z-z_r)}{1+c^0T^0(z)} & k\left(T^0\right)\dfrac{d^2[\ ]}{dz^2}\end{pmatrix}\begin{pmatrix}\Psi_{1,3}^{(2)}(z)\\ \Psi_{2,3}^{(2)}(z)\end{pmatrix}=\begin{pmatrix}0\\0\end{pmatrix}, \tag{8.93}$$

$$\left.\frac{d\left[k\left(T^0\right)\Psi_{1,3}^{(2)}\right]}{dz}\right|_{z=\ell/2} = 0, \ at\ z=\ell/2, \tag{8.94}$$

$$\Psi_{1,3}^{(2)}(z)=0, \ at\ \ z=-\ell/2. \tag{8.95}$$

$$\left.\frac{d\Psi_{2,3}^{(2)}}{dz}\right|_{z=\ell/2} = 0, \; at \; z = \ell/2, \tag{8.96}$$

$$\Psi_{2,3}^{(2)}(z) = 0, \; at \; z = -\ell/2. \tag{8.97}$$

Solving Equations 8.93 through 8.97 yields the following expressions for the components of the second-level adjoint function $\Psi_3^{(2)} \triangleq \left(\Psi_{1,3}^{(2)}, \Psi_{2,3}^{(2)}\right)$:

$$\Psi_{1,3}^{(2)}(z) = \frac{k(T_a)}{k\left[T^0(z)\right]}, \tag{8.98}$$

and

$$\Psi_{2,3}^{(2)}(z) = C\left(T^0, z_r; \boldsymbol{\alpha}^0\right) k(T_a)\left[z + \frac{\ell}{2} - (z - z_r)H(z - z_r)\right]. \tag{8.99}$$

Adding Equations 8.88 and 8.92 and identifying the coefficients multiplying the respective parameter variations yields the following expressions for the second-order sensitivities $R_{3i}^{(2)}\left(\boldsymbol{\alpha}^0\right) \triangleq \partial^2 T(z_r)/(\partial T_a \, \partial \alpha_i)$:

$$R_{31}^{(2)}\left(\boldsymbol{\alpha}^0, T^0; z_r\right) \triangleq \frac{\partial^2 T(z_r)}{\partial T_a \, \partial Q} = -\int_{-\ell/2}^{\ell/2} \Psi_{2,3}^{(2)}(z)dz = \frac{\partial^2 T(z_r)}{\partial Q \, \partial T_a} = -\left[k\left(T^0\right)\frac{d\Psi_{2,1}^{(2)}(z)}{dz}\right]_{z=-\ell/2}$$
$$= -C\left(T^0, z_r; \boldsymbol{\alpha}^0\right)\frac{k(T_a^0)}{2}\left(\frac{3\ell^2}{4} + z_r\ell - z_r^2\right), \tag{8.100}$$

$$R_{32}^{(2)}\left(\boldsymbol{\alpha}^0, T^0; z_r\right) \triangleq \frac{\partial^2 T(z_r)}{\partial T_a \, \partial q} = \left[\Psi_{2,3}^{(2)}(z)\right]_{z=\ell/2} = \frac{\partial^2 T(z_r)}{\partial q \, \partial T_a} = -\left[k\left(T^0\right)\frac{d\Psi_{2,2}^{(2)}(z)}{dz}\right]_{z=-\ell/2}$$
$$= C\left(T^0, z_r; \boldsymbol{\alpha}^0\right) k\left(T_a^0\right)\left(z_r + \frac{\ell}{2}\right), \tag{8.101}$$

$$R_{33}^{(2)}\left(\boldsymbol{\alpha}^0, T^0; z_r\right) \triangleq \frac{\partial^2 T(z_r)}{\partial T_a^2} = -\left[c^0 k_0^0 \frac{d\Psi^0(z)}{dz} + k\left(T^0\right)\frac{d\Psi_{2,3}^{(2)}(z)}{dz}\right]_{z=-\ell/2} \tag{8.102}$$
$$= C\left(T^0, z_r; \boldsymbol{\alpha}^0\right)\left\{k^2\left[T^0(z_r)\right] - k^2\left(T_a^0\right)\right\},$$

$$R_{34}^{(2)}\left(\alpha^0,T^0;z_r\right)\triangleq\frac{\partial^2 T(z_r)}{\partial T_a\,\partial k_0}=\frac{1+c^0T_a^0}{k\left[T^0(z_r)\right]}+\frac{Q^0}{k_0^0}\int_{-\ell/2}^{\ell/2}\Psi_{2,3}^{(2)}(z)dz$$
$$-\frac{1}{k_0^0}\int_{-\ell/2}^{\ell/2}\Psi_{1,3}^{(2)}(z)\delta(z-z_r)dz-\frac{q^0}{k_0^0}\left[\Psi_{2,3}^{(2)}(z)\right]_{z=\ell/2} \quad (8.103)$$
$$=C\left(T^0,z_r;\alpha^0\right)k\left(T_a^0\right)\tau^0(z_r),$$

$$R_{35}^{(2)}\left(\alpha^0,T^0;z_r\right)\triangleq\frac{\partial^2 T(z_r)}{\partial T_a\,\partial c}$$
$$=\frac{k_0^0T_a^0}{k\left[T^0(z_r)\right]}+C\left(T^0,z_r;\alpha^0\right)\frac{k_0^0k\left(T_a^0\right)}{c^0}\left\{\tau^0(z_r)+T_a^0-c^0\left[T^0(z_r)\right]^2-2T^0(z_r)\right\}. \quad (8.104)$$

The symmetry of the second-order sensitivity $\partial^2T(z_r)/\partial T_a\partial Q$ implies the equality between the relations expressed in Equations 8.100 and 8.65. This equality provides an independent verification of the correctness of the respective expressions as well as a verification of the solution accuracy of computing the second-level adjoint functions $\Psi_1^{(2)}\triangleq\left(\Psi_{1,1}^{(2)},\Psi_{2,1}^{(2)}\right)$ and $\Psi_3^{(2)}\triangleq\left(\Psi_{1,3}^{(2)},\Psi_{2,3}^{(2)}\right)$. Furthermore, the symmetry of the second-order sensitivity $\partial^2T(z_r)/\partial T_a\partial q$ implies the equality between the relations expressed in Equations 8.101 and 8.79. This equality provides an independent verification of the correctness of the respective expressions as well as a verification of the solution accuracy of computing the second-level adjoint functions $\Psi_2^{(2)}\triangleq\left(\Psi_{1,2}^{(2)},\Psi_{2,2}^{(2)}\right)$ and $\Psi_3^{(2)}\triangleq\left(\Psi_{1,3}^{(2)},\Psi_{2,3}^{(2)}\right)$. The second-order relative sensitivities $\left(R_{3i}^{(2)}\right)_{rel}\triangleq\left[\partial^2T(z_r)/\partial T_a\,\partial\alpha_i\right]\left[T_a\alpha_i/T(z_r)\right]$ are depicted in Figures 8.16 through 8.18 as functions of the arbitrary location $z_r\in[-\ell/2,\ell/2]$.

As Figures 8.16 through 8.18 indicate, the relative importance (magnitude) of the second-order relative sensitivities $\left(R_{3i}^{(2)}\right)_{rel}\triangleq\left[\partial^2T(z_r)/\partial T_a\,\partial\alpha_i\right]\left[T_a\alpha_i/T(z_r)\right]$ depends on the position z_r. Ranking them by the largest attainable maximum absolute values of these relative sensitivities, the importance of the model parameters is as follows: the boundary heat flux q, as previously depicted in Figure 8.13, and the heat source Q, as previously depicted in Figure 8.9. On the contrary, Figures 8.16 through 8.18 indicate that the remaining parameters (namely the linear heat conductivity coefficient k_0, the ambient temperature T_a, and the nonlinear heat conductivity coefficient c) are much less important than Q and q. The specific values of the second-order relative sensitivities $\left(R_{3i}^{(2)}\right)_{rel}\triangleq\left[\partial^2T(z_r)/\partial T_a\,\partial\alpha_i\right]\left[T_a\alpha_i/T(z_r)\right]$ at selected locations z_r are presented in Table 8.4. Ranking them by the largest attainable absolute values of these second-order relative sensitivities, the most important parameters are the heat source Q and the boundary heat flux q (see also Tables 8.2 and 8.3, respectively).

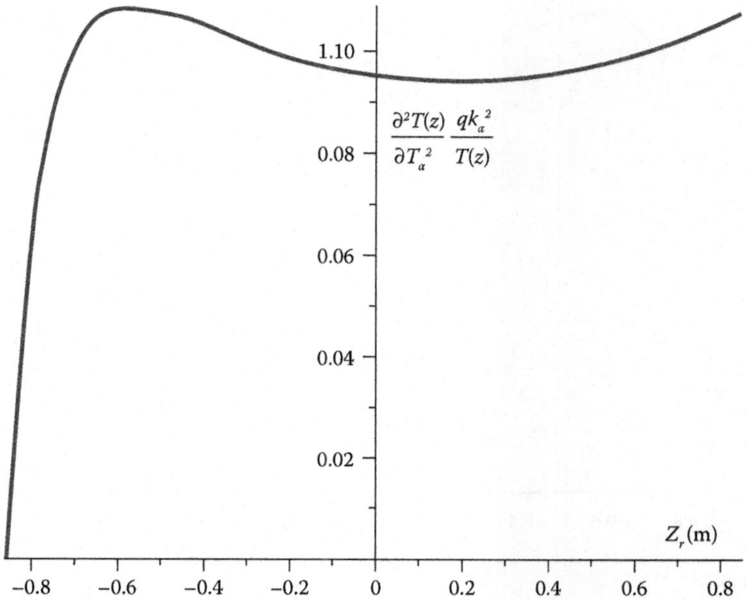

FIGURE 8.16

Relative sensitivity $\frac{\partial^2 T(z_r)}{\partial T_a^2} \frac{T_a^2}{T(z_r)}$.

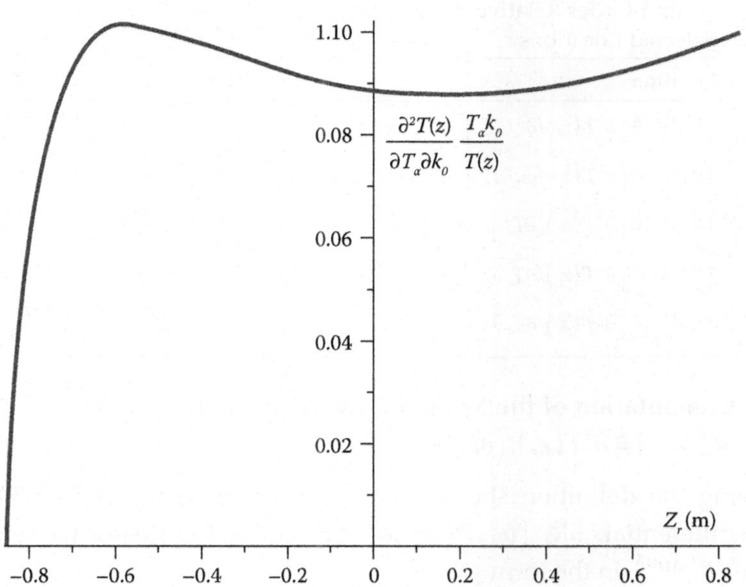

FIGURE 8.17

Relative sensitivity $\frac{\partial^2 T(z_r)}{\partial T_a \partial k_0} \frac{T_a k_0}{T(z_r)}$.

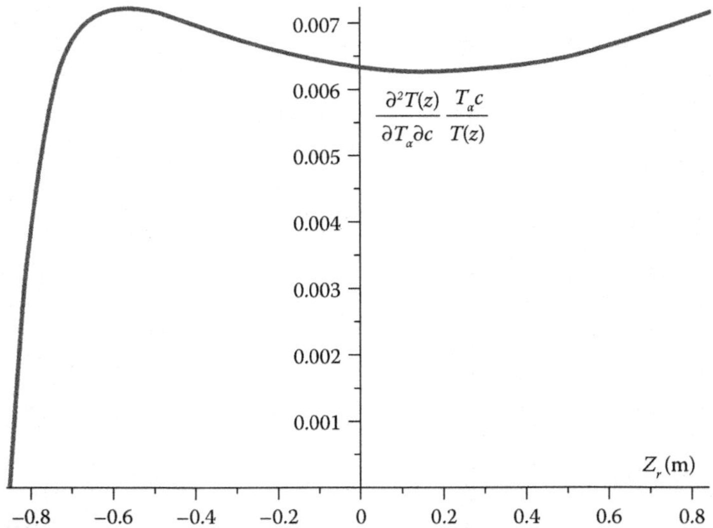

FIGURE 8.18
Relative sensitivity $\dfrac{\partial^2 T(z_r)}{\partial T_a \partial c}\dfrac{T_a c}{T(z_r)}$.

TABLE 8.4

Second-Order Relative Sensitivities $\left(R_{3i}^{(2)}\right)_{rel} \triangleq \dfrac{\partial^2 T(z_r)}{\partial T_a \partial \alpha_i}\dfrac{T_a \alpha_i}{T(z_r)}$ at Selected Locations z_r

Location	$z_r = -\ell/2$	$z_r = z_{max}$	$z_r = \ell/2$
$\left(R_{31}^{(2)}\right)_{rel} \triangleq \left[\partial^2 T(z_r)/\partial Q \partial T_a\right]\left[Q T_a/T(z_r)\right]$	0.0	−0.2	−0.5
$\left(R_{32}^{(2)}\right)_{rel} \triangleq \left[\partial^2 T(z_r)/\partial q \partial T_a\right]\left[q T_a/T(z_r)\right]$	0.0	0.1	0.4
$\left(R_{33}^{(2)}\right)_{rel} \triangleq \left[\partial^2 T(z_r)/\partial T_a^2\right]\left[T_a^2/T(z_r)\right]$	0.0	0.095	0.12
$\left(R_{34}^{(2)}\right)_{rel} \triangleq \left[\partial^2 T(z_r)/\partial T_a \partial k_0\right]\left[T_a k_0/T(z_r)\right]$	0.0	0.088	0.01
$\left(R_{35}^{(2)}\right)_{rel} \triangleq \left[\partial^2 T(z_r)/\partial T_a \partial c\right]\left[T_a c/T(z_r)\right]$	0.0	0.006	0.007

8.3.4 Computation of the Second-Order Sensitivities $R_{4i}^{(2)}\left(\alpha^0\right) \triangleq \partial^2 T(z_r)/\left(\partial k_0 \partial \alpha_i\right)$

Applying the definition shown in Equation 8.37 to Equation 8.30 yields the G-differential, $\delta R_4^{(1)}\left(\alpha^0, T^0, \Psi^0; \delta T, \delta\Psi, \delta\alpha\right)$, of the first-order sensitivity $R_4^{(1)}\left(\alpha^0, T^0, \Psi^0\right)$, in the form

$$\delta R_4^{(1)}\left(\alpha^0, T^0, \Psi^0; \delta T, \delta\Psi, \delta\alpha\right) \triangleq \sum_{i_2=1}^{5} R_{4i_2}^{(2)}\left(\alpha^0, T^0\right)\delta\alpha_{i_2}$$
$$\left[\delta R_4^{(1)}\left(\alpha^0, T^0, \Psi^0; \delta T, \delta\Psi, \delta\alpha\right)\right]_{direct} + \left[\delta R_4^{(1)}\left(\alpha^0, T^0, \Psi^0; \delta T, \delta\Psi, \delta\alpha\right)\right]_{indirect}, \tag{8.105}$$

where the direct-effect term $\left[\delta R_4^{(1)}\left(\alpha^0,T^0,\Psi^0;\delta T,\delta\Psi,\delta\alpha\right)\right]_{direct}$ is defined as

$$\left[\delta R_4^{(1)}\left(\alpha^0,T^0,\Psi^0;\delta T,\delta\Psi,\delta\alpha\right)\right]_{direct} \triangleq \left[\frac{(\delta Q)}{k_0^0}-(\delta k_0)\frac{Q^0}{\left(k_0^0\right)^2}\right]\int_{-\ell/2}^{\ell/2}\Psi(z)dz - \left[\frac{(\delta q)}{k_0^0}-(\delta k_0)\frac{q^0}{\left(k_0^0\right)^2}\right]\Psi(\ell/2), \tag{8.106}$$

while the indirect-effect term $\left[\delta R_4^{(1)}\left(\alpha^0,T^0,\Psi^0;\delta T,\delta\Psi,\delta\alpha\right)\right]_{indirect}$ is defined as

$$\left[\delta R_4^{(1)}\left(\alpha^0,T^0,\Psi^0;\delta T,\delta\Psi,\delta\alpha\right)\right]_{indirect} \triangleq \frac{Q^0}{k_0^0}\int_{-\ell/2}^{\ell/2}\delta\Psi(z)dz-\frac{q^0}{k_0^0}\delta\Psi(\ell/2). \tag{8.107}$$

The direct-effect term defined in Equation 8.106 can be evaluated by using Equation 8.23 to obtain

$$\left[\delta R_4^{(1)}\left(\alpha^0,T^0,\Psi^0;\delta T,\delta\Psi,\delta\alpha\right)\right]_{direct} = \left[(\delta k_0)\frac{Q^0}{\left(k_0^0\right)^2}-\frac{(\delta Q)}{k_0^0}\right]\frac{1}{2k\left[T^0(z_r)\right]}\left(\frac{3\ell^2}{4}+z_r\ell-z_r^2\right) + \left[\frac{(\delta q)}{k_0^0}-(\delta k_0)\frac{q^0}{\left(k_0^0\right)^2}\right]\frac{z_r+\ell/2}{k\left[T^0(z_r)\right]}. \tag{8.108}$$

The indirect-effect term defined in Equation 8.107 needs to be evaluated by constructing the corresponding 2nd-LASS for a second-level adjoint function, denoted as $\Psi_4^{(2)} \triangleq \left(\Psi_{1,4}^{(2)},\ \Psi_{2,4}^{(2)}\right)$, by following the general principles of the 2nd-ASAM presented in Chapter 7. Applying these principles leads to the following expression for the indirect-effect term defined in Equation 8.107:

$$\delta R_4^{(1)}\left(\alpha^0,T^0,\Psi^0;\delta T,\delta\Psi,\delta\alpha\right)=\int_{-\ell/2}^{\ell/2}\left[\Psi_{1,4}^{(2)}(z)Q^{(2)}(\alpha^0,T^0;\delta\alpha)+\Psi_{2,4}^{(2)}(z)Q^{(1)}(\alpha^0,T^0;\delta\alpha)\right]dz - q_1\left(\alpha,T^0;\delta\alpha\right)\left[\Psi_{2,4}^{(2)}(z)\right]_{z=\ell/2}-(\delta T_a)\left[k\left(T^0\right)\frac{d\Psi_{2,4}^{(2)}(z)}{dz}\right]_{z=-\ell/2}, \tag{8.109}$$

where the second-level adjoint function $\Psi_4^{(2)} \triangleq \left(\Psi_{1,4}^{(2)}, \Psi_{2,4}^{(2)}\right)$ is the solution of the following 2nd-LASS:

$$\begin{pmatrix} \dfrac{d^2}{dz^2}\left\{k\left(T^0\right)[\ ]\right\} & 0 \\ \dfrac{c^0\delta(z-z_r)}{1+c^0T^0(z)} & k\left(T^0\right)\dfrac{d^2[\ ]}{dz^2} \end{pmatrix}\begin{pmatrix} \Psi_{1,4}^{(2)}(z) \\ \Psi_{2,4}^{(2)}(z) \end{pmatrix} = \begin{pmatrix} \dfrac{Q^0}{k_0^0} - \dfrac{q^0}{k_0^0}\delta\left(z-\dfrac{\ell}{2}\right) \\ 0 \end{pmatrix}, \quad (8.110)$$

$$\left.\frac{d\left[k\left(T^0\right)\Psi_{1,4}^{(2)}\right]}{dz}\right|_{z=\ell/2} = 0,\ at\ z = \ell/2, \quad (8.111)$$

$$\Psi_{1,4}^{(2)}(z) = 0,\ at\ \ z = -\ell/2. \quad (8.112)$$

$$\left.\frac{d\Psi_{2,4}^{(2)}}{dz}\right|_{z=\ell/2} = 0,\ at\ z = \ell/2, \quad (8.113)$$

$$\Psi_{2,4}^{(2)}(z) = 0,\ at\ \ z = -\ell/2. \quad (8.114)$$

Solving Equations 8.110 through 8.114 yields the following expressions for the components of the second-level adjoint function $\Psi_4^{(2)} \triangleq \left(\Psi_{1,4}^{(2)}, \Psi_{2,4}^{(2)}\right)$:

$$\Psi_{1,4}^{(2)}(z) = \frac{1}{k\left[T^0(z)\right]}\left\{-\frac{Q^0}{2k_0^0}\left(\frac{3\ell^2}{4} + z_r\ell - z_r^2\right) + \frac{q^0}{k_0^0}\left[z + \frac{\ell}{2} - \left(z - \frac{\ell}{2}\right)H\left(z - \frac{\ell}{2}\right)\right]\right\}, \quad (8.115)$$

and

$$\Psi_{2,4}^{(2)}(z) = -C\left(T^0, z_r; \alpha^0\right)\tau(z_r)\left(\frac{3\ell^2}{4} + z_r\ell - z_r^2\right)\left[z + \frac{\ell}{2} - (z - z_r)H(z - z_r)\right]. \quad (8.116)$$

Replacing Equations 8.115 and 8.116 in Equation 8.109, carrying out the integrations over z and performing the ensuing algebra yields the following expression for the indirect-effect term $\left[\delta R_4^{(1)}\left(\alpha^0, T^0, \Psi^0; \delta T, \delta\Psi, \delta\alpha\right)\right]_{indirect}$:

$$\begin{aligned} \left[\delta R_4^{(1)}\left(\alpha^0, T^0, \Psi^0; \delta T, \delta\Psi, \delta\alpha\right)\right]_{indirect} &= \left[(\delta Q) - (\delta k_0)\frac{Q^0}{k_0^0}\right]\frac{c^0k_0^0\tau(z_r)}{2k^3\left[T^0(z_r)\right]}\left(\frac{3\ell^2}{4} + z_r\ell - z_r^2\right) \\ &+ (\delta c)\frac{c^0\left(k_0^0\right)^2\tau(z_r)}{2k^3\left[T^0(z_r)\right]}\left\{T_a^2 - \left[T^0(z_r)\right]^2\right\} + \left[\frac{(\delta k_0)}{k_0^0} + (\delta c)\frac{T^0(z_r)}{1+c^0T^0(z_r)}\right]\frac{\tau(z_r)}{k\left[T^0(z_r)\right]} \quad (8.117) \\ &+ (\delta T_a)k(T_a)\frac{c^0k_0^0\tau(z_r)}{k^3\left[T^0(z_r)\right]} + \left[(\delta k_0)\frac{q^0}{k_0^0} - (\delta q)\right]\frac{c^0k_0^0\tau(z_r)}{k^3\left[T^0(z_r)\right]}\left(z_r + \frac{\ell}{2}\right). \end{aligned}$$

Summing the expression for the indirect-effect term given in Equation 8.117 with the expression of the direct-effect term given in Equation 8.108 and identifying the coefficients of the respective parameter variations yields the following expressions for the second-order derivatives $R_{4i}^{(2)}(\boldsymbol{\alpha}^0) \triangleq \partial^2 T(z_r)/(\partial k_0\, \partial \alpha_i)$:

$$R_{41}^{(2)}(\boldsymbol{\alpha}^0, T^0; z_r) \triangleq \frac{\partial^2 T(z_r)}{\partial k_0\, \partial Q}$$
$$= C(T^0, z_r; \boldsymbol{\alpha}^0)\frac{1}{2}\left\{\tau^0(z_r) - \frac{1}{c^0}\left[1 + c^0 T^0(z_r)\right]^2\right\}\left(\frac{3\ell^2}{4} + z_r\ell - z_r^2\right), \tag{8.118}$$

$$R_{42}^{(2)}(\boldsymbol{\alpha}^0, T^0; z_r) \triangleq \frac{\partial^2 T(z_r)}{\partial k_0\, \partial q} = C(T^0, z_r; \boldsymbol{\alpha}^0)\left\{\frac{1}{c^0}\left[1 + c^0 T^0(z_r)\right]^2 - \tau^0(z_r)\right\}\left(z_r + \frac{\ell}{2}\right), \tag{8.119}$$

$$R_{43}^{(2)}(\boldsymbol{\alpha}^0, T^0; z_r) \triangleq \frac{\partial^2 T(z_r)}{\partial k_0\, \partial T_a} = C(T^0, z_r; \boldsymbol{\alpha}^0)\, k(T_a^0)\, \tau^0(z_r), \tag{8.120}$$

$$R_{44}^{(2)}(\boldsymbol{\alpha}^0, T^0; z_r) \triangleq \frac{\partial^2 T(z_r)}{\partial k_0^2} = \tau^0(z_r)\left\{\frac{2}{k_0^0 k\left[T^0(z_r)\right]} - C(T^0, z_r; \boldsymbol{\alpha}^0)\,\tau^0(z_r)\right\}, \tag{8.121}$$

$$R_{45}^{(2)}(\boldsymbol{\alpha}^0, T^0; z_r) \triangleq \frac{\partial^2 T(z_r)}{\partial k_0\, \partial c}$$
$$= C(T^0, z_r; \boldsymbol{\alpha}^0)\frac{k_0^0 \tau^0(z_r)}{c^0}\left\{c^0\left[T^0(z_r)\right]^2 + 2T^0(z_r) - \tau^0(z_r) - T_a^0\right\}. \tag{8.122}$$

The symmetry of the second-order sensitivity $\partial^2 T(z_r)/\partial k_0 \partial Q$ implies the equality between the relations expressed in Equations 8.118 and 8.66. This equality provides an independent verification of the correctness of the respective expressions as well as a verification of the solution accuracy of computing the second-level adjoint functions $\boldsymbol{\Psi}_1^{(2)} \triangleq \left(\Psi_{1,1}^{(2)}, \Psi_{2,1}^{(2)}\right)$ and $\boldsymbol{\Psi}_4^{(2)} \triangleq \left(\Psi_{1,4}^{(2)}, \Psi_{2,4}^{(2)}\right)$. Furthermore, the symmetry of the second-order sensitivity $\partial^2 T(z_r)/\partial k_0 \partial q$ implies the equality between the relations expressed in Equations 8.119 and 8.85. This equality provides an independent verification of the correctness of the respective expressions as well as a verification of the solution accuracy of computing the second-level adjoint functions $\boldsymbol{\Psi}_2^{(2)} \triangleq \left(\Psi_{1,2}^{(2)}, \Psi_{2,2}^{(2)}\right)$ and $\boldsymbol{\Psi}_4^{(2)} \triangleq \left(\Psi_{1,4}^{(2)}, \Psi_{2,4}^{(2)}\right)$. Finally, the symmetry of the second-order sensitivity $\partial^2 T(z_r)/\partial k_0 \partial T_a$ implies the equality between the relations expressed in Equations 8.120 and 8.103. This equality provides an independent verification of the correctness of the respective expressions as well as a verification of the solution accuracy of computing the second-level adjoint functions $\boldsymbol{\Psi}_3^{(2)} \triangleq \left(\Psi_{1,3}^{(2)}, \Psi_{2,3}^{(2)}\right)$ and $\boldsymbol{\Psi}_4^{(2)} \triangleq \left(\Psi_{1,4}^{(2)}, \Psi_{2,4}^{(2)}\right)$. The second-order

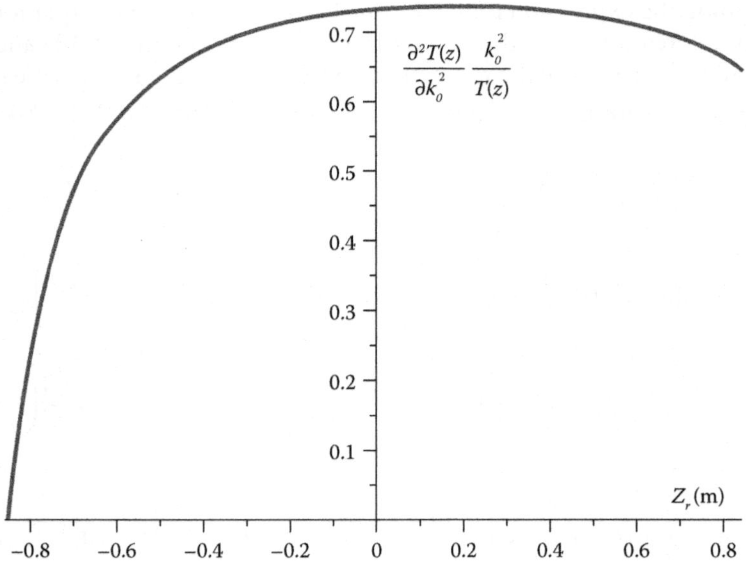

FIGURE 8.19
Relative sensitivity $\frac{\partial^2 T(z_r)}{\partial k_0^2}\frac{k_0^2}{T(z_r)}$.

relative sensitivities $\left[\partial^2 T(z_r)/\partial k_0^2\right]\left[k_0^2/T(z_r)\right]$ and $[\partial^2 T(z_r)/\partial k_0 \partial c][k_0 c/T(z_r)]$ are depicted in Figures 8.19 and 8.20 as functions of the arbitrary location $z_r \in [-\ell/2, \ell/2]$.

8.3.5 Computation of the Second-Order Sensitivities $R_{5i}^{(2)}(\alpha^0) \triangleq \partial^2 T(z_r)/(\partial c\, \partial \alpha_i)$

Applying the definition shown in Equation 8.37 to Equation 8.31 yields the G-differential, $\delta R_5^{(1)}(\alpha^0, T^0, \Psi^0; \delta T, \delta\Psi, \delta\alpha)$, of the first-order sensitivity $R_5^{(1)}(\alpha^0, T^0, \Psi^0)$, in the form

$$
\begin{aligned}
&\delta R_5^{(1)}(\alpha^0, T^0, \Psi^0; \delta T, \delta\Psi, \delta\alpha) \triangleq \sum_{i_2=1}^{5} R_{5i_2}^{(2)}(\alpha^0, T^0)\,\delta\alpha_{i_2} \\
&= \frac{1}{2}\frac{d}{d\varepsilon}\left\{\int_{-\ell/2}^{\ell/2} \frac{\left(T_a^0 + \varepsilon\delta T_a\right)^2 - \left[T^0 + \varepsilon\delta T(z)\right]^2}{1 + \left(c^0 + \varepsilon\delta c\right)\left[T^0 + \varepsilon\delta T(z)\right]}\,\delta(z - z_r)\,dz\right\}_{\varepsilon=0} \qquad (8.123) \\
&= \left[\delta R_5^{(1)}(\alpha^0, T^0, \Psi^0; \delta T, \delta\Psi, \delta\alpha)\right]_{direct} + \delta R_5^{(1)}(\alpha^0, T^0, \Psi^0; \delta T, \delta\Psi, \delta\alpha)_{indirect},
\end{aligned}
$$

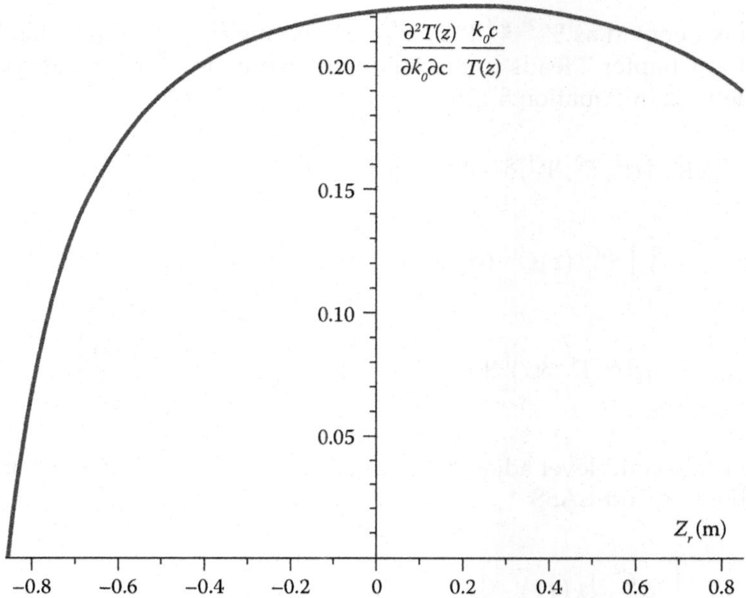

FIGURE 8.20
Relative sensitivity $\frac{\partial^2 T(z_r)}{\partial k_0 \partial c}\frac{k_0 c}{T(z_r)}$.

where

$$\left[\delta R_5^{(1)}\left(\alpha^0, T^0, \Psi^0; \delta T, \delta\Psi, \delta\alpha\right)\right]_{direct} \triangleq (\delta T_a)\frac{T_a^0}{1+c^0 T^0(z_r)} + (\delta c)\frac{\left[T^0(z_r)\right]^2 - \left(T_a^0\right)^2}{2\left[1+c^0 T^0(z_r)\right]^2} T^0(z_r), \tag{8.124}$$

and

$$\left[\delta R_5^{(1)}\left(\alpha^0, T^0, \Psi^0; \delta T, \delta\Psi, \delta\alpha\right)\right]_{indirect} \triangleq -\int_{-\ell/2}^{\ell/2} \frac{c^0\left(T_a^0\right)^2 + 2T^0(z) + c^0\left[T^0(z)\right]^2}{2\left[1+c^0 T^0(z)\right]^2}\delta T(z)\delta(z-z_r)\,dz. \tag{8.125}$$

The indirect-effect term defined in Equation 8.125 needs to be evaluated by constructing the corresponding 2nd-LASS for a second-level adjoint function, denoted as $\Psi_5^{(2)} \triangleq \left(\Psi_{1,5}^{(2)}, \Psi_{2,5}^{(2)}\right)$. Applying the general principles presented in Chapter 7 leads to the following expression for the indirect-effect term defined in Equation 8.125:

$$
\begin{aligned}
&\left[\delta R_5^{(1)}\left(\alpha^0, T^0, \Psi^0; \delta T, \delta\Psi, \delta\alpha\right)\right]_{indirect} \\
&= \int_{-\ell/2}^{\ell/2} \left[\Psi_{1,5}^{(2)}(z) Q^{(2)}(\alpha^0, T^0; \delta\alpha) + \Psi_{2,5}^{(2)}(z) Q^{(1)}(\alpha^0, T^0; \delta\alpha)\right] dz \\
&- q_1\left(\alpha, T^0; \delta\alpha\right)\left[\Psi_{2,5}^{(2)}(z)\right]_{z=\ell/2} - (\delta T_a)\left[k\left(T^0\right)\frac{d\Psi_{2,5}^{(2)}(z)}{dz}\right]_{z=-\ell/2},
\end{aligned}
\tag{8.126}
$$

where the second-level adjoint function $\Psi_5^{(2)} \triangleq \left(\Psi_{1,5}^{(2)}, \Psi_{2,5}^{(2)}\right)$ is the solution of the following 2nd-LASS:

$$
\begin{pmatrix}
\dfrac{d^2}{dz^2}\left\{k\left(T^0\right)[\ ]\right\} & 0 \\
\dfrac{c^0 \delta(z - z_r)}{1 + c^0 T^0(z)} & k\left(T^0\right)\dfrac{d^2[\ ]}{dz^2}
\end{pmatrix}
\begin{pmatrix}
\Psi_{1,5}^{(2)}(z) \\
\Psi_{2,5}^{(2)}(z)
\end{pmatrix}
=
\begin{pmatrix}
0 \\
-\left(k_0^0\right)^2 \dfrac{c^0\left(T_a^0\right)^2 + 2T^0(z) + c^0\left[T^0(z)\right]^2}{2k^3\left[T^0(z)\right]} \delta(z - z_r)
\end{pmatrix},
\tag{8.127}
$$

$$
\left.\frac{d\left[k\left(T^0\right)\Psi_{1,5}^{(2)}\right]}{dz}\right|_{z=\ell/2} = 0, \ at\ z = \ell/2, \Psi_{1,5}^{(2)}(z) = 0, \ at\ \ z = -\ell/2,
\tag{8.128}
$$

$$
\left.\frac{d\Psi_{2,5}^{(2)}}{dz}\right|_{z=\ell/2} = 0, \ at\ z = \ell/2, \Psi_{2,5}^{(2)}(z) = 0, \ at\ \ z = -\ell/2.
\tag{8.129}
$$

Solving Equations 8.127 through 8.129 yields

$$
\begin{aligned}
&\Psi_{1,5}^{(2)}(z) = 0, \\
&\Psi_{2,5}^{(2)}(z) = -C\left(T^0, z_r; \alpha^0\right)\frac{k_0^0}{2c^0} \\
&\left\{c^0\left(T_a^0\right)^2 + 2T^0(z_r) + c^0\left[T^0(z_r)\right]^2\right\}\left[(z - z_r)H(z - z_r) - z - \frac{\ell}{2}\right].
\end{aligned}
\tag{8.130}
$$

Note that the equations satisfied by $\Psi_{2,5}^{(2)}(z)$ have the same form as the 1st-LASS (cf. Equations 8.20 through 8.22), except for a different factor multiplying the source term $\delta(z-z_r)$ in Equation 8.127. Therefore, the solution for $\Psi_{2,5}^{(2)}(z)$ can be written down by simply multiplying the solution of the first-level adjoint function $\Psi(z)$ in Equation 8.23 by the respective factor.

Replacing Equation 8.130 in Equation 8.126, carrying out the integrations over z, and performing the ensuing algebra yields the following expression for the indirect-effect term $\left[\delta R_5^{(1)}\left(\alpha^0,T^0,\Psi^0;\delta T,\delta\Psi,\delta\alpha\right)\right]_{indirect}$:

$$\begin{aligned}
&\left[\delta R_5^{(1)}\left(\alpha^0,T^0,\Psi^0;\delta T,\delta\Psi,\delta\alpha\right)\right]_{indirect} = \\
&-C\left(T^0,z_r;\alpha^0\right)\frac{k_0^0}{2c^0}\left\{c^0\left(T_a^0\right)^2+2T^0(z_r)+c^0\left[T^0(z_r)\right]^2\right\}\left\{(\delta Q)\frac{1}{2}\left(\frac{3\ell^2}{4}+z_r\ell-z_r^2\right)\right. \\
&-(\delta q)\left(z_r+\frac{\ell}{2}\right)+(\delta T_a)k(T_a)-(\delta k_0)\left[\frac{Q^0}{2k_0^0}\left(\frac{3\ell^2}{4}+z_r\ell-z_r^2\right)-\frac{q^0}{k_0^0}\left(z_r+\frac{\ell}{2}\right)\right] \\
&\left.+(\delta c)\frac{\left(T_a^0\right)^2-\left[T^0(z_r)\right]^2}{2}\right\}.
\end{aligned} \tag{8.131}$$

Adding Equations 8.131 and 8.124 and identifying the coefficients multiplying the respective parameter variations yields the following expressions for the second-order sensitivities $R_{5i}^{(2)}\left(\alpha^0\right)\triangleq\partial^2 T(z_r)/(\partial c\,\partial\alpha_i)$:

$$\begin{aligned}
R_{51}^{(2)}\left(\alpha^0,T^0;z_r\right)&\triangleq\frac{\partial^2 T(z_r)}{\partial c\,\partial Q}=C\left(T^0,z_r;\alpha^0\right)\frac{k_0^0}{2c^0} \\
&\times\left\{T_a^0+\tau^0(z_r)-2T^0(z_r)-c^0\left[T^0(z_r)\right]^2\right\}\left(\frac{3\ell^2}{4}+z_r\ell-z_r^2\right),
\end{aligned} \tag{8.132}$$

$$\begin{aligned}
R_{52}^{(2)}\left(\alpha^0,T^0;z_r\right)&\triangleq\frac{\partial^2 T(z_r)}{\partial c\,\partial q} \\
&=C\left(T^0,z_r;\alpha^0\right)\frac{k_0^0}{c^0}\left\{c^0\left[T^0(z_r)\right]^2+2T^0(z_r)-\tau^0(z_r)-T_a^0\right\}\left(z_r+\frac{\ell}{2}\right),
\end{aligned} \tag{8.133}$$

$$\begin{aligned}
R_{53}^{(2)}\left(\alpha^0,T^0;z_r\right)&\triangleq\frac{\partial^2 T(z_r)}{\partial c\,\partial T_a} \\
&=\frac{k_0^0T_a^0}{k\left[T^0(z_r)\right]}+C\left(T^0,z_r;\alpha^0\right)\frac{k_0^0k\left(T_a^0\right)}{c^0}\left\{\tau^0(z_r)+T_a^0-c^0\left[T^0(z_r)\right]^2-2T^0(z_r)\right\},
\end{aligned} \tag{8.134}$$

$$R_{54}^{(2)}\left(\boldsymbol{\alpha}^0, T^0; z_r\right) \triangleq \frac{\partial^2 T(z_r)}{\partial c\, \partial k_0}$$
$$= C\left(T^0, z_r; \boldsymbol{\alpha}^0\right) \frac{k_0^0 \tau^0(z_r)}{c^0} \left\{ c^0 \left[T^0(z_r) \right]^2 + 2T^0(z_r) - \tau^0(z_r) - T_a^0 \right\}, \tag{8.135}$$

$$R_{55}^{(2)}\left(\boldsymbol{\alpha}^0, T^0; z_r\right) \triangleq \frac{\partial^2 T(z_r)}{\partial c^2} = C\left(T^0, z_r; \boldsymbol{\alpha}^0\right) \left(\frac{k_0^0}{c^0} \right)^2 \left[T_a^0 + \tau^0(z_r) - T^0(z_r) \right]$$
$$\times \left\{ c^0 \left[T^0(z_r) \right]^2 + c^0 \left(T_a^0 \right)^2 + T^0(z_r) + \tau^0(z_r) + T_a^0 \right\}. \tag{8.136}$$

The symmetry of the second-order sensitivity $\partial^2 T(z_r)/\partial c \partial Q$ implies the equality between the relations expressed in Equations 8.132 and 8.67. This equality provides an independent verification of the correctness of the respective expressions as well as a verification of the solution accuracy of computing the second-level adjoint functions $\boldsymbol{\Psi}_1^{(2)} \triangleq \left(\Psi_{1,1}^{(2)}, \Psi_{2,1}^{(2)}\right)$ and $\boldsymbol{\Psi}_5^{(2)} \triangleq \left(\Psi_{1,5}^{(2)}, \Psi_{2,5}^{(2)}\right)$. The symmetry of the second-order sensitivity $\partial^2 T(z_r)/\partial c \partial q$ implies the equality between the relations expressed in Equations 8.133 and 8.86. This equality provides an independent verification of the correctness of the respective expressions as well as a verification of the solution accuracy of computing the second-level adjoint functions $\boldsymbol{\Psi}_2^{(2)} \triangleq \left(\Psi_{1,2}^{(2)}, \Psi_{2,2}^{(2)}\right)$ and $\boldsymbol{\Psi}_5^{(2)} \triangleq \left(\Psi_{1,5}^{(2)}, \Psi_{2,5}^{(2)}\right)$. Furthermore, the symmetry of the second-order sensitivity $\partial^2 T(z_r)/\partial c \partial T\alpha$ implies the equality between the relations expressed in Equations 8.134 and 8.104. This equality provides an independent verification of the correctness of the respective expressions as well as a verification of the solution accuracy of computing the second-level adjoint functions $\boldsymbol{\Psi}_3^{(2)} \triangleq \left(\Psi_{1,3}^{(2)}, \Psi_{2,3}^{(2)}\right)$ and $\boldsymbol{\Psi}_5^{(2)} \triangleq \left(\Psi_{1,5}^{(2)}, \Psi_{2,5}^{(2)}\right)$. Finally, the symmetry of the second-order sensitivity $\partial^2 T(z_r)/\partial c \partial k_0$ implies the equality between the relations expressed in Equations 8.135 and 8.122. This equality provides an independent verification of the correctness of the respective expressions as well as a verification of the solution accuracy of computing the second-level adjoint functions $\boldsymbol{\Psi}_4^{(2)} \triangleq \left(\Psi_{1,4}^{(2)}, \Psi_{2,4}^{(2)}\right)$ and $\boldsymbol{\Psi}_5^{(2)} \triangleq \left(\Psi_{1,5}^{(2)}, \Psi_{2,5}^{(2)}\right)$. The second-order relative sensitivity $\partial^2 T(z_r)/\partial c^2 [c^2/T(z_r)]$ is depicted in Figure 8.21 as a function of the arbitrary location $z_r \in [-\ell/2, \ell/2]$.

Table 8.5 presents the values of the second-order relative sensitivities $\left(R_{44}^{(2)}\right)_{rel}$, $\left(R_{45}^{(2)}\right)_{rel}$, and $\left(R_{55}^{(2)}\right)_{rel}$ at selected locations z_r.

8.3.6 Computation of Standard Deviation and Skewness of the Temperature Distribution

The first- and second-order temperature sensitivities will be used in this section to illustrate their essential role for quantifying standard deviations and non-Gaussian features (e.g., asymmetries) of the various response distributions.

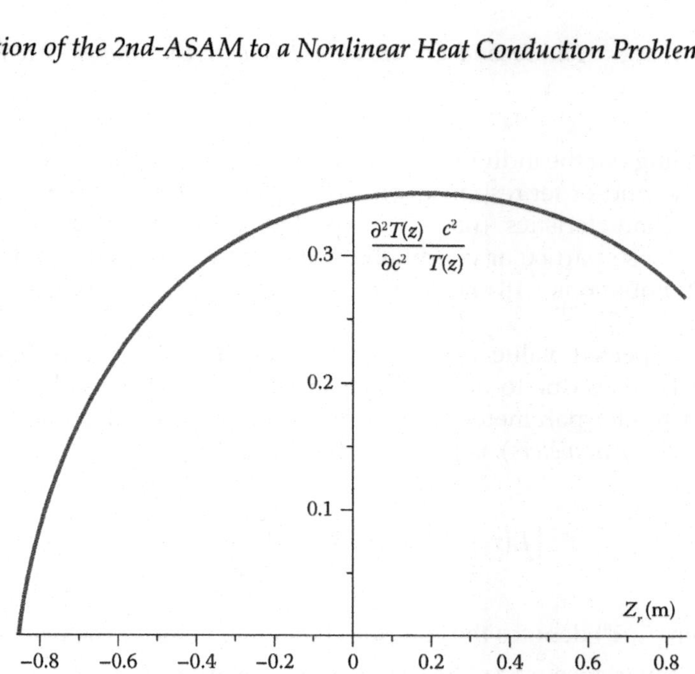

FIGURE 8.21

Relative sensitivity $\frac{\partial^2 T(z_r)}{\partial c^2}\frac{c^2}{T(z_r)}$.

TABLE 8.5

Second-Order Relative Sensitivities $\left(R_{44}^{(2)}\right)_{rel}$, $\left(R_{45}^{(2)}\right)_{rel}$, and $\left(R_{55}^{(2)}\right)_{rel}$ at Selected Locations z_r

Location	$z_r = -\ell/2$	$z_r = z_{\max}$	$z_r = \ell/2$
$\left(R_{44}^{(2)}\right)_{rel} \triangleq \left[\partial^2 T(z_r)/\partial k_0^2\right]\left[k_0^2/T(z_r)\right]$	0.0	0.74	0.65
$\left(R_{45}^{(2)}\right)_{rel} \triangleq \left[\partial^2 T(z_r)/\partial k_0 \partial c\right]\left[k_0 c/T(z_r)\right]$	0.0	0.22	0.18
$\left(R_{55}^{(2)}\right)_{rel} \triangleq \left[\partial^2 T(z_r)/\partial c^2\right]\left[c^2/T(z_r)\right]$	0.0	0.35	0.27

To quantify asymmetries in distribution, at least the third-order ("skewness") response correlations need to be computed, which require, in turn, the exact computation of (at least) the first- and second-order response sensitivities to model parameters. To illustrate the impact of first-order versus second-order sensitivities in contributing to the uncertainty in the temperature response $T(z_r)$ at an arbitrary location z_r, consider that the model parameters Q, q, T_a, k_0, and c are uncorrelated and normally distributed, with absolute standard deviations denoted as $\sigma(Q)$, $\sigma(q)$, $\sigma(T_a)$, $\sigma(k_0)$, and $\sigma(c)$, respectively, and all having relative standard deviations of 10%, that is, $\sigma(Q)/Q^0 = \sigma(q)/q^0 = \sigma(T_a)/T_a^0 = \sigma(T_a)/T_a^0 = \sigma(k_0)/k_0^0 = \sigma(c)/c^0 = 10\%$. These parameter standard deviations are deliberately selected to all be equal in order to equalize their respective effects on the response's expected value, variance, and skewness, while singling out the individual and, subsequently, cumulative effects of the first- and second-order response sensitivities, respectively, on the response's statistical characteristics (namely: expected value, variance, and skewness). For these particular conditions, the response correlations derived in Chapter 1, Equations 1.18 and 1.19, reduce to the following expressions:

1. The expected value of a response r_k, denoted here as $[E(r_k)]^{UG}$, which arises due to uncertainties in uncorrelated normally distributed model parameters (the superscript *UG* indicates *uncorrelated Gaussian parameters*), is given by the expression

$$\left[E(r_k)\right]^{UG} = r_k\left(\boldsymbol{\alpha}^0\right) + \frac{1}{2}\sum_{i=1}^{N_\alpha} \frac{\partial^2 r_k}{\partial \alpha_i^2}\sigma_i^2 \tag{8.137}$$

   where $r_k(\boldsymbol{\alpha}^0)$ denotes the computed nominal value of the response.

2. The covariance, $cov(r_k, r_l)$, between two responses r_k and r_l arising from normally distributed uncorrelated parameters is given by

$$\left[cov\left(r_k, r_\ell\right)\right]^{UG} = \sum_{i=1}^{N_\alpha}\left(\frac{\partial r_k}{\partial \alpha_i}\frac{\partial r_\ell}{\partial \alpha_i}\right)\sigma_i^2 + \frac{1}{2}\sum_{i=1}^{N_\alpha}\left(\frac{\partial^2 r_k}{\partial \alpha_i^2}\right)\left(\frac{\partial^2 r_\ell}{\partial \alpha_i^2}\right)\sigma_i^4. \tag{8.138}$$

3. The variance, $\mathrm{var}(r_k)$, of a response r_k is obtained by setting $r_k \equiv r_l$ in the above expression to obtain

$$\left[\mathrm{var}(r_k)\right]^{UG} = \sum_{i=1}^{N_\alpha}\left(\frac{\partial r_k}{\partial \alpha_i}\right)^2 \sigma_i^2 + \frac{1}{2}\sum_{i=1}^{N_\alpha}\left(\frac{\partial^2 r_k}{\partial \alpha_i^2}\right)^2 \sigma_i^4 . \tag{8.139}$$

It follows from Equation 8.139 that the contributions stemming from the first-order parameter derivatives (in the absence of second- and higher-order derivatives) yield the following *first-order standard deviation*, denoted as $\mathrm{Std.Dev}_{FO}[T(z_r)]$, of $T(z_r)$:

$$\begin{aligned}\mathrm{Std.Dev}_{FO}\left[T(z_r)\right] = \Bigg\{&\left[\frac{\partial T(z_r)}{\partial Q}\sigma(Q)\right]^2 + \left[\frac{\partial T(z_r)}{\partial q}\sigma(q)\right]^2 \\ &+ \left[\frac{\partial T(z_r)}{\partial T_a}\sigma(T_a)\right]^2 + \left[\frac{\partial T(z_r)}{\partial k_0}\sigma(k_0)\right]^2 + \left[\frac{\partial T(z_r)}{\partial c}\sigma(c)\right]^2\Bigg\}^{1/2}.\end{aligned} \tag{8.140}$$

The magnitude of each of the terms on the right side of Equation 8.140 quantifies the contribution made by each of the respective model parameters, considered to be uncorrelated and normally distributed, to the first-order standard deviation, Std.Dev$_{FO}[T(z_r)]$. These magnitudes are plotted, as a function of z_r, in Figures 8.22 through 8.26. Using the results plotted in Figures 8.22 through 8.26 in Equation 8.140 yields the first-order standard deviation, denoted as Std.Dev$_{FO}[T(z_r)]$, of $T(z_r)$, which is plotted in Figure 8.27.

As shown in Figures 8.22 through 8.26, the quantities $|\sigma(Q)\partial T(z_r)/\partial Q|$ and $|\sigma(q)\partial T(z_r)/\partial q|$ are monotonically increasing, the quantity $|\sigma(T_a)\partial T(z_r)/\partial T_a|$ displays a minimum, while the quantities $|\sigma(k_0)\partial T(z_r)/\partial k_0|$ and $|\sigma(c)\partial T(z_r)/\partial c|$ display maxima, as functions of z_r, corresponding to the behavior of the respective first-order derivatives, of course. Since the derivatives $|\partial T(z_r)/\partial Q|$ and $|\partial T(z_r)/\partial q|$ are the largest, the respective contributions cause the quantity Std.Dev$_{FO}[T(z_r)]$ to increase monotonically as a function of z_r. The smallest contribution to Std.Dev$_{FO}[T(z_r)]$ stems from $|\sigma(c)\partial T(z_r)/\partial c|$, indicating that the parameter c, which actually controls the strength of the nonlinearity in the conduction equation, is not very important in the temperature range under consideration (400–800 K).

The quantities $\left(\sigma_i^2\,\partial^2 r_k/\partial\alpha_i^2\right)^2/2$ contain the contributions involving the second-order derivatives to the total standard deviation, denoted as Std.Dev$[T(z_r)]$, of $T(z_r)$. The magnitudes of each of these second-order contributions are plotted, as a function of z_r, in Figures 8.28 through 8.32.

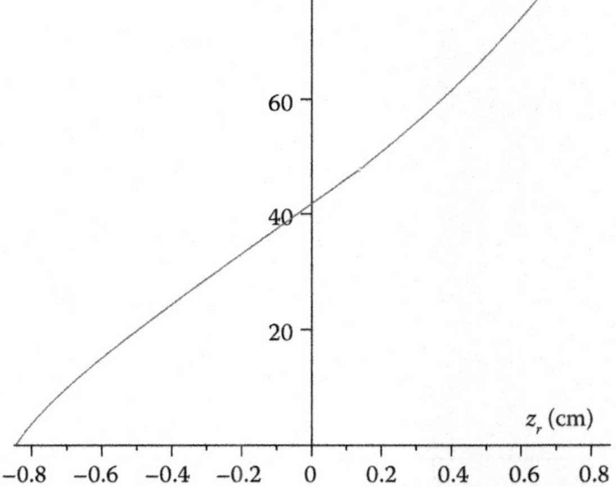

FIGURE 8.22
Contributions from first-order sensitivities of Q to Std.Dev. in $T(z_r)$.

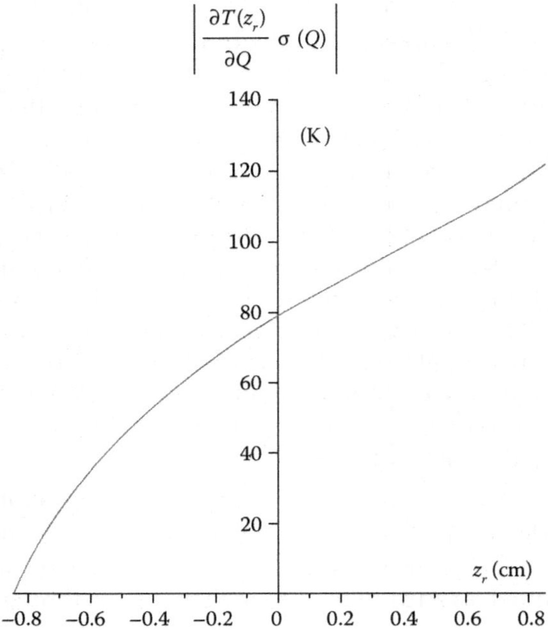

FIGURE 8.23
Contributions from first-order sensitivities of q to Std.Dev. in $T(z_r)$.

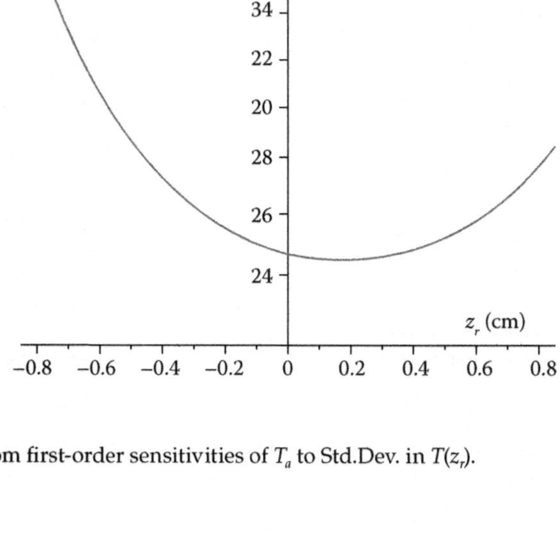

FIGURE 8.24
Contributions from first-order sensitivities of T_a to Std.Dev. in $T(z_r)$.

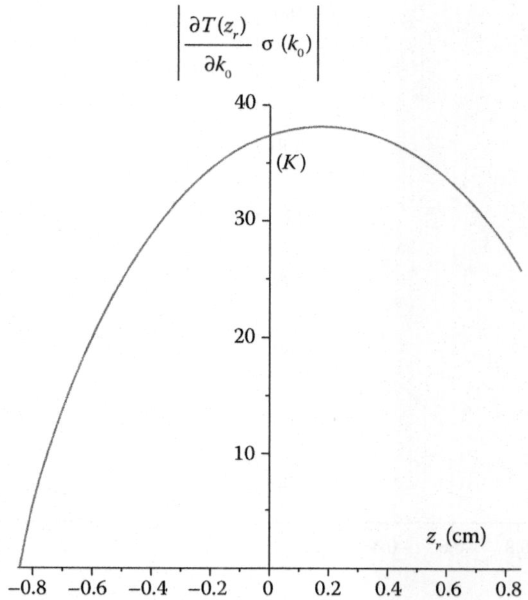

FIGURE 8.25
Contributions from first-order sensitivities of k_0 to Std.Dev. in $T(z_r)$.

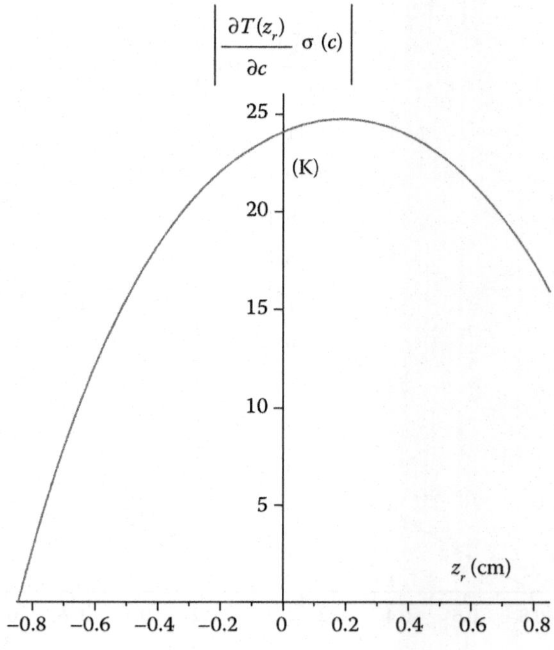

FIGURE 8.26
Contributions from first-order sensitivities of c to Std.Dev. in $T(z_r)$.

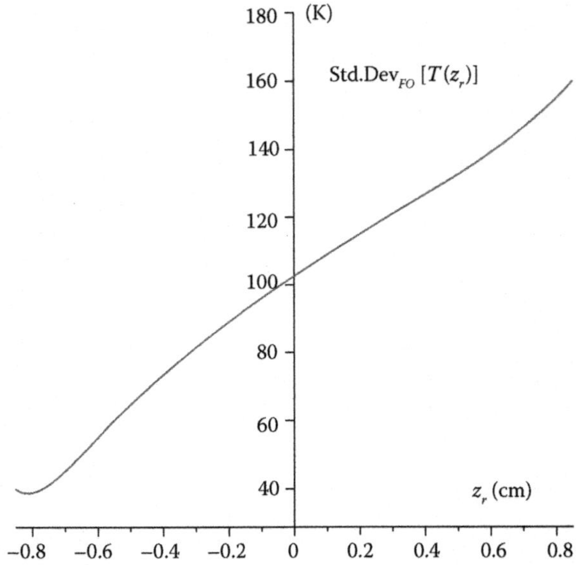

FIGURE 8.27
Contributions from all first-order sensitivities to Std.Dev. in $T(z_r)$.

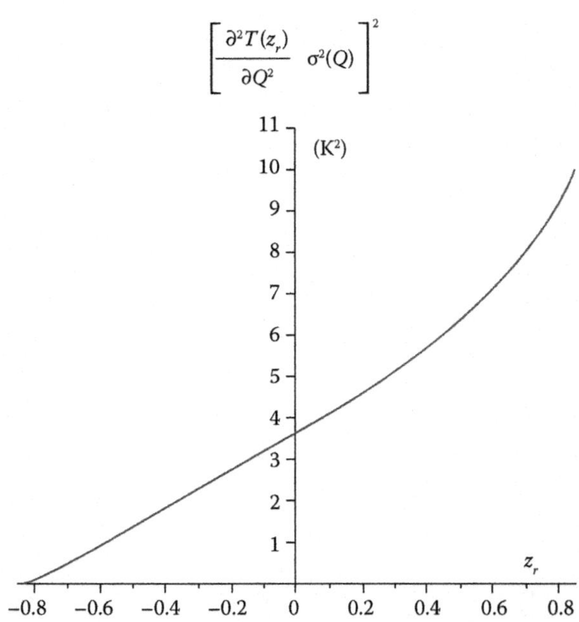

FIGURE 8.28
Contributions from second-order sensitivities of Q to Std.Dev. in $T(z_r)$.

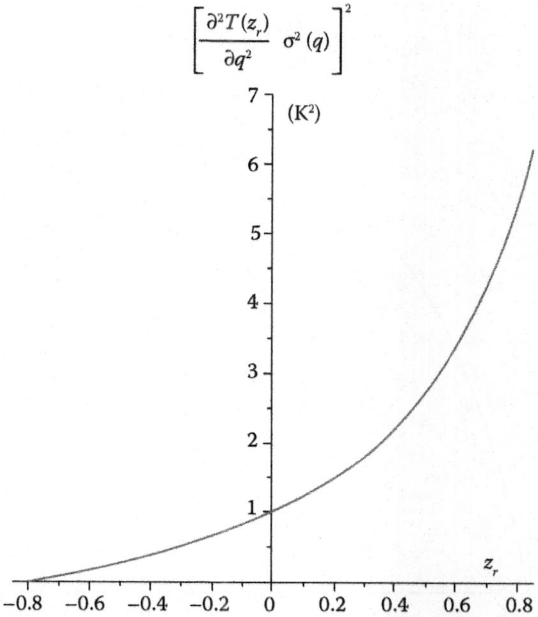

FIGURE 8.29
Contributions from second-order sensitivities of q to Std.Dev. in $T(z_r)$.

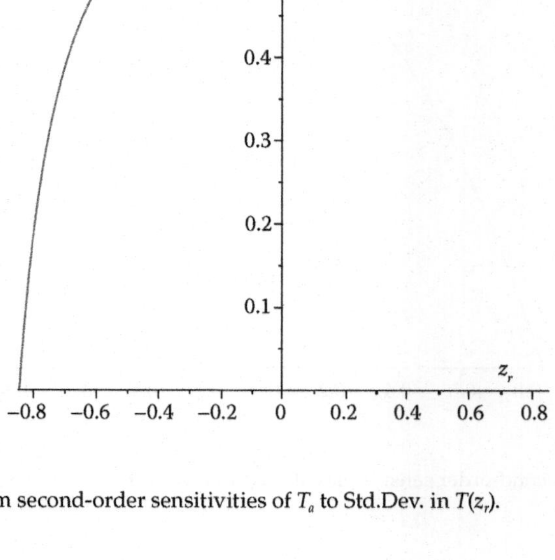

FIGURE 8.30
Contributions from second-order sensitivities of T_a to Std.Dev. in $T(z_r)$.

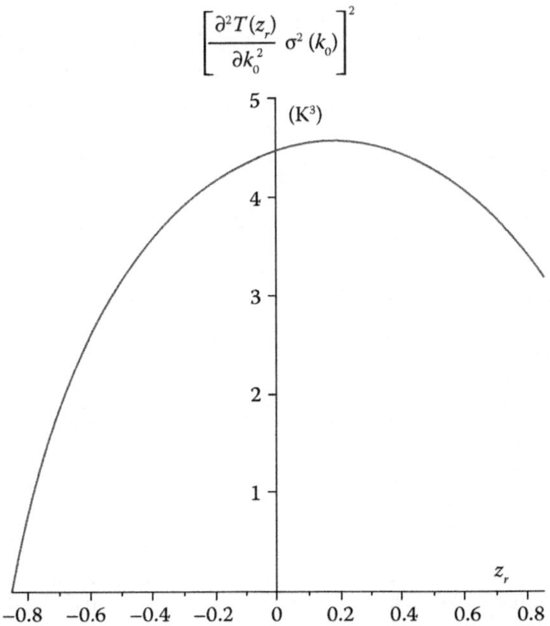

FIGURE 8.31
Contributions from second-order sensitivities of k_0 to Std.Dev. in $T(z_r)$.

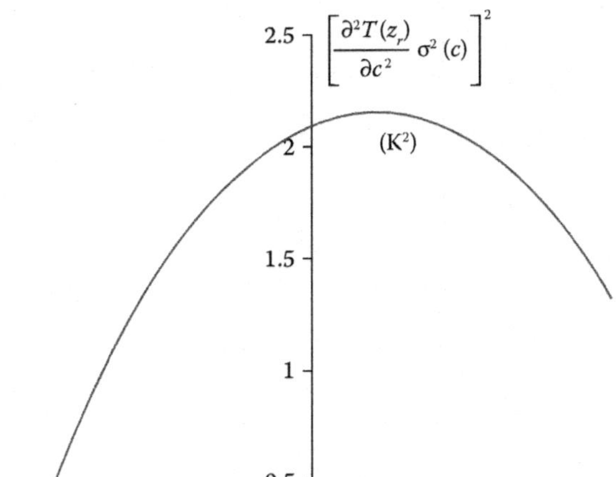

FIGURE 8.32
Contributions of second-order sensitivities of c to Std.Dev. in $T(z_r)$.

Comparing the results plotted in Figures 8.22 through 8.26 to the corresponding results plotted in Figures 8.28 through 8.32 indicate that although some of the second-order sensitivities have magnitudes comparable to the first-order ones, the contributions stemming from the second-order sensitivities are much smaller than those stemming from the first-order sensitivities. This is because the contributions stemming from the second-order sensitivities are multiplied by the fourth power of the respective standard deviations, rather than by the second power (of the respective standard deviation), as are the first-order sensitivities. The contributions to the total standard deviation in $T(z_r)$ stem predominantly from the heat source Q and the boundary heat flux q, which cause this standard deviation to increase monotonically as a function of the location z_r, from the inlet to the outlet, reaching its maximum value of 170 K at the outlet. Equivalently, the relative standard deviation of $T(z_r)$ increases from 10% at the inlet to 24% at the outlet. In particular, Table 8.6 presents the specific values of the various contributions to $\text{Std.Dev}_{FO}[T(z_r)]$ and $\text{Std.Dev}[T(z_r)]$, at the test section's inlet and outlet, as well as at $z_r = z_{\max} =$ 18 cm, which denotes the location where the temperature distribution $T(z_r)$ reaches its maximum value, $T_{max}(z_{max}) = 874.7$ K. The respective parameter contributions are also separated in contributions stemming from the first-order sensitivities (designated using a subscript *1*) and those stemming from the second-order sensitivities (designated using a subscript 2). Notably, the terms involving the second-order sensitivities do not contribute at all at the bottom of the test section, but their contributions increase monotonically from the bottom to the test section's top. The spatial variation of the total standard deviation of the temperature distribution $T(Z_r)$ is depicted in Figure 8.33.

TABLE 8.6

Individual and Total Contributions of Parameters' Standard Deviations to the First-Order and Total Standard Deviation in $T(z_r)$ at Selected Locations

Location	$z_r = -\ell/2$	$z_r = z_{\max}$	$z_r = \ell/2$
$\text{Std.Dev}_{Q1}[T(z_r)]$, (K)	0	88	122
$\text{Std.Dev}_{Q2}[T(z_r)]$, (K)	0	5	10
$\text{Std.Dev}_{q1}[T(z_r)]$, (K)	0	50	96
$\text{Std.Dev}_{q2}[T(z_r)]$, (K)	0	1.4	6
$\text{Std.Dev}_{Ta1}[T(z_r)]$, (K)	40	24	29
$\text{Std.Dev}_{Ta2}[T(z_r)]$, (K)	0	0.5	0.6
$\text{Std.Dev}_{k1}[T(z_r)]$, (K)	0	38	26
$\text{Std.Dev}_{k2}[T(z_r)]$, (K)	0	3	5
$\text{Std.Dev}_{c1}[T(z_r)]$, (K)	0	25	16
$\text{Std.Dev}_{c2}[T(z_r)]$, (K)	0	1.3	2
$\text{Std.Dev}_{FO}[T(z_r)]$, (K)	40	117	160
$\text{Std.Dev}[T(z_r)]$, (K)	40	125	170

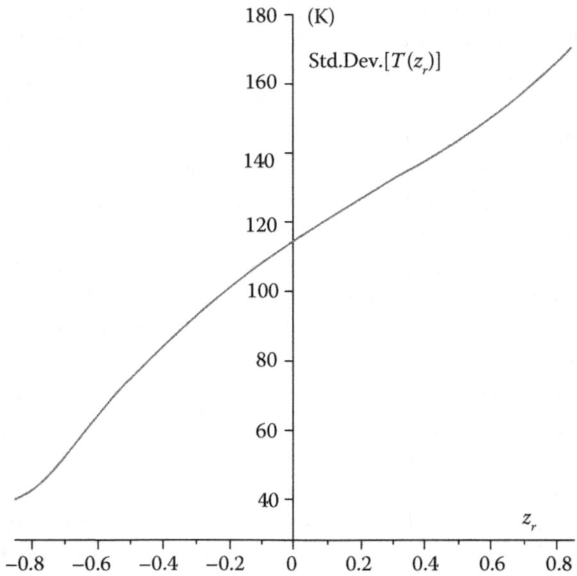

FIGURE 8.33
Spatial variation of the standard deviation of the temperature distribution $T(z_r)$.

As the results in Table 8.6 indicate, the first-order standard deviation, Std.Dev$_{FO}[T(z_r)]$, and the total standard deviation, Std.Dev$[T(z_r)]$, increase from the test section inlet, at $z_r = \ell/2$, and attain their respective maximum values at the test section's outlet, at $z_r = \ell/2$. As the absolute value of the difference {Std.Dev$[T(z_r)]$–Std.Dev$_{FO}[T(z_r)]$} indicates, the largest contribution to the total standard deviation, Std.Dev$[T(z_r)]$, stems from the first-order sensitivities, which contribute to Std.Dev$_{FO}[T(z_r)]$. The absolute value of the difference {Std.Dev$[T(z_r)]$–Std.Dev$_{FO}[T(z_r)]$} increases from 0(K) at the test section's inlet to 10(K) at the test section's outlet, which indicates that the importance of the contributions stemming from the second-order sensitivities increases from the inlet (bottom) to outlet (top) of the test section. Recalling that the temperature attains $T^0(\ell/2) =$ 700K at the top of the test section, it follows from the values presented in Table 8.6 at $z_r = \ell/2$ that the total relative standard deviation is 24.3%, of which 23% stems from the first-order sensitivities and 1.3% stems from the second-order ones.

For normally distributed uncorrelated parameters, the third-order response correlation given in Equation 1.20 reduces to the following expression:

$$\left[\mu_3\left(r_k, r_l, r_m\right)\right]^{UG} = \sum_{i=1}^{N_\alpha}\left(\frac{\partial^2 r_k}{\partial \alpha_i^2}\frac{\partial r_l}{\partial \alpha_i}\frac{\partial r_m}{\partial \alpha_i} + \frac{\partial r_k}{\partial \alpha_i}\frac{\partial^2 r_l}{\partial \alpha_i^2}\frac{\partial r_m}{\partial \alpha_i} + \frac{\partial r_k}{\partial \alpha_i}\frac{\partial r_l}{\partial \alpha_i}\frac{\partial^2 r_m}{\partial \alpha_i^2}\right)\sigma_i^4. \quad (8.141)$$

In particular, the third-order central moment, $\mu_3(r_k)$, of the response r_k is determined by setting $k = l = m$ in Equation 8.141 to obtain

$$\left[\mu_3(r_k)\right]^{UG} = 3\sum_{i=1}^{N_\alpha}\left(\frac{\partial r_k}{\partial \alpha_i}\right)^2 \frac{\partial^2 r_k}{\partial \alpha_i^2}\sigma_i^4. \tag{8.142}$$

The *skewness*, $\gamma_1(r_k)$, of a response r_k can be computed using the customary definition

$$\gamma_1(r_k) \equiv \mu_3(r_k)/\left[\text{var}(r_k)\right]^{3/2}. \tag{8.143}$$

Recall that the *skewness* of a distribution quantifies the departure of the subject distribution from symmetry. Symmetric univariate distributions (e.g., the Gaussian) are characterized by $\gamma_1(r_k) = 0$. A distribution with a long right tail would have a positive skewness, while a distribution with a long left tail would have a negative skewness. In other words, if $\gamma_1(r_k) < 0$, then the respective distribution is skewed towards the left of the mean $[E(r_k)]^{UG}$, favoring lower values of r_k relative to $[E(r_k)]^{UG}$. On the contrary, if $\gamma_1(r_k) > 0$, then the respective distribution is skewed towards the right of the mean $[E(r_k)]^{UG}$, favoring higher values of r_k relative to $[E(r_k)]^{UG}$. As Equation 8.143 indicates, neglecting the second-order sensitivities for normally distributed model parameters would nullify the third-order response correlations and hence would nullify the *skewness*, $\gamma_1(r_k)$, of a response r_k.

The skewness induced in the temperature distribution $T(z_r)$ by each model parameter considered separately will be denoted as $\gamma_Q[T(z_r)]$, $\gamma_q[T(z_r)]$, $\gamma_{Ta}[T(z_r)]$, $\gamma_k[T(z_r)]$, and $\gamma_c[T(z_r)]$, respectively. In other words, the quantity $\gamma_Q[T(z_r)]$ denotes the skewness that would be induced in the temperature distribution $T(z_r)$ if only the heat source Q were a normally distributed variate with a relative standard deviation of 10%. The quantity $\gamma_q[T(z_r)]$ denotes the skewness that would be induced in the temperature distribution $T(z_r)$ if only the boundary heat flux q were a normally distributed variate with a relative standard deviation of 10%, and so on. Figures 8.34 through 8.43 depict the third-order moment and the skewness of the temperature distribution $T(z_r)$ stemming from each individual variate, as functions of the location z_r, from the bottom to the top of the test section, respectively.

Figures 8.35 and 8.37 indicate that $\gamma_Q[T(z_r)]<0$ and $\gamma_q[T(z_r)]<0$. Hence, if only the heat source Q and/or the boundary heat flux q were normally distributed variates, each having a relative standard deviation of 10%, they would cause the temperature distribution $T(z_r)$ to be skewed significantly towards values *lower* than the mean temperature (i.e., the distribution would be skewed to the left of the mean value). On the contrary, Figures 8.41 and 8.43 indicate that $\gamma_k[T(z_r)]<0$ and $\gamma_c[T(z_r)]<0$. Hence, if only the model parameters k_0 and/or c were normally distributed variates, each having a relative standard deviation of 10%, they would cause the temperature distribution $T(z_r)$

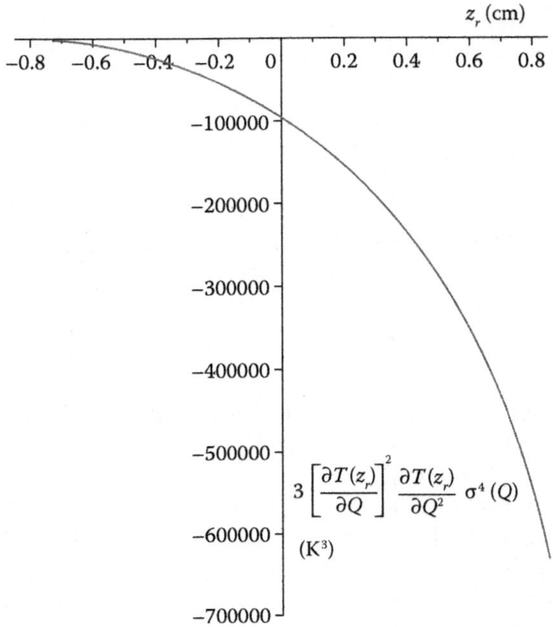

FIGURE 8.34
Third-order moment of $T(z_r)$ stemming solely from the variate Q.

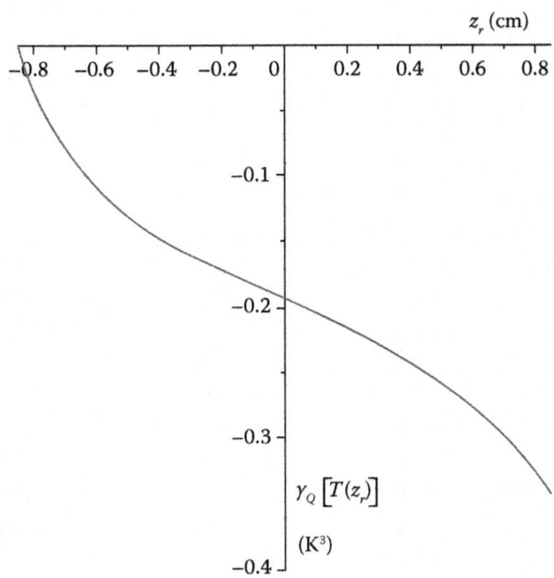

FIGURE 8.35
Skewness of $T(z_r)$ stemming solely from the variate Q.

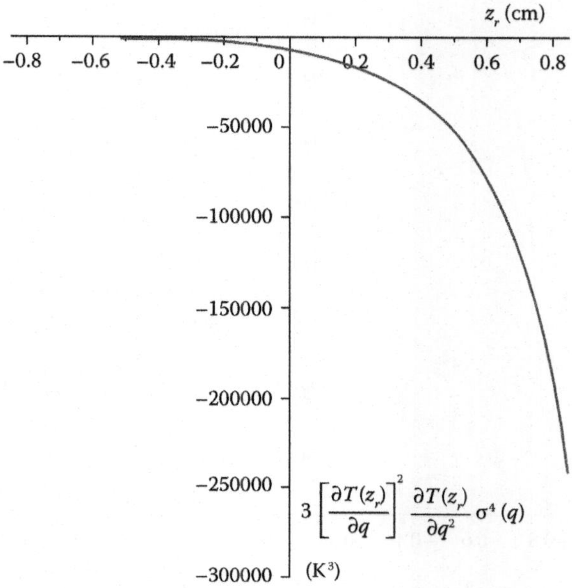

FIGURE 8.36
Third-order moment of $T(z_r)$ stemming solely from the variate q.

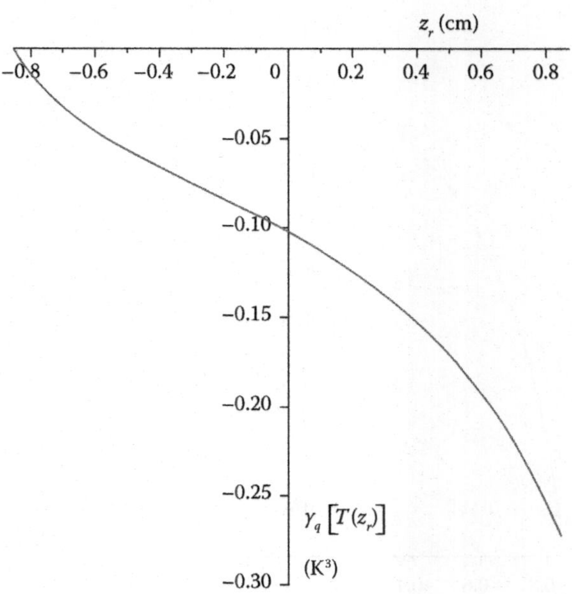

FIGURE 8.37
Contributions to skewness in $T(z_r)$ stemming solely from the variate q.

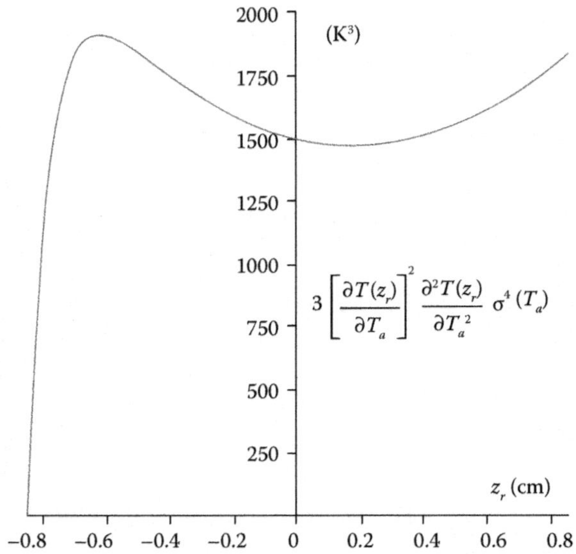

FIGURE 8.38
Third-order moment of $T(z_r)$ stemming solely from the variate T_a.

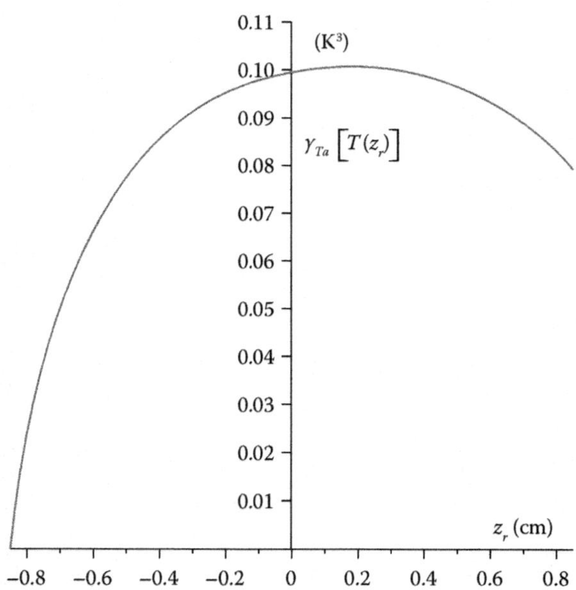

FIGURE 8.39
Contributions to skewness in $T(z_r)$ stemming solely from the variate T_a.

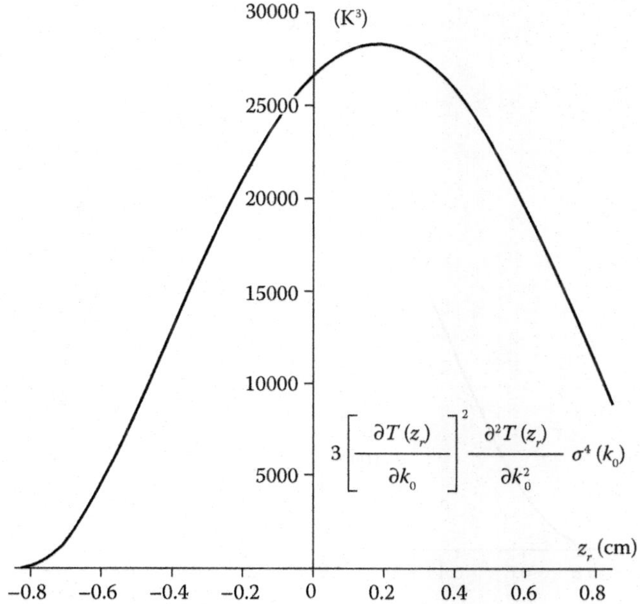

FIGURE 8.40
Third-order moment of $T(z_r)$ stemming solely from the variate k_0.

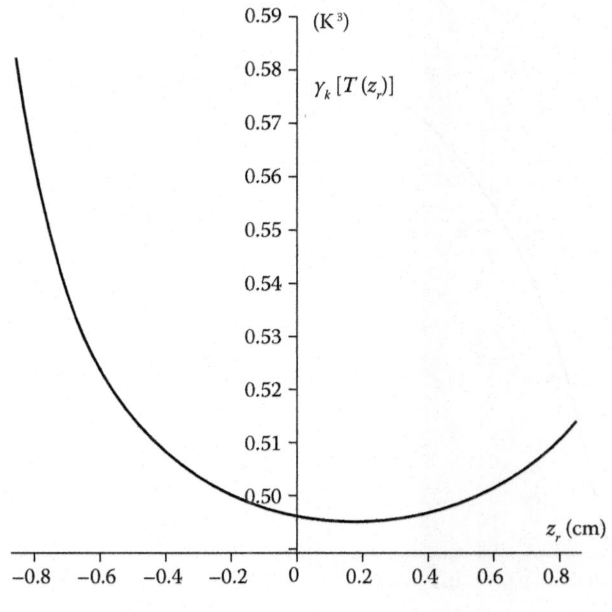

FIGURE 8.41
Contributions to skewness in $T(z_r)$ stemming solely from the variate k_0.

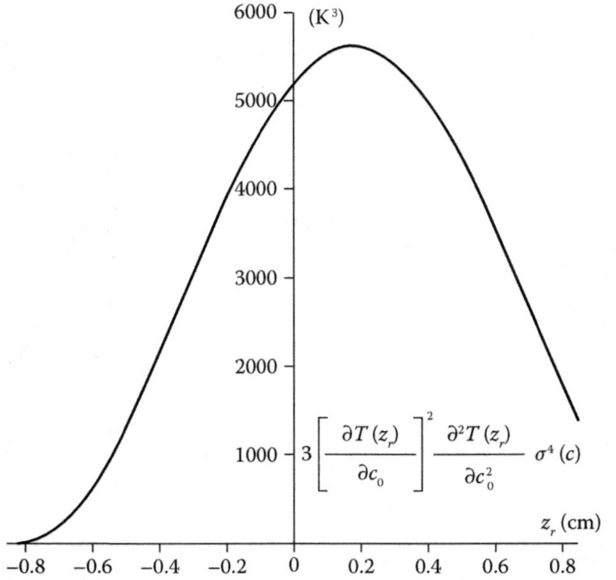

FIGURE 8.42
Third-order moment of $T(z_r)$ stemming solely from the variate c.

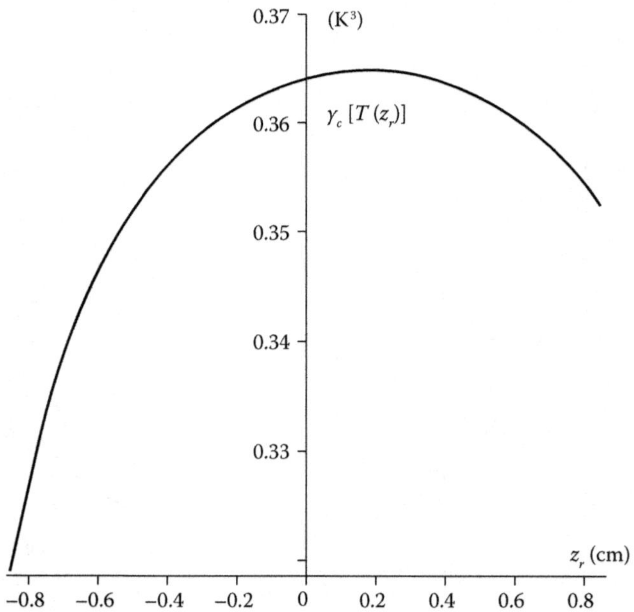

FIGURE 8.43
Contributions to skewness in $T(z_r)$ stemming solely from the variate c.

to be skewed significantly towards values *higher* than the mean temperature (i.e., the distribution would be skewed to the right of the mean value). Finally, Figure 8.39 indicates that $\gamma_{Ta}[T(z_r)]<0$ but has rather small positive values. Hence, if the temperature T_a at the bottom of the test section were the sole variate, normally distributed and having a relative standard deviation of 10% (and all other parameters were perfectly known, taking on exactly their nominal values), then the temperature distribution $T(z_r)$ would display a slight asymmetry towards temperatures higher than the mean temperature.

The total skewness of the temperature distribution $T(z_r)$, reflecting the cumulative effects of the model parameters (assumed to be normally distributed and having all relative standard deviations of 10%), is depicted in Figure 8.44. This figure indicates that the negative contributions stemming from the third moments of the heat source, Q, and the boundary flux, q, are initially small, so the positive contributions from the third-order moments of the other model parameters dominate at the bottom of the test section. Therefore, the temperature distribution $T(z_r)$ is skewed slightly toward higher temperatures in the region extending about 20 cm from the bottom of the test section. For the remainder of the test section, however, the negative contributions stemming from the third moments of the heat source, Q, and the boundary flux, q, become dominant, so the temperature distribution $T(z_r)$ becomes increasingly skewed towards temperatures lower than (i.e., to the left of) the mean temperature towards the top of the test section. Notably, the influence of the model parameter c, which controls the strength of the nonlinearity in this illustrative benchmark problem, would be strong if it were

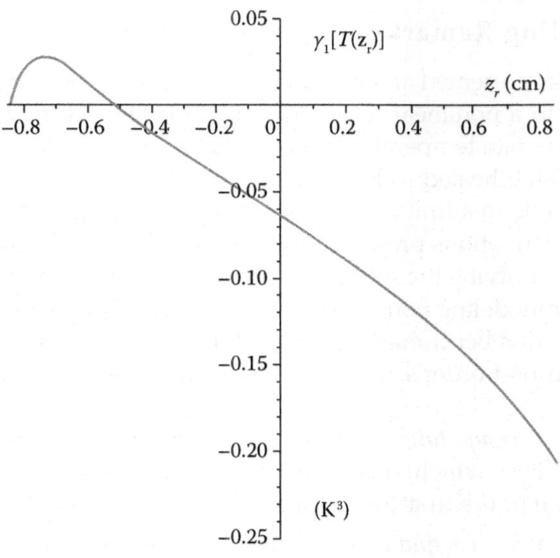

FIGURE 8.44
Spatial variation of the skewness in $T(z_r)$.

TABLE 8.7

Individual and Total Contributions Stemming from Second-Order Sensitivities to the Skewness, $\gamma_1\ [T(z_r)]$, in $T(z_r)$ at Selected Locations z_r

Location	$z_r = -\ell/2$	$z_r = z_{max}$	$z_r = \ell/2$
$\gamma_Q[T(z_r)]$	0	−0.22	−0.35
$\gamma_q[T(z_r)]$	0	−0.12	−0.27
$\gamma_{Ta}[T(z_r)]$	0	0.1	0.08
$\gamma_k[T(z_r)]$	0	0.49	0.51
$\gamma_c[T(z_r)]$	0	0.36	0.35
$\gamma_1[T(z_r)]$	0	−0.08	−0.2

the only uncertain model parameter. However, if all of the other parameters are also uncertain, all having equal relative standard deviations, the uncertainties in the heat source Q and boundary heat flux q mitigate the impact uncertainties in c, for the range of temperatures (400–900 K) considered for this benchmark problem.

The numerical values of the impact of the individual parameters, as well as their cumulative impact on the temperature distribution $T(z_r)$ at the bottom and top of the test section, as well as at the location $z_r = z_{max} = 18$ cm, are presented in Table 8.7.

8.4 Concluding Remarks

This chapter has presented an illustrative application of the 2nd-ASAM for *nonlinear* systems to a nonlinear heat conduction benchmark problem that simulates the steady-state temperature distribution in a conceptual design of a test section comprising heated rods immersed in LBE. This benchmark problem is sufficiently simple to admit an exact solution, thereby making transparent the mathematical derivations presented in Chapter 7, yet it provides an important benchmark for verifying the solutions provided by commercially available computer codes for modeling nonlinear heat conduction. For the illustrative nonlinear heat conduction benchmark considered in this chapter, the computations of the first- and second-order sensitivities have underscored the following:

1. *One adjoint computation* sufficed to determine the first-level adjoint function $\Psi(z)$, which, in turn, sufficed to compute, using just quadratures, all of the first-order sensitivities $S_i \triangleq \partial R/\partial \alpha_i$, $i = 1, 2, 3, 4, 5$.
2. *Five adjoint computations* were performed, one each for computing the respective second-level adjoint functions $\boldsymbol{\Psi}_i^{(2)} \triangleq \left(\Psi_{1,i}^{(2)}, \Psi_{2,i}^{(2)}\right)$, $i = 1,2,3,4,5$, and subsequently determining

all of the second-order sensitivities $S_{ij} \triangleq \partial^2 R/\partial\alpha_i$, i = 1, 2, 3, 4, 5 $S_{ij} \triangleq \partial^2 R/\partial\alpha_i\,\partial\alpha_j;\; i,j=1,2,3,4,5$. These computations also provide independent solution verifications of the second-level adjoint functions $\mathbf{\Psi}_i^{(2)} \triangleq \left(\Psi_{1,i}^{(2)}, \Psi_{2,i}^{(2)}\right),\; i=1,2,3,4,5$, due to the symmetry of the second-order sensitivities. Notably, similar differential operators appear on the left sides of the 1st- and 2nd-LASS, which therefore would use the same solvers of differential equations; *only the sources on the right sides of these systems differ from each other.*

3. The magnitudes of the second-order sensitivities are generally of the same order as the magnitudes of the first-order sensitivities.

   The second-order sensitivities play the following important roles:

   a. They cause the expected value of the response to differ from the computed nominal value of the response, but for the nonlinear heat conduction benchmark these differences were insignificant.
   b. They contribute to increasing the response variances and modifying the response covariances, but for the nonlinear heat conduction benchmark their contributions are much smaller than the contributions stemming from the first-order response sensitivities.
   c. They provide the leading contributions to causing asymmetries in the response distribution. For the benchmark test section considered in this work, the heat source, the boundary heat flux, and the temperature at the bottom boundary of the test section would cause the temperature distribution in the test section to be skewed significantly toward values *lower* than the mean temperature. On the contrary, the model parameters entering the nonlinear temperature-dependent expression of the LBE conductivity would cause the temperature distribution in the test section to be skewed significantly toward values *higher* than the mean temperature. These opposite effects partially cancel each other. At the locations considered in this illustrative problem, the temperature distribution was skewed toward temperatures lower than the respective mean temperatures. Notably, the influence of the model parameter that controls the strength of the nonlinearity in the heat conduction coefficient for this benchmark would be significant if it were the only uncertain model parameter. However, if all of the model parameters are uncertain, all having equal relative standard deviations, the uncertainties in the heat source and boundary heat flux compensate for the impact of uncertainties in the nonlinear heat conduction coefficient for the range of temperatures (400–900 K) considered for this benchmark.

The illustrative examples presented in this chapter, as well as the other illustrative examples presented in this book, are intended to underscore the fact that erroneous conclusions are likely to be drawn based on partial information, that is, conclusions based on only some, rather than all, sensitivities. This is because the sensitivities have varying magnitudes and signs (negative and positive), so cancellations and/or enhancement of contributions to the total uncertainties (covariance and variances, skewness, etc.) in the respective response distributions are likely to occur. It is therefore very important to compute *all*, rather than just some, response sensitivities. This argument underscores, once again, the practical importance and usefulness of the 2nd-ASAM, which is the only practical methodology for computing exactly all of the first- and second-order response sensitivities for physical systems involving many parameters.

References

ANSYS Academic Research. 2015. *Release 16.0, FLUENT Adjoint Solver, ANSYS®, Inc.*

Cacuci, D. G. 1981a. Sensitivity theory for nonlinear systems: I. Nonlinear functional analysis approach. *J. Math. Phys.* 22: 2794–2802.

Cacuci, D. G. 1981b. Sensitivity theory for nonlinear systems: II. Extensions to additional classes of responses. *J. Math. Phys.* 22: 2803–2812.

Cacuci, D. G. 1988. The forward and the adjoint methods of sensitivity analysis. Chapter 3 in *Uncertainty Analysis*, ed. Y. Ronen, pp. 71–144. Boca Raton, FL: CRC Press.

Cacuci, D. G. 2003. *Sensitivity and uncertainty analysis: Theory*. Volume 1. Boca Raton, FL: Chapman & Hall/CRC.

Cacuci, D. G. 2014. Predictive modeling of coupled multi-physics systems: I. Theory. *Ann. Nucl. Energ.* 70: 266–278.

Cacuci, D. G. 2015a. Second-order adjoint sensitivity analysis methodology (2nd-ASAM) for computing exactly and efficiently first- and second-order sensitivities in large-scale linear systems: I. Computational methodology. *J. Comp. Phys.* 284: 687–699.

Cacuci, D. G. 2015b. Second-order adjoint sensitivity analysis methodology (2nd-ASAM) for computing exactly and efficiently first- and second-order sensitivities in large-scale linear systems: II. Illustrative application to a paradigm particle diffusion problem. *J. Comp. Phys.* 284: 700–717.

Cacuci, D. G. 2016a. Second-order adjoint sensitivity and uncertainty analysis of a benchmark heat transport problem: I. Analytical results. *Nucl. Sci. Eng.* 183: 1–21.

Cacuci, D. G. 2016b. Second-order adjoint sensitivity analysis methodology (2nd-ASAM) for large-scale nonlinear systems: I. Theory. *Nucl. Sci. Eng.* 184: 16–30.

Cacuci, D. G. 2016c. Second-order adjoint sensitivity analysis methodology (2nd-ASAM) for large-scale nonlinear systems: II. Illustrative application to a paradigm nonlinear heat conduction benchmark. *Nucl. Sci. Eng.* 184: 31–52.

Cacuci, D. G. 2016d. A heat transport benchmark problem for predicting the impact of measurements on thermal-hydraulics experimental facility design. *Nucl. Eng. Design* 300: 12–27.

Cacuci, D. G. 2016e. Second-order adjoint sensitivity and uncertainty analysis of a benchmark heat transport problem: I. Analytical results, *Nucl. Sci. Eng.* 183: 1–21. doi: 10.13182/NSE15-80.

Cacuci, D. G. 2017. Inverse predictive modeling of radiation transport through optically thick media in the presence of counting uncertainties, *Nucl. Sci. Eng.* 186: 199–223. doi: 10.1080/00295639.2017.1305244.

Cacuci, D. G. and M. Ionescu-Bujor. 2010. Model calibration and best-estimate prediction through experimental data assimilation: I. Mathematical framework. *Nucl. Sci. Eng.* 165: 18–44.

Cacuci, D. G., E. Greenspan, J. H. Marable, and M. L. Williams. 1980. Developments in sensitivity theory, In: *ANS Topical Conference "1980 Advances in Reactor Physics and Shielding*, pp. 692–704. NAS/70048, 14–17 September 1980. Sun Valley, ID.

Cacuci, D. G., M. Ionescu-Bujor, and M. I. Navon. 2005. *Sensitivity and uncertainty analysis: Applications to large scale systems*. Volume 2. Boca Raton, FL: Chapman & Hall/CRC.

Cacuci, D. G., M. I. Navon, and M. Ionescu-Bujor. 2014. *Computational methods for data evaluation and assimilation*. Boca Raton, FL: Chapman & Hall/CRC.

Cacuci, D. G., R. Fang, M. Ilic, and M. C. Badea, 2016a. A Heat Conduction and Convection Analytical Benchmark for Adjoint Solution Verification of CFD Codes Used in Reactor Design, *Nucl. Sci. Eng.*, 182: 452–480. doi: 10.13182/NSE15-69.

Cacuci, D. G., M. Ilic, M. C. Badea, and R. Fang. 2016b. Second-order adjoint sensitivity and uncertainty analysis of a benchmark heat transport problem. II: Computational results for G4M reactor thermal-hydraulics parameters. *Nucl. Sci. Eng*. 183: 22–38.

Daescu, D. N. and I. M. Navon. 2014. Sensitivity analysis in nonlinear variational data assimilation: Theoretical aspects and applications. in *Advanced numerical methods for complex environmental models: Needs and availability*, pp. 276–300, eds. I. Faragó, Á. Havasi, and Z. Zlatev, Bentham Science, Oak Park, IL.

Dems, K. and Z. Mroz. 1985. Variational approach to first- and second-order sensitivity analysis of elastic structures. *Int. J. Numer. Methods Eng*. 21: 637–661.

Gandini, A. 1978. Implicit and explicit higher-order perturbation methods for nuclear reactor analysis. *Nucl. Sci. Eng*. 67: 347–355.

GEN4 ENERGY, INC. 2012. *Reactor Core Design*. Denver, CO.

Greenspan, E. 1982. New developments in sensitivity theory, in *Advances in nuclear science and technology*, Volume 14. eds. J. Lewins and M. Becker, Plenum Press, New York.

Greenspan, E., D. Gilai, and E. M. Oblow. 1978. Second-order generalized perturbation theory for source-driven systems. *Nucl. Sci. Eng*. 68: 1–9.

Haftka, R. T. 1982. Second-order sensitivity derivatives in structural analysis. *AIAA J*. 20: 1765–1766.

Haftka, R. T. and Z. Mroz. 1986. First-and second-order sensitivity analysis of linear and nonlinear structures, *AIAA J*. 24: 1187–1192.

Haug, E. J. 1981. Second-order design sensitivity analysis of structural systems. *AIAA J*. 19: 1087–1088.

Haug, E. J. and P. E. Ehle. 1982. Second-order design sensitivity analysis of mechanical system dynamics. *Int. J. Numer. Meth. Eng*. 18: 1699–1717.

Haug, E., J. V. Komkov, and K. K. Choi. 1986. *Design sensitivity analysis of structural systems*. New York: Academic Press.

Lahoz, W., B. Khattatov, and R. Ménard. (eds.) 2010. *Data Assimilation: Making Sense of Observations*, Heidelberg, Springer Verlag.

Le Dimet, F. X., I. M. Navon, and D. N. Daescu. 2002. Second-order information for data assimilation. *Mon Weather Rev*. 130: 629–648.

Mitani, H. 1973. Higher-order perturbation method in reactor calculation. *Nucl. Sci. Eng*. 51: 180–188.

Negm, M. H., M. H. Bakr, O. S. Ahmed, N. K. Nikolova, and J. W. Bandler. 2014. Wideband second-order adjoint sensitivity analysis exploiting TLM. *IEEE Trans. Microw. Theor. Tech*. 62: 389–398.

Peltz, J. J. and D. G. Cacuci. 2016. Predictive modeling applied to a paradigm spent fuel dissolver model: I. Adjoint sensitivity analysis. *Nucl. Sci. Eng*. 183: 305–331.

Prágcr, T. and F. D. Kelemen. 2014. Adjoint methods and their application in earth sciences, in *Advanced numerical methods for complex environmental models: Needs and availability*, eds. I. Faragó, Á. Havasi, and Z. Zlatev, pp. 203–275, Bentham Science.

Saltelli, A., K. Chan, and E. M. Scott. Ed. 2000. *Sensitivity analysis*. Chichester: John Wiley & Sons.

Sandu, A. and L. Zhang. 2008. Discrete second order adjoints in atmospheric chemical transport modeling. *J. Comput. Phys.* 227: 5949–5983.

Seki, Y. 1973. Evaluation of the second-order perturbation terms by the generalized perturbation method. *Nucl. Sci. Eng.* 51: 243–251.

Wang, Z., I. M. Navon, F. X. Le Dimet, and X. Zou. 1992. The second order adjoint analysis: Theory and applications. *Meteor. Atmos. Phys.* 50: 3–20.

Wang, Z., I. M. Navon, X. Zou, and F. X. Le Dimet. 1995. A truncated Newton optimization algorithm in meteorology applications with analytic Hessian/vector products. *Comput. Optim. Appl.* 4: 241–262.

Wigner, E. P. 1945. *Effect of small perturbations on pile period*. Chicago Report CP-G-3048.

Ye, X., P. Li, and F. Y. Liu. 2010. Exact time-domain second-order adjoint-sensitivity computation for linear circuit analysis and optimization. *IEEE Trans. Circ Syst.* 57: 236–248.

Index

H

I

K

L

R

S